Carbon Nanomaterials in Clean Energy Hydrogen Systems

Carbon Nanomaterials in Clean Energy Hydrogen Systems

Contributors

Liga Grinberga and Janis Kleperis et al.

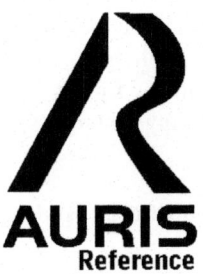

AURIS
Reference

www.aurisreference.com

Carbon Nanomaterials in Clean Energy Hydrogen Systems

Contributors: Liga Grinberga and Janis Kleperis et al.

Published by Auris Reference Limited

www.aurisreference.com

United Kingdom

Carbon Nanomaterials in Clean Energy Hydrogen Systems

ISBN: 978-1-78154-864-6

British Library Cataloguing in Publication Data
A CIP record for this book is available from the British Library

Printed in the United Kingdom

Exclusively distributed by CBS Publishers & Distributors Pvt. Ltd.

Sales & Distribution Rights only for India, Pakistan, Bangladesh, Sri Lanka, Nepal and Bhutan.This book is not to be sold outside these territories.

Contents

List of Abbreviations

CBB	Calvin-Benson-Bassham
CNC	Carbon nanocone
CNH	Carbon nanohorn
CNT	Carbon nanotube
CCCP	Carbonyl-cyanide m-chlorophenylhydrazone
CCVD	Catalytic chemical vapor deposition
CIEMAT	Centre of Energy, Environment and Technology Research
CLC	Chemical looping combustion
CLH	Chemical looping hydrogen production
CVD	Chemical vapor decomposition
CCS	Circular consensus sequencing
CDCL	Coal direct chemical looping
COF	Covalent-organic frameworks
DOE	Department of Energy
DSSC	Dye-sensitized solar cells
FHL	Formate hydrogen lyase
GTO	Gaussian type function
HGAp	Hierarchical genome assembly algorithm
ICEs	Internal combustion engines
LCA	Life-cycle assessment
MBH	Membrane-bound hydrogenases
MOF	Metal-organic framework
MFC	Microbial fuel cells
MPECVD	Microwave plasma-enhanced chemical vapor deposition
MMT	Million metric tons
MMTCE	Million metric tons of carbon equivalent
PDOS	Partial density of states
PEC	Photoelectrochemical
PNS	Purple non-sulfur
SWCNC	Single walled carbon nanocone
SWCNT	Single-wall carbon nanotube
SMR	Steam reforming of methane
SCL	Syngas chemical looping
TPD	Temperature programmed desorption
TGA	Thermogravimetric analysis
VGCF	Vapour grown carbon fibres
VLS	Vapour-liquid-solid
VOC	Volatile organic component
WGS	Water-gas shift reaction

List of Contributors

Liga Grinberga
Institute of Solid State Physics, University of Latvia Latvia

Janis Kleperis
Institute of Solid State Physics, University of Latvia Latvia

Karen Wawrousek
Department of Chemical and Petroleum Engineering, University of Wyoming, Laramie, Wyoming, United States of America

Scott Noble
Biosciences Center, National Renewable Energy Laboratory, Golden, Colorado, United States of America

Jonas Korlach
Pacific Biosciences, Menlo Park, California, United States of America

Jin Chen
Department of Energy Plant Research Laboratory, Michigan State University, East Lansing, Michigan, United States of America

Carrie Eckert
Department of Chemical and Petroleum Engineering, University of Wyoming, Laramie, Wyoming, United States of America

Jianping Yu
Department of Chemical and Petroleum Engineering, University of Wyoming, Laramie, Wyoming, United States of America

Pin-Ching Maness
Department of Chemical and Petroleum Engineering, University of Wyoming, Laramie, Wyoming, United States of America

Leonid Bazyma
National Aerospace University "Kharkov Aviation Institute", Ukraine

Andrew Basteev
National Aerospace University "Kharkov Aviation Institute", Ukraine

Michail Obolensky
V.N. Karazin Kharkiv National University, Ukraine

Andrew Kravchenko
V.N. Karazin Kharkiv National University, Ukraine

Vladimir Beletsky
V.N. Karazin Kharkiv National University, Ukraine

Yuri Petrusenko
National Science Center - Kharkov Institute of Physics and Technology, Ukraine

Valeriy Borysenko
National Science Center - Kharkov Institute of Physics and Technology, Ukraine

Sergey Lavrynenko
National Science Center - Kharkov Institute of Physics and Technology, Ukraine

Oleg Kravchenko
A.N. Podgorny Institute for Mechanical Engineering Problems, Ukraine

Irina Suvorova
A.N. Podgorny Institute for Mechanical Engineering Problems, Ukraine

Vladimir Golovanevskiy
Western Australian School of Mines, Curtin University, Australia

S. Abdel Aal
Department of Chemistry, Faculty of Science, Benha University, Benha, Egypt

A. S. Shalabi
Department of Chemistry, Faculty of Science, Benha University, Benha, Egypt

K. A. Soliman
Department of Chemistry, Faculty of Science, Benha University, Benha, Egypt

Cesar Mota
Centro Nacional de Investigaciones Metalúrgicas (CSIC), Madrid, Spain

Antonio Madroñero
Centro Nacional de Investigaciones Metalúrgicas (CSIC), Madrid, Spain

Jose María Amo
Centro Nacional de Investigaciones Metalúrgicas (CSIC), Madrid, Spain

Jose Ignacio Robla
Centro Nacional de Investigaciones Metalúrgicas (CSIC), Madrid, Spain

Mario Culebras
Materials Science Institute, University of Valencia, Valencia, Spain

Andrés Cantarero
Materials Science Institute, University of Valencia, Valencia, Spain

Clara Maria Gómez,
Materials Science Institute, University of Valencia, Valencia, Spain

Vladimir A. Blagojević
University of Belgrade, Faculty for Physical Chemistry, Serbia

Dragica M. Minić
University of Belgrade, Faculty for Physical Chemistry, Serbia

Dejan G. Minić
Kontrola LLC, Austin, TX, USA

Jasmina Grbović Novaković
Laboratory for Material Sciences, Institute for Nuclear Science Vinča, University of Belgrade, Belgrade, Serbia

Doki Yamaguchi
CSIRO Earth Science and Resource Engineering,, Australia

Liangguang Tang
CSIRO Earth Science and Resource Engineering,, Australia

Nick Burke
CSIRO Earth Science and Resource Engineering,, Australia

Ken Chiang
CSIRO Earth Science and Resource Engineering,, Australia

Lucas Rye
CSIRO Marine and Atmospheric Research,, Australia

Trevor Hadley
CSIRO Process Science and Engineering,, Australia

Seng Lim
CSIRO Process Science and Engineering,, Australia

Michael U. Niemann
Clean Energy Research Center, College of Engineering, University of South Florida, 4202 East Fowler Avenue, Tampa, FL 33620, USA

Sesha S. Srinivasan
Clean Energy Research Center, College of Engineering, University of South Florida, 4202 East Fowler Avenue, Tampa, FL 33620, USA

Ayala R. Phani
Nano-RAM Technologies, 98/2A Anjanadri, 3rd Main, Vijayanagar, Bangalore 5600040, Karnataka, India

Ashok Kumar
Clean Energy Research Center, College of Engineering, University of South Florida, 4202 East Fowler Avenue, Tampa, FL 33620, USA

D. Yogi Goswami
Clean Energy Research Center, College of Engineering, University of South Florida, 4202 East Fowler Avenue, Tampa, FL 33620, USA

Elias K. Stefanakos
Clean Energy Research Center, College of Engineering, University of South Florida, 4202 East Fowler Avenue, Tampa, FL 33620, USA

Yit Thai Ong
School of Chemical Engineering, Engineering Campus,, Universiti Sains Malaysia, Seri Ampangan, 14300, Nibong Tebal, SPS, Pulau Pinang, Malaysia

Abdul Latif Ahmad
School of Chemical Engineering, Engineering Campus,, Universiti Sains Malaysia, Seri Ampangan, 14300, Nibong Tebal, SPS, Pulau Pinang, Malaysia

Sharif Hussein Sharif Zein
School of Chemical Engineering, Engineering Campus,, Universiti Sains Malaysia, Seri Ampangan, 14300, Nibong Tebal, SPS, Pulau Pinang, Malaysia

Soon Huat Tan
School of Chemical Engineering, Engineering Campus,, Universiti Sains Malaysia, Seri Ampangan, 14300, Nibong Tebal, SPS, Pulau Pinang, Malaysia

Peter A Schultz
Multiscale Science Department, Sandia National Laboratories, Albuquerque, NM 87185, USA

Clark S Snow
Applied Science and Technology Maturation Department, Sandia National Laboratories, Albuquerque, NM 87185, USA

Renju Zacharia
Institut de Recherche sur l'Hydrogène, Université du Québec à Trois-Rivières, Trois-Rivieres, QC, Canada G9A 5H7
Gas Processing Center, College of Engineering, Qatar University, Doha, Qatar

Sami ullah Rather[3]
Chemical and Materials Engineering Department, King Abdulaziz University, Jeddah 21589, Saudi Arabia

Yury S. Nechaev
Bardin Institute for Ferrous Metallurgy, Moscow, Russia

Alp Yürüm
Nanotechnology Research and Application Center, Sabanci University, Istanbul, Turkey

Adem Tekin
Informatics Institute, Istanbul Technical University, Istanbul, Turkey

Nilgün Karatepe Yavuz
Energy Institute, Istanbul Technical University, Istanbul, Turkey

Yuda Yürüm
Faculty of Engineering and Natural Sciences, Sabanci University, Istanbul, Turkey

T. Nejat Veziroglu[6]
International Association for Hydrogen Energy, Miami, USA

Preface

This book, Carbon Nanomaterials in Clean Energy Hydrogen Systems, aims to provide the wide overview of the latest scientific results on basic research and technological applications of hydrogen interactions with carbon materials. Unlike the conventional graphite phase, carbon nanostructures possess metallic or semiconductor properties that can induce catalysis by participating directly in the charge transfer process. Further, the electrochemical properties of these materials facilitate modulation of their charge transfer properties and aid in the design of catalysts for hydrogenation, sensors, and fuel cells. Analyzing today's situation and tracing tendencies, it is clear that the primary energy consumption is increasing but reserves are running out very rapidly. Meanwhile, global utilization of fossils is causing environmental problems throughout the world. As a consequence, investigations of alternative energy strategies have recently become important, particularly for future world stability. In first chapter a short description of water photocatalysis and photosynthetic hydrogen production and work of our laboratory on these subjects are given. In second chapter, we report the sequencing and analysis of the genome of the purple non-sulfur photosynthetic bacterium Rubrivivax gelatinosus CBS. High capacity hydrogen storage in Ni decorated carbon nanocone has been described in third chapter. The aim of fourth chapter is to examine hydrogen storage capacity and the possibility of hydride formation upon hydrogen storage operation and to determine hydrogen storage capacity in the presence of oxygen molecules at the Ni decorated CNC. The fifth chapter deals with the possibility of improving the hydrogen storage capacity using an activation process consisting of γ rays irradiation. The goal of sixth chapter is to use water thermolysis either in solar concentrators or in nuclear power plants to produce hydrogen directly using thermal energy. Solar concentrators can produce very high temperatures (over 1800K) by concentrating sunlight using a system of mirrors. Seventh chapter reviews the existing hydrogen production technologies then highlights the recent progress made on hydrogen production from small scale CL processes. Eighth chapter focuses the application of nanostructured materials for storing atomic or molecular hydrogen. The synergistic effects of nanocrystalinity and nanocatalyst doping on the metal or complex hydrides for improving the thermodynamics and hydrogen reaction kinetics are discussed. In ninth chapter, the contribution of CNTs is addressed in terms of sustainable environment and green technologies perspective, such as waste water treatment, air pollution monitoring, biotechnologies, renewable energy technologies, supercapacitors and green nanocomposites. Tenth chapter presents about mechanical properties of metal dihydrides. Eleventh chapter presents review of solid state hydrogen storage methods adopting different kinds of novel materials.

Chapter 1

COMPOSITE NANOMATERIALS FOR HYDROGEN TECHNOLOGIES

Liga Grinberga and Janis Kleperis
Institute of Solid State Physics, University of Latvia Latvia

INTRODUCTION

Analyzing today's situation and tracing tendencies, it is clear that the primary energy consumption is increasing but reserves are running out very rapidly. Meanwhile, global utilization of fossils is causing environmental problems throughout the world. As a consequence, investigations of alternative energy strategies have recently become important, particularly for future world stability. The most important property of alternative energy sources is their environmental compatibility. One such new energy carrier currently being investigated is hydrogen. Many hold the hopes that it could maintain mankind's growing need for energy. However, the hydrogen alternative has both positive and negative aspects. The main advantage of hydrogen as a fuel is sustainable development. It is a non-toxic energy carrier and holds a higher energy content: 9.5 kg of hydrogen is equivalent to that of 25 kg of gasoline (Midilli, 2005). Hydrogen can be produced in many ways, however the most environmentally friendly and the less fossils consuming are processes driven directly by sunlight – photocatalysis and photo-biological methods. Studies demonstrate that solid materials can be utilized to solve the storage problem by reversible absorption and desorption of large amounts of hydrogen (Schlapbach & Zuttel, 2001). Although several metal hydrides and composite materials are capable of meeting this target, the high desorption temperatures, slow absorption/ desorption rates and small cycling capacity limit the widespread application of current metal hydrides. Nanostructuring of materials and enhancement of surface absorption capability are two main factors to increase the amount of sorbed hydrogen. One way to combine the effectiveness of hydrogen absorption in metal hydrides and the desirable weight/volume proportion is to make composite material from alloy forming hydride and appropriate support material. In this chapter a short description of water photocatalysis and photosynthetic hydrogen production and work of our laboratory on these subjects are given. There are described various composite materials for solid hydrogen storage and compared their characteristics. The chapter is concluded with the main results of our work on hydrogen storage in the modified AB_5 type metal hydride where the idea of the possible gain of using the spill-over effect to enhance catalytic activity and the amount of absorbed hydrogen was explored.

HYDROGEN PRODUCTION

Hydrogen production is the first step toward the transition to a hydrogen economy. Fossil fuel systems for hydrogen production are the oldest technologies and tend to be the cheapest. Because fossil fuels are carbon-based, carbon dioxide is produced as a by-product when they are decomposed to release energy. Biomass pyrolysis and gasification processes are very similar to fossil fuel reforming and gasification processes. The water splitting methods can use nuclear heat or alternative energy sources do not produce harmful emissions but they are more expensive than fossil fuel processes. Photocatalytic and photobiological processes use solar energy, and the sulphur iodine process uses nuclear heat. In the case of electrolysis, electricity supplies the energy required and there could be used alternative energy sources like wind, solar, or water power. These processes are attractive because the water feedstock contains only hydrogen and oxygen, so no carbon dioxide is released in its decomposition.

Photocatalysis

Photocatalytic water splitting using solar energy could be one of solutions of environmentally friendly and clean ways of hydrogen production. The basic material for the production of 'solar hydrogen' is water that is a renewable resource and on the earth it is enough and easy to access. However, there are many problems that must be solved before this technology become economically feasible. One of tasks is the development of efficient nano sized photocatalyst that works in the visible light. The other task is to increase the efficiency of solar energy utilization. The photochemical water splitting concept lies on the materials (photocatalysts) that can produce chemical reactions by absorption a quantum of light. The photocatalysts mainly are semiconductors that use photons to excite an electron from the valence band to the conduction band. The excited electrons by 'moving' to the conduction band and 'leaving' holes in the valence band cause reduction - oxidation reactions similarly to electrolysis (1). Electron is creating hydrogen by water molecule reduction (2) while holes form oxygen by oxidation (3) (Fig.1.). Energy of absorbed photons must be greater than band-gap energy of a semiconductor, although the band-gap should be higher than 1.23 eV, which is the energy needed for to split water. Due to the orbital configuration of the oxide semiconductor metal cations the energy levels are more positive than oxidation potential of water and, consequently, the band-gap become wider. For ultra violet light absorption the band-gap exceeds 3 eV, however for visible light absorption the band-gap should be around 2 – 2,2 eV (Navarro et. al., 2009). Photocatalysis reactions are considered as 'up-hill' reactions because back reactions can proceed very easy. There are several important requirements that must be fulfilled for successful photocatalytic water decomposition. Generated electrons and holes have to be separated and before recombination reaction takes place they should migrate to the surface. In this process the bulk properties of semiconductors are determinative; tough the reduction and oxidation reactions are dependent of a surface area and active sites on the surface of a photocatalyst. Summarizing, functional photocatalyst is a complex material that should provide the right band structure and suitable bulk and surface properties.

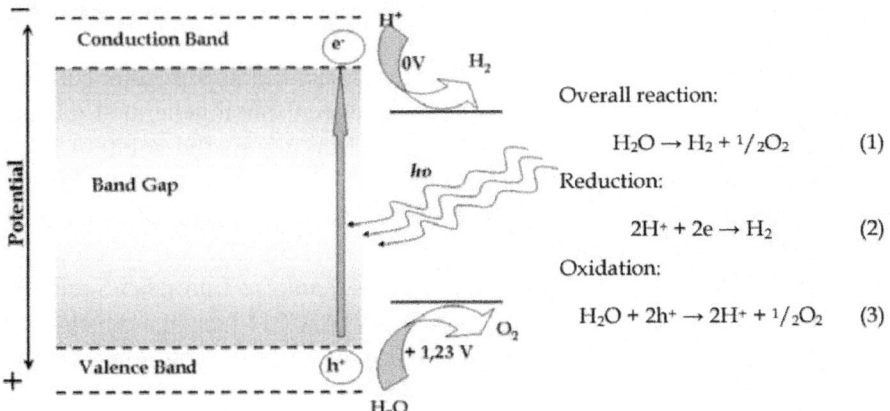

Overall reaction:

$$H_2O \rightarrow H_2 + 1/2O_2 \qquad (1)$$

Reduction:

$$2H^+ + 2e \rightarrow H_2 \qquad (2)$$

Oxidation:

$$H_2O + 2h^+ \rightarrow 2H^+ + 1/2O_2 \qquad (3)$$

Figure. 1: Schematic description of water decomposition on semiconductor photocatalysts

Table 1: Selection of photocatalysts developed for water splitting reaction under visible light (Navarro et al., 2008)

Photocatalysts	Band-gap energy (eV)	Cocatalyst	Sacrificial reagent	Reference
TiO_2-Cr-Sb	2,2	-	$AgNO_3$	Kato &Kudo, 2002
$SrTiO_3$ - Cr - Ta	2,3	Pt	CH_3OH	Ishii et al., 2004
$SrTiO_3$ - Cr - Sb	2,4	Pt	CH_3OH	Kato &Kudo, 2002
$La_2Ti_2O_7$ - Cr	1,8 - 2,3	Pt	CH_3OH	Hwang et al., 2005
TaON	2,5	Pt	CH_3OH	Hitoki et al., 2002
		-	$AgNO_3$	
$CaTaO_2N$	2,4	Pt	CH_3OH	Yamashita et al., 2004
		-	$AgNO_3$	
$SrTaO_2N$	2,1	Pt	CH_3OH	Yamashita et al., 2004
		-	$AgNO_3$	
$Sr_2Nb_2O_{7-x}N_x$	2,1	Pt	CH_3OH	Ji et al., 2005
		-	$AgNO_3$	
$BiVO_4$	2,4	-	CH_3OH	Kudo et al., 1998
		-	$AgNO_3$	
$(Ga_{1-x}Zn_x)(N_{1-x}O_x)$	2,4 - 2,8	Cr/Rh	-	Maeda et al., 2006
$(Zn_{1+x}Ge)(N_2O_x)$	2,7	RuO_2	-	Lee et al., 2007
CdS	2,4	-	S^{2-}/SO^{2-}_3	Navarro et al., 2008
CdS - CdO - ZnO	2,3	-	S^{2-}/SO^{2-}_3	Navarro et al., 2008
$Cd_{0.7}Zn_{0.3}S$	2,68	-	S^{2-}/SO^{2-}_3	del Valle et al., 2008
ZnS - Cu	2,5	-	SO^{2-}_3	Kudo and Sekizawa, 1999
$(AgIn)_xZn_2(1-x)S_2$	2,4	Pt	S^{2-}/SO^{2-}_3	Kudo et al., 2002
$(CuAg\ In)_xZn_2(1-x)S_2$	2,4	Ru	S^{2-}	Tsuji et al., 2005
$Na_{14}In_{17}Cu_3S_{35}$	2,0	-	-	Zheng et al., 2005

To develop suitable visible light photocatalysts the band-gap tuning by doping of transitionmetal cations has often been used. Unfortunately photocatalytic activity significantly decreases because doped cations inducing a formation of recombination centres between photogenerated electrons and holes and impend a migration of holes. Making a solid solution is other way of preparation of photocatalysts. There the different width band-gap semiconductor ratio can change energy levels in the composite material in total. Table 1 shows several photocatalyst materials that are developed and investigated lately. Our group's investigations thanks to European Social Fund project 2009/0202/1DP/ 1.1.1.2.0/09/APIA/VIAA/141 lies on a developing of a new photocatalysts where the method of doping and solid solution preparation is combined.

Bio-Hydrogen

A number of technologies for biological H2 production are available, but they are not established for significant amount of hydrogen production yet. Methods for engineering and manufacturing these systems have not been fully evaluated. Biological processes of hydrogen recovery and collection from organic resources such as municipal wastewater and sludge facilitate recycling of sewage are environmentally benign and necessary for alternative independent household support. Nowadays, many institutions and universities worldwide are involved in the research of hydrogen production by microorganisms and algae (Das & Veziroglu, 2008, Kotay Meher & Das, 2008, Donohue & Cogdell, 2006). Microorganisms are capable of producing H2 via two main pathways: fermentation and photosynthesis. The processes of bio-hydrogen production include: 1. direct biophotolysis by green algae – the photosynthetic production of hydrogen by splitting water into molecular hydrogen and oxygen using sunlight under specific conditions; 2. indirect biophotolysis by cyanobacteria with specialized cells (heterocysts) that perform nitrogen fixation and contain enzymes (nitrogenase and hydrogenase) directly involved in hydrogen metabolism and synthesis of molecular H_2; 3. photo-fermentation by purple non-sulfur bacteria that evolve molecular H_2 catalyzed by nitrogenase enzyme under nitrogen-deficient conditions using the energy of light and organic acids; 4. dark-fermentation by anaerobic bacteria grown in the dark on carbohydrate-rich substrates (Holladay et. Al., 2009, Das & Veziroglu, 2001, Levin, 2004). Bacterial hydrogen production by fermentation of carbohydrate-containing substrates (glucose, cellulose, starch and organic waste materials) is frequently preferred to photolysis, because it does not rely on the availability of light sources. In the fermentation of glucose by enterobacteria, e.g. Escherichia coli, one of the pyruvate oxidation products, alongside with acetyl-CoA, is formate, which is produced by pyruvate formate lyase and is the sole source of hydrogen in these bacteria. The formate is split into CO_2 and H_2 by formate hydrogen lyase (FHL) complex, which comprises seven proteins, six of them being encoded hyc operon. Five hyc operon encoded proteins are membrane-embedded electron transporters. The hycE protein is one of the three E.coli NiFe hydrogenases. The hycE and FDH-H components of FHL complex are soluble peri¬plasmic proteins. The hydrogen evolved from FHL is consumed by E.coli uptake hydrogenases Hyd-1 and Hyd-2. In contrast to enterobacteria, strictly anaerobic fermenters, e.g. Clostridia, use a reduced ferredoxin (required to oxidize pyruvate to acetyl-CoA) for H2 production by the hydrogenase that generates ferredoxin in the oxidized form and releases electrons as molecular hydrogen (Nath & Das, 2004, Hallenbeck & Benemann, 2002, Maeda et. al., 2008). Glucose fermentation by enteric bacteria yields the maximum of 2 mol H2/ mol glucose (Wang & Wan, 2009). To enhance the hydrogen production and utilize the substrate in full measure for complete conversion, the synergy of biological processes (two-stage/hybrid ones) should be applied. Gaseous hydrogen is formed in liquid media during bacterial fermentation. Hydrogen gas hardly dissolves in aquatic solutions, but special methods are required to discharge hydrogen into atmosphere and in so doing to escape oversaturation

(Mandelis & Christofides, 1993). In the atmosphere the parameters of hydrogen gas are measured by classical volumetric, mass-spectrometric and chromatography methods, or using chemical gas sensors. Wilkins (Wilkins et. al., 1974) described a method for measuring gas production by microorganisms using a platinum electrode and a reference Calomel ($Hg–Hg_2Cl_2$) electrode. To measure hydrogen gas concentration in liquid a hydrogen electrode is usually used (Pt or another noble metal − gold, rhodium, palladium, etc.). The hydrogen H+ ions and the molecular hydrogen H2 set the equilibrium potential in compliance with the reaction: H2 ⇔ 2H+ + 2e−. This reaction proceeds very fast, so in its course the equilibrium state remains stable; in electrochemistry this electrode is adopted as zero reference (with zero potential). In microbiology, to measure dissolved oxygen and hydrogen gases the micro-respiration Clark electrodes are used (Ghirardi e.al., 1994). In a Clark's electrode the cathode polarized versus an internal Ag/AgCl anode is placed behind an electrically insulating silicone rubber membrane, which is extremely permeable to oxygen. The flow of electrons from the anode to the oxygen-reducing cathode reflects linearly the partial oxygen pressure around the sensor tip and is in the pA range. The same principle holds for a hydrogen Clark-type sensor: the environmental hydrogen is driven by the external partial pressure and penetrates through the sensor tip membrane to be oxidized at the platinum anode surface. Flynn et al. [Flynn et. al., 2002) used chemochromic sensors for screening in order to identify positive (i.e. hydrogen-producing) algal colonies. A chemochromic sensor film, which is normally transparent, turns blue in the presence of hydrogen gas. Hydrogen gas is produced during the bacterial fermentation process in anaerobic conditions. In practice, hydrogen is collected in the gaseous state, since dissolved hydrogen tends to become gas. To optimize the hydrogen collection methods it is necessary to study properly the hydrogen production kinetics in liquid phase during the fermentation process. The experimental test system for bacterial hydrogen production and micro-sensors were used to determine the hydrogen gas concentrations in liquid; the mass-spectrometry method was employed for measurements in the hydrogen-containing head space. In our experiments Escherichia coli strain MSCL 332 (i.e. from Microbial Strain Collection of Latvia) was grown on Luria-Bertani (LB) nutrient agar plates (5 g/l yeast extract, 10 g/l tryptone, 10 g/l sodium chloride, 15 g/l Bacto agar). E.coli from single colonies on the agar plates were inoculated in 2 x 150 ml flasks containing LB liquid medium. The flasks were aerobically shaken at 37°C for 12 hours at 120 rpm using a multi-shaker PSU-20. The bacteria cell number in the overnight culture was titrated at 10-6 dilution. The amount of bacterial cell protein was calculated assuming that one E.coli cell contains 1.54 × 10–13 g of protein. The overnight culture in LB liquid medium was mixed (1:1) with phosphate buffer saline (PBS) pH 7.3 (0.8

g/l NaCl, 0.2 g/l KCl, 1.43 g/l Na_2HPO_4, 0.2 g/l KH_2PO_4) in a vessel sterilized for measurements. The PBS contained a complex trace element medium pH 6.5 (0.039g/l $Fe(NH4)_2 \cdot SO4 \cdot 6H_2O$, 0.172 mg/l $Na2SeO3$, 0.02 mg/l $NiCl_2$, 0.4 mg/l $(NH_4)6Mo7O_24$). Glucose (3.3mM, final concentration, sterilized through membrane 0.2μm filter) was added at the start of experiment. The hydrogen and oxygen concentrations were measured with Clark-type microsensors in the sample liquid phase. The microsensors were connected with the signal amplifier – a pico-ammeter and an A/D current converter connected to PC using USB port. Before the measurements, both oxygen and hydrogen microsensors were calibrated in a liquid culture medium (similar to the sample measured by 15 min bubbling Ar) for zero concentrations and hydrogen gas and clean air for 100% dissolved H_2 and O_2 concentrations (730 and 760 μmol/l, accordingly). The system is able to work independently when measurements are made in one sample. If there are several samples at a time, it is necessary to move microsensors manually and to sterilize the sensor tip using 96% ethanol, 0.1 M NaOH and distilled water every time when it is taken out from the sample. The gas from the headspace of liquid bacterial culture in the test vessel was taken to an RGAPro-100 mass-spectrometer to analyze its components. The gas from an argon balloon through a diffuser was let in the test vessel with bacteria culture to sustain the anaerobic environment (see Fig.2) and put in a water bath to maintain a temperature of 37±2 °C.

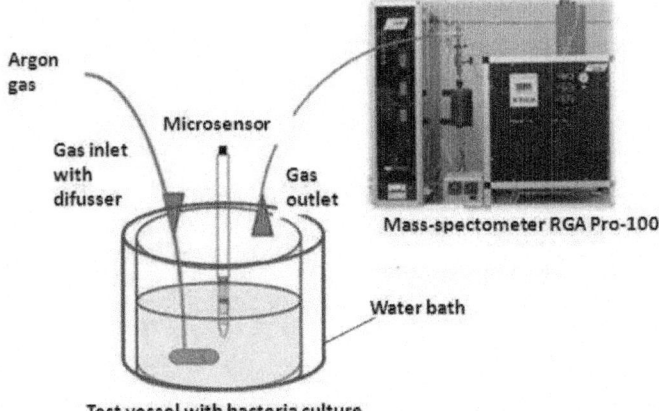

Figure. 2: Experimental test system for H2 concentration measurements with mass-spectrometer

Argon gas was bubbled through the liquid for 15 min (flow 13 l/h) and gas measurements were made with 30 min and an hour intervals. The total time of measurements was six hours. The gas volume taken for each analysis was

20 cm3. During the mass-spectrometric analysis, simultaneous measurements with a hydrogen microsensor were taken in order to make unbiased comparison of mass-spectrometric and hydrogen microsensor analyses. The concentration of dissolved hydrogen gas was measured with a microsensor; the massspectrometric analyses were made for the atmospheric composition in the headspace of the sample bacteria culture and nutrients.

The experimental results of microsensoric measurements were analyzed using Sensor Trace Basic and MicOX (A/S Unisense) programs, and processed by Microsoft Office Excel 2007. The mass-spectrometric data were analyzed by RGA 3.0 Software for SR Residual Gas Analyzers program. Summarizing experiments: The hydrogen output was measured for seven hours after the beginning of fermentation process; increase in the hydrogen concentration was observed starting from the second hour after adding glucose. The constancy of oxygen concentration in the measurements evidences that the system had reliable anaerobic conditions (Fig. 3.). As is seen from this figure, the concentration of dissolved hydrogen stopped increasing after 5–10 h as glucose exhausted. The maximum rate of hydrogen formation in the test system was 612 μmol/ l/20 min or 1.4 mmol [2.4 mg] /l per h for 43 mg protein mass (i.e. 32.6 μmol/mg protein mass). The maximum concentration of dissolved hydrogen (2481 μmol/l or 2.5 mmol/l) is reached in the fourth hour of fermentation as is seen in Fig. 2. This concentration at least three times exceeds the maximum thermodynamically allowed concentration of dissolved hydrogen in water.

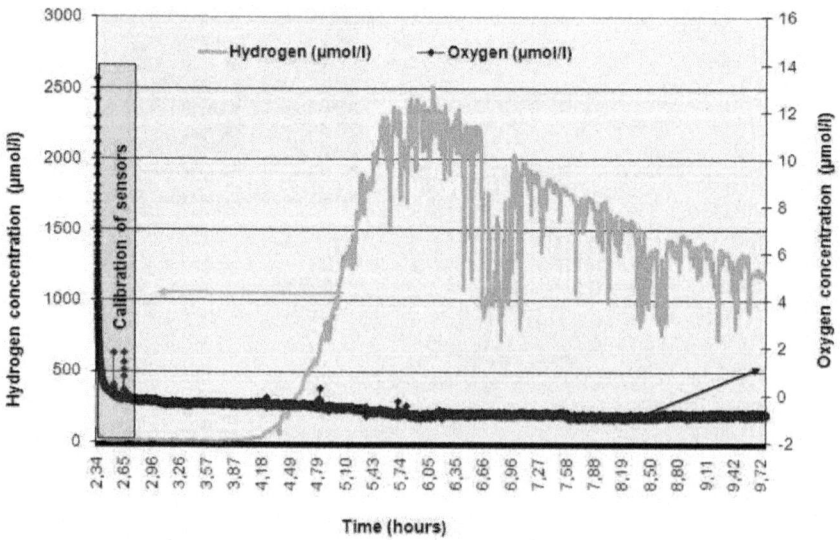

Figure. 3: Microsensoric fermentation measurements on the sample with E.coli

To demonstrate that the hydrogen production began only after glucose had been added various glucose concentrations were tested. The correlation between the glucose concentration and the hydrogen output is shown in Fig. 4. To calculate the partial pressure of hydrogen in the headspace Henry's law was used. The calculations were done using the measured dissolved hydrogen concentrations in the test system (2481 μmol/l after 4 h fermentation). At room temperature and normal atmospheric pressure the Henry constant is k_H = 1282,05 (l/atm·mol), therefore pH = 1282.05·cH (atm) and in our case pH = 1282.05·2481·10–6 = 3.18 atm, which obviously does not fit the experimental results obtained in the mass-spectrometric analysis.

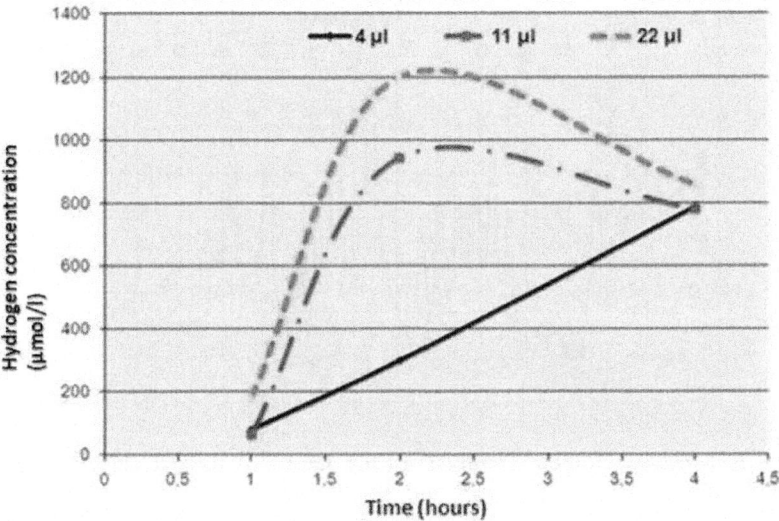

Figure. 4: Time dependence of hydrogen output at different glucose concentrations

To calculate the partial pressure of hydrogen in the headspace Henry's law was used. The calculations were done using the measured dissolved hydrogen concentrations in the test system (2481 μmol/l after 4 h fermentation). At room temperature and normal atmospheric pressure the Henry constant is kH = 1282,05 (l/atm·mol), therefore pH = 1282,05·cH (atm) and in our case pH = 1282,05·2481·10^{-6} = 3.18 atm, which obviously does not fit the experimental results obtained in the mass-spectrometric analysis. As is seen from Fig. 5, the partial pressure of hydrogen of 3.18 atm above the test system's headspace is inadequate to the concentrations determined by mass-spectrometric analysis – only 6·10^{-3} atm or 0.6% vol. The mass-spectrometric analysis has revealed the presence of different volatile substances – the end products of bacterial formation: acetate, carbon dioxide, ethanol, acteone, and hydrogen gas. In three separate measurements (without Ar bubbling and liquid mixing, with argon bubbling only, and with argon bubbling & liquid mixing) the massspectrometric analysis showed a hydrogen concentration increase from 0% to 0.4% after 6 h fermentation only in one measurement when no argon bubbling and mixing was applied. Such an increase in the hydrogen concentration (only 3.6·10–3 atm partial pressure) is not convincing as compared with the concentrations measured in liquid phase. A reason for that could be the limited hydrogen migration from liquid to gaseous phase (dissolved hydrogen oversaturation); besides, the mass-spectrometry measurement method was not perfect (it is to be optimized in the future experiments).

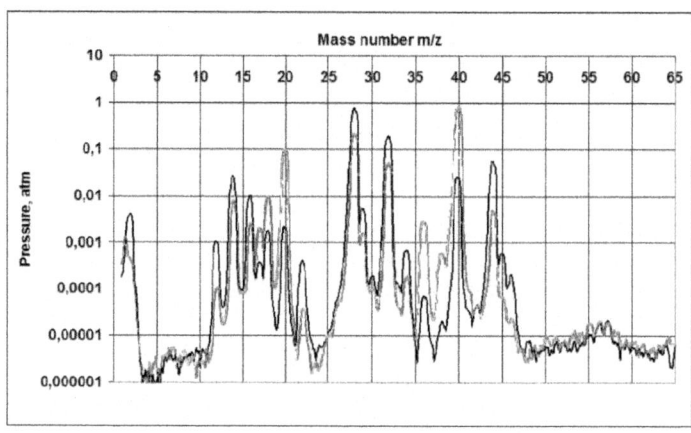

Figure. 5: Mass-spectometric analysis of the sample with bacteria E.coli before fermentation (light gray curve) and after 6h (black curve)

The mass-spectrometric analysis has revealed the presence of different volatile substances – the end products of bacterial formation: acetate, carbon dioxide, ethanol, acteone, and hydrogen gas. In three separate measurements (without Ar bubbling and liquid mixing, with argon bubbling only, and with argon bubbling & liquid mixing) the massspectrometric analysis showed a hydrogen concentration increase from 0% to 0.4% after 6 h fermentation only in one measurement when no argon bubbling and mixing was applied. Such an increase in the hydrogen concentration (only 3.6·10–3 atm partial pressure) is not convincing as compared with the concentrations measured in liquid phase. A reason for that could be the limited hydrogen migration from liquid to gaseous phase (dissolved hydrogen oversaturation); besides, the mass-spectrometry measurement method was not perfect (it is to be optimized in the future experiments). Concentrations of dissolved gas are higher than theoretically possible in the anaerobic processes where gases are formed in the liquid phase and tend to reach the gas phase. Such over-saturation could be associated with biological processes: lower pH due to the formation of gases (e.g. CO_2, H_2S) in anaerobic processes; besides, a negative thermodynamic effect is caused by inhibitor gases – e.g. H_2, since the hydrogen synthesising enzymes are sensitive to H_2 concentrations and are subject to the end-product inhibition. As concentrations of hydrogen increase, its synthesis rate decreases: the evolved H_2 is consumed by E.coli uptake hydrogenases Hyd-1 and Hyd-2. As mentioned, the hydrogen synthesising enzymes are sensitive to the end product – the hydrogen gas concentration. As this concentration increases the synthesis rate decreases, with formation of mixed-acid hydrogen-containing fermentation products (ethanol, acetate, butane). To enhance the hydrogen gas output, the bacterial metabolism has to be switched from alchocol and acid formation to volatile fatty acids. This can be facilitated by the system's optimization, for example, using continuous bubbling with inert gas to reduce the partial pressure of hydrogen in the liquid phase thereby increasing its formation in the gaseous phase. To enhance hydrogen formation, very delicate bubbling/mixing procedures should be applied, since in our measurements, with intense bubbling and mixing by a magnetic stirrer, the least hydrogen increase in the headspace was observed.

Conclusions

Escherichia coli wild-type strain MSCL 332 from Microbial Strain Collection of Latvia was successfully used. The hydrogen concentration in the headspace was analyzed by mass spectrometry. Due to incompleteness of the test system the hydrogen concentration in the gaseous phase was not detected, which therefore remains to be a subject of future activities. Also, the gas measuring system should be improved based on the mass-spectrometry. To enhance the transfer of dissolved hydrogen into the headspace, continuous gas bubbling and mixing are required. The dissolved hydrogen and oxygen concentrations in liquid phase during the fermentation process were measured using microsensors. The maximum of dissolved hydrogen concentration (2481 $\mu mol/l$) was reached by the fourth hour of fermentation, which is markedly higher than predicted by Henry's law (730 $\mu mol/l$). Also, alternative methods should be employed for the hydrogen collection directly from the nutritional broth thus making it possible to develop commercial hydrogen production.

HYDROGEN STORAGE

Hydrogen storage is clearly one of the key challenges in developing hydrogen economy and hydrogen storage for vehicle applications is one of the most important challenges. Hydrogen storage basically implies the reduction of the huge volume of the hydrogen gas because a 1kg of hydrogen at ambient temperature and atmospheric pressure takes a volume of 11 m3. To increase a hydrogen density the compressing of hydrogen or decreasing of temperature below critical must be performed. Available technologies permit to store hydrogen directly by modifying its physical state in gaseous or liquid form in pressurized or in cryogenic tanks. Storage by absorption as chemical compounds or by adsorption on carbon materials have definite advantages from the safety perspective such that some form of conversion or energy input is required to release the hydrogen for use. A great deal of effort has been made on new hydrogen-storage systems, including metal, chemical or complex hydrides and carbon nanostructures. Hydrogen interaction with other elements depends from material, it occurs as anion (H-) or cation (H$^+$) in ionic compounds, it participates with its electron to form covalent bonds, and it can even behave like a metal and form alloys at ambient temperature. The hydrogen molecule H$_2$ can be found in various forms depending on the temperature and the pressure which are shown in the phase diagram (Fig. 5.). The phase diagram shows that the liquid hydrogen with a density of 70.8 kg·m^{-3} at -253°C only exists between the solid line and the line from the triple point at 21.2K and the critical point at 32K. At low temperature hydrogen is a solid with a density of 70.6 kg·m^{-3} at -262°C and is a gas at higher temperatures with a density of 0.089886 kg·m^{-3} at 0°C and a pressure of 1 bar (Zuttel, 2004).

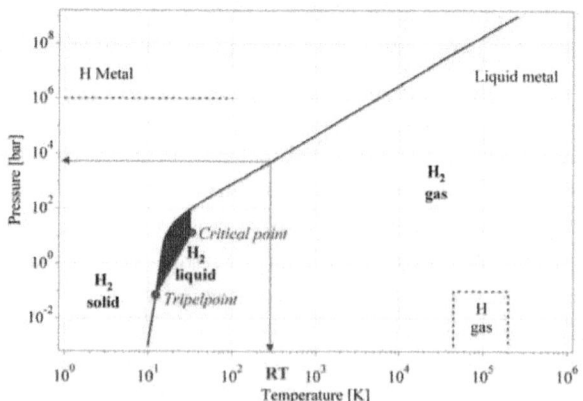

Figure. 5: Primitive phase diagram for hydrogen (Zuttel, 2004)

Metal Hydrides

Hydrogen is a highly reactive element and forms hydrides and solid solutions with thousands of metals and alloys that will release hydrogen at elevated temperatures. Metal hydrides are arranged of metal atoms that form a host lattice, and hydrogen atoms that are installed in the interstitial sites. The absorption process of hydrogen in the metal can be described using one-dimensional Lennard-Jones potential scheme for hydrogen molecule and 2 hydrogen atoms (Fig. 6) (Zuttel, 2004). The potential of molecule and both atoms are separated by the heat of dissociation energy ED=435.99 kJ/mol in some distance from the metal surface. Approximately one hydrogen molecule radius from the metal surface the hydrogen molecule interacts with metal surface and due to the Van der Waals forces physisorption happens that is illustrated as the flat minimum in the H2+M curve. Closer to the surface the hydrogen has to overcome an activation barrier for dissociation and formation of the hydrogen metal bond and hydrogen becomes chemisorbed that is showed as deep minimum of the 2H+M curve. If the both curves cross above zero energy level for the chemisorption activation energy is needed and the kinetics of adsorption is getting more slowly. Depending on the surface elements the activation barrier height can be changed and chemisorbed hydrogen atoms may have a high surface mobility, interact with each other and form surface phases. Furthermore the chemisorbed hydrogen atom can jump in the subsurface layer and finally diffuse on the interstitial sites through the host metal lattice (Schlapbach, 1992). Hydride formation from gaseous phase can be described by pressurecomposition isotherms (Fig. 7.). The host metal dissolves some hydrogen and solid solution phase or α-phase is formed. The metal lattice expands proportionally to the hydrogen concentration by approximately 2 – 3Å3 per hydrogen atom.

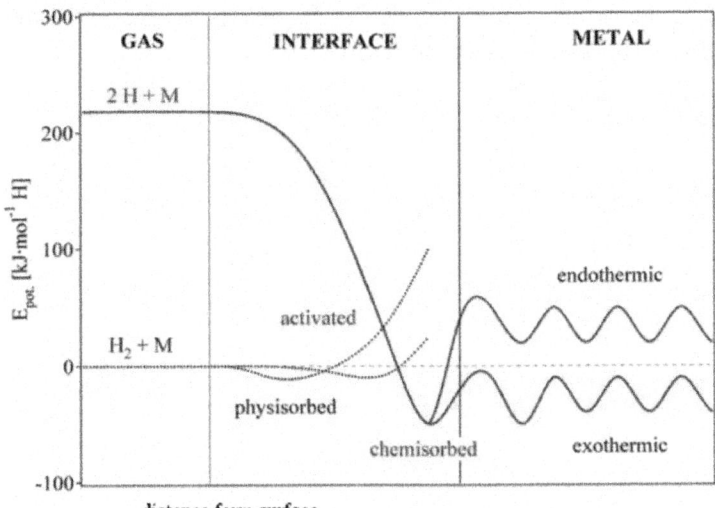

Figure. 6: Potential energy curves for activated and non-activated chemisorption of hydrogen on metal surface (Zuttel, 2003)

If the hydrogen pressure and concentration Hydrogen/Metal exceeds ratio 0,1 a H–H interaction becomes significant and the β-phase nucleates and grows. While the α-phase and α-phase coexists, the isotherms show a flat plateau, the length of which determines how much H_2 can be reversibly stored. When the α-phase completely transfers to the β-phase, the H_2 pressure rises steeply with the concentration. Further enlargements of hydrogen pressure can cause formation of other plateaux and hydride phases. The two-phase region ends in a critical point TC, above which the transition from α- to β-phase, is continuous (Fig. 7.). The hydrogen concentration in the hydride phase is often found to be H/M=1. The volume expansion between the coexisting α- and the β-phase corresponds in many cases to 10–20% of the metal lattice. Therefore, at the phase boundary a large amount of stress is built up and often leads to a decrepitation of brittle host metals such as intermetallic compounds. The final hydride is a powder with a typical particle size of 10–100 μm (Zuttel, 2004; Schlapbach, 1998; Schlapbach, 1992). The plateau pressure strongly depends on temperature that is related to the changes of enthalpy and entropy (4). Solving the Van't Hoff equation and from the gained slope where the pressure is a function of temperature, the heat of hydride formation can be evaluated.

$$\ln\left(\frac{p_{eq}}{p_{eq}^0}\right) = \frac{\Delta H}{R} \cdot \frac{1}{T} \cdot \frac{\Delta S}{R},$$

(4)

where p_{eq} is plateau pressure at equilibrium state, p^0_{eq} stands for the plateau pressure at standard conditions, R is the universal gas constant, T is the temperature at p_{eq} and Δ_H is the enthalpy change.

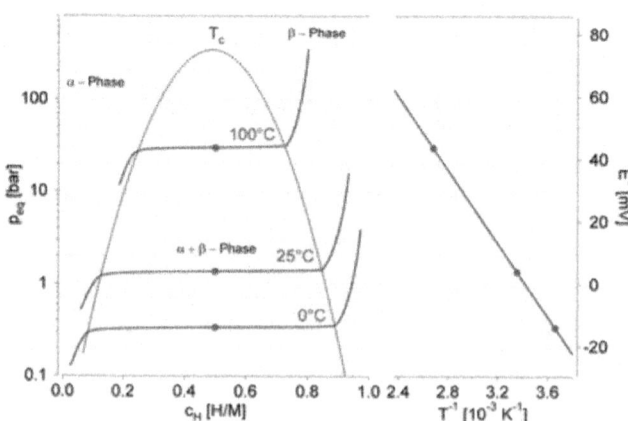

Figure. 7: Pressure composition isotherms for typical intermetallic compound is shown on the left side. The construction of the Van t Hoff plot is shown on the right hand side (Zuttel, 2004)

The enthalpy term characterizes the stability of the metal hydrogen bond. A hydride forming entropy changes leads to great heat generation during the hydrogen absorption. The same heat has to be provided to the metal hydride to desorb the hydrogen. Hydrogen absorption is very much involved with a phase transition. Pressure does not increase with the amount of absorbed hydrogen as long as the phase transition takes place therefore a, metal hydrides can absorb large amounts of hydrogen at a constant pressure. Hydrogen sorption characteristics can be changed by partial substitution of the hydride forming elements, thereby it is possible to form some metal hydrides that works at ambient temperature and close to atmospheric pressure (Schlapbach, 1992) A particular interest is about intermetallic hydrides because the variations of the elements allow modification of the properties of the hydrides (Table 2). The element A usually is a rare earth or an alkaline earth metal and tends to form a stable hydride or has a high affinity to hydrogen. The B element is often a transition metal and forms unstable hydrides or has a low affinity to hydrogen. The combination of the both elements A and B gives as alloys suitable for practical applications.

AB5 Intermetallic Compounds

The on of the classical alloys is a combination of the La and Ni, where La individually forms LaH2 at 25oC, $p=3 \cdot 10^{-29}$ atm, $\Delta H_f = -208$ kJ/mol H_2, with Ni that alone forms NiH at 25oC, p=3400atm, $\Delta H_f = -8,8$ kJ/mol H_2. Although the alloy LaNi5 forms a hydride at 25oC, p=1,6 atm, $\Delta H_f = -30,8$ kJ/mol H_2, that can be considered as a interpolation between boundaries of elemental hydride forming activities (Sandrock, 1999).

Table 2: The most important hydride forming intermetallic compounds (Zuttel, 2004)

Intermetallic compound	Prototype	Hydrides	Structure
AB_5	$LaNi_5$	$LaNi_5H_6$	Hexagonal
AB_2	ZrV_2, $ZrMn_2$, $TiMn_2$	$ZrV_2H_{3,5}$	Hexagonal or cubic
AB_3	$CeNi_3$, YFe_3	$CeNi_3H_4$	Hexagonal
A_2B_7	Y_2Ni_7, Th_2Fe_7	$Y_2Ni_7H_3$	Hexagonal
A_6B_{23}	Y_6F_{23}, Ho_6Fe_{23}	$Ho_6Fe_{23}H_{12}$	Cubic
AB	TiFe, ZrNi	$TiFeH_2$	Cubic
A_2B	Mg_2Ni, Ti_2Ni	Mg_2NiH_4	Cubic

LaNi$_5$ has a CaCu$_5$-type, structure containing three octahedral and three tetragonal sites per elemental cell unit. The alloy forms at least two hydrides: α phase - LaNi$_5$H0.3 - with low hydrogen content and β phase - LaNi$_5$H$_{5.5}$ - with high hydrogen–content. Both hydrides differ significantly in the specific lattice volume; β phase has a 25% larger lattice expansion as α phase that causes a crumbling of the alloy particles on hydriding - dehydriding cycles. The alloy of rare earth elements in composition with nickel was used for the first time by Lindholm (Lindholm, 1996) as the electrode in the fuel cells in 1966 and then by Dilworth and Wunderlin (Dilworth & Wunderlin, 1968) in 1968. The term MmNi$_5$ was used where Mm (mischmetal)-represents a natural mixture of rare earth elements, mostly consisting of Ce (30-52 wt%), La (13 to 25 wt%), Nd, Pr and Sm (13-57 wt%) where an amount and elements of additives depends on the place of origin.

Though, the element substitutions in the AB$_5$-type alloys have been made also artificially to get better alloys for practical use. La can be replaced with Mm, Ce, Pr, Nd, Zr, Hf and Ni can be exchanged with Al, Mn, Si, Zn, Cr, Fe, Cu, Co; thereby altering the hydrogen storage capacity, the stability of the hydride phase or the corrosion resistance. For example, a partial replacement of the A and B components significantly changes macrostructure of an alloy and other properties (Table 3).

The stoichiometry of an alloy influences its durability in the long-term hydriding - dehydriding cycles, and typical commercial AB5-type alloys consists of at least 5-6 different metals, for example La0.64Ce0.36Nd0.46Ni 0.95Cr0.19Mn0.41Co0.15 (Bernd, 1992). Commonly metal hydrides are very effective for storing large amounts of hydrogen in a safe and compact way but they are mostly heavy or working in the not suitable conditions for vehicle applications. The transition hydrides are reversible and works around ambient temperature and atmospheric pressure but the gravimetric hydrogen density is limited to to <3 mass%. It is still a challenge to explore the properties of the lightweight metal hydrides or investigate new hydride composite materials.

Our laboratory at the Institute of Solid State Physics works on the classical AB$_5$ hydride type material LaNi5 modification. As hydrogen absorber, LaNi$_5$ has been one of the most investigated intermetallic compounds during the last decades. Despite of its high hydrogen capacity of a one hydrogen atom to the each metal atom and easy activation, the binary compound is not suitable for applications due to its high plateau pressure and short lifecycle (Bittner & Badcock, 1983). However, modification of the physical and chemical properties of LaNi5 can be achieved by substituting lanthanum atom with a rare earth metal (e.g. Ce, Pr, Nd, Er) or nickel with a transition metal (Al, Mn, Co, Cr). Specific sample preparation methods, such as melt–spinning, sputtering and mechanical milling have been used to improve the hydrogenation kinetics of intermetallic compounds.

Table 3: Effect on composition on properties of AB5-type alloys

Composition	Elements and their role
Substitutions of A in AB5 $La_{1-y}M_yB_5$	Zr, Ce, Pr, Nd decrease the unit cell volume, improve activation, high-rate discharge and cycle life, but increase the self-discharge due to a higher dissociation pressure of the metal hydride. The use of Mm instead of La reduces the alloy costs.
Substitutions of B in AB5 $A(Ni_{1-z}M_z)_5$	A= La, Mm; M= Co, Cu, Fe, Mn, Al; $0<z<0.24$ Ni $(1-z)>2.2$ is indispensable to prevent the decrease of the amount of absorbed hydrogen and the electrode capacity Co decreases the volume expansion upon hydriding, retards an increase of the internal cell pressure, decreases the corrosion rate and improves the cycle life of the electrode, especially at elevated temperature (40 C), but increases the alloy costs **Substitution of Co by Fe** allows cost reduction without affecting cell performance, decreases decrepitation of alloy during hydriding. **Al.** increases hydride formation energy, prolongs cyclic life. **Mn** decreases equilibrium pressure without decreasing the amount of stored hydrogen. **V** increases the lattice volume and enhances the hydrogen diffusion. **Cu** increases high rate discharge performance.
Special additions to B in AB5 $A(Ni,M)_{5-x}B_x$	A= La, Mm; M = Co, Cu, Fe, Mn, Al; B= Al, Si, Sn, Ge, In, Tl, **Al, Si, Sn and Ge** – minimise corrosion of the hydride electrode. Ge-substituted alloys exhibit facilitated kinetics of hydrogen absorption/desorption in comparison with Sn-containing alloys. **In, Tl, Ga** increase overvoltage of hydrogen evolution (prevent generation of gaseous hydrogen).
Nonstoichiometric alloys $AB_{5\pm x}$	A= La, Mm; B=(Ni,Mn,Al,Co,V,Cu) Additional Ni forms separate finely dispersed phase. In **MmB5.12** the Ni3Al-type second phase with high electrocatalytic activity is formed. **Alloys poor in Mm** are destabilised and the attractive interaction between the dissolved hydrogen atoms increases. Second phase (Ce2Ni7), which forms very stable hydride is present in MmB4.88. When (5-x)<4.8, the hydrogen gas evolution during overcharge decreases.
Addition of alloys with increased catalytic activity $AB_5 + DE_3$	D= Mo, W, Ir; E= Ni, Co DE3 is a catalyst for hydrogen sorption-desorptionreactions.
Mixture of two alloys $A^1B^1_5 + A^2B^2_5$	Mixing of two alloys characterised by various hydrogen equilibrium absorption pressures increases the electrode performance.

The resulting alloys exhibit particular structural characteristics as nano-crystalline grains with a high density of grain boundaries and lack a long-range order (similar to an amorphous state). These microstructures currently provide fast hydrogenation kinetics and better lifecycle behaviour. Mechanical milling has become a popular technique because of its simplicity, relative inexpensive equipment and applicability to most intermetallic compounds. This technique has been used for several hydrogen storage alloys there was observed a good improvement in hydrogen activation and kinetics (Ares et. al., 2004, Zaluska et. al., 2001). The diffusion of active species on the surface may play an important role in reactions on multifunctional catalyst transport phenomena. Especially, migration of hydrogen atoms from a metal to an oxide or carbon surface that by itself has no activity for dissociate hydrogen adsorption is important. It is well known that noble metals, like Pt and Pd, can adsorb and diffuse hydrogen in reactive forms over relatively large distances. This property named as spill-over effect, is widely exploited in catalysis (Scarano et. al., 2006). The spill-over of hydrogen involves a transfer of electrons to acceptors within the support; this process modifies the chemical nature of the support and can also activate a previously inactive material and/or induce subsequent hydrogen physisorption (Roland et. al., 1997). Dissociation of hydrogen molecule on a metal and subsequent spill-over of atomic hydrogen to its support is highly dependent upon the chemical bridges formed at the interface. Hydrogen spill-over can be assessed in a number of ways, but perhaps the most common is simple calculation of the hydrogen to metal ratio, either the surface metal or total metal content. When spill-over occurs, the relation $H:M_{surface}$ will typically exceed unity. In the case of materials that form hydrides, this relation will exceed the stoichiometric ratio of the hydride.

An AB_5 type alloy with a trade name 7-10 produced by the company Metal Rare Earth Limited of China was chosen for experiments. To study this material the measurements were carried out using Scanning Electron Microscope (SEM) of Carl Zeiss brand, model EVO 50 XVP located at the Institute of Solid State Physics. The SEM images were taken in secondary electrons (SE); the acceleration voltage was equal to 30 kV, and the emission current was between 0,5pA and 500nA. The energy dispersive detector for X-rays (EDX) was used for composite determination in the alloy 7-10. Structural properties of the samples were studied by X-ray Diffractometer System X-STOE Theta/theta, using K Cu radiation but the diffraction patterns were analyzed by appropriate software of STOE system at the DTU RISOE National Laboratory (RNL). Tungsten – carbide crucibles with 2 balls from the same material and a high energetic ball mill Retsch® MM200 was used for grinding the raw material 7-10 for 30 minutes at frequency 25 Hz per min as well as for preparing a composite with a glass. The composite consists of 3,7 weight parts of an alloy 7-10 and of a 1 part of the Pyrex glass.

A thermogravimetric technique has been used to study the hydrogen sorption on prepared samples. The measurements were performed by equipment based on the Sartorius high pressure balance (HPB) combined with pressure, temperature and gas supplying systems. The sample is placed in the steel container that can be sealed to provide a vacuum or gas atmosphere and the pressure and temperature ranges of 10^{-3} - 30 bar and from room temperature to 300°C, respectively. The studied sample 340±2 mg was initially degassed under vacuum down to 10^{-3} bars at the ambient temperature and flushed with helium gas. That was prolonged by heating up the system until 260°C with followed vacuuming, activating at 10^{-3} bar hydrogen pressure and cooling down. Subsequently the previous actions were repeated two times and finished with the vacuuming of system. Afterwards, stepwise changes of the pressure inside the measuring device were applied. At a constant temperature the increase of the sample weight, as a function of time was measured for each pressure step. The same weight and treatment procedures was chosen for a composite material, though, for calculations one have to remember that in composite an amount of hydride forming alloy is less than that for pure alloy. SEM and EDX results (Fig. 8, Table 4) shows that the composition of the alloy 7-10 corresponds to formula $A_{0.96}B_{5.04}$ (A=La, Ce, Nd, Pr; B=Ni, Co, Mn, Al, Cr) that is close to AB5 stoichiometry and the molecular mass of this sample becomes 435,74 g/mol. For determination of the molecular mass of the alloy and further calculations there was assumed that after the treatment a metalhydride material contains a diminutive amount of oxygen. The XRD pattern shows that alloy 7-10 belongs to a single phase LaNi5 hexagonal CaCu5- type structure in the space group P6/mmm.

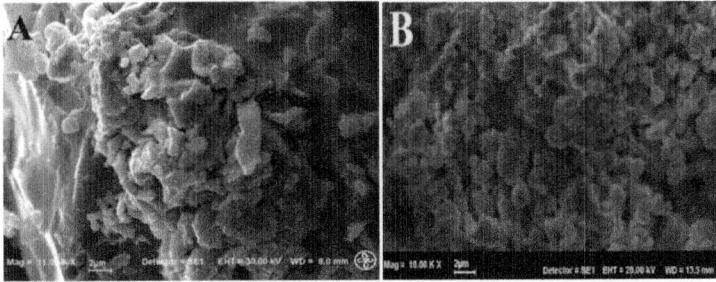

Figure. 8: SEM images of A – an alloy 7-10, grain size 50 μm and smaller, magnification 11 000; B – ball milled alloy 7-10, grain size 2 μm and smaller but agglomerated in clusters, magnification 10 000

Analysing a treated data it was figured out that a raw hydride material 7-10 absorbing more hydrogen than a classical $LaNi_5$ by itself. The HPB plots clearly shows that a pure $LaNi_5$ absorbing and desorbing hydrogen when the pressure is changed. There just a 0.26 w% of hydrogen is remaining in the pure $LaNi_5$ sample at the one atmosphere. Though, the data plot of a sample 7-10 displays opposite tendency – the absorbed hydrogen amount just slightly decreasing after lowering the pressure to the one atmosphere (Fig. 9.).

Table 4: Quantitative EDX data for the total surface of the sample 7-10

Element	Weight %	Atomic %
La	17,51	8,24
Ce	9,88	4,61
Nd	3,07	1,39
Pr	0,98	0,45
Ni	52,89	58,89
Co	6,23	6,91
Mn	5,08	6,05
Al	1,93	4,67
Cr	0,39	0,49
O	2,03	8,30

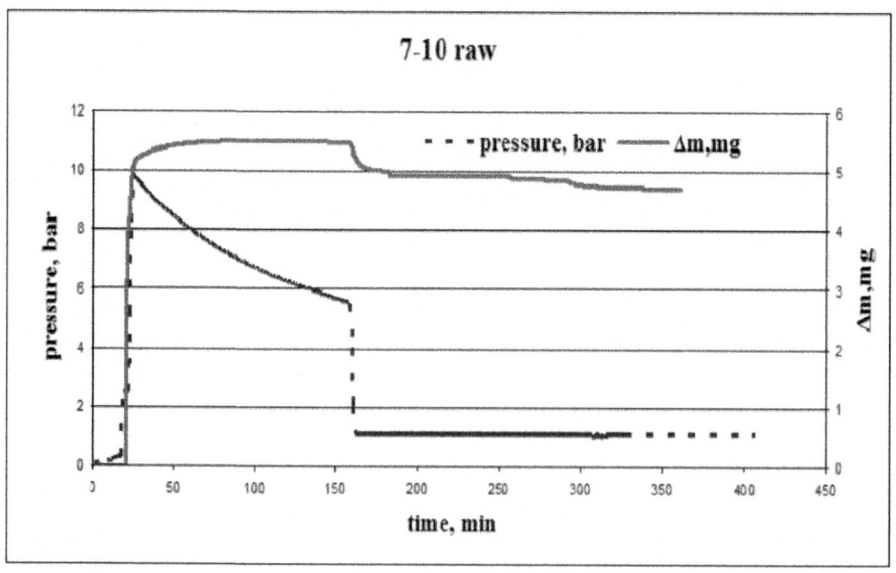

Figure. 9: HPB data plot of the raw sample 7-10 at the room temperature

Comparison of HPB results of both materials with the glass phase additives confirmed the same tendency – the composite of the alloy 7-10 and a glass absorbed more hydrogen than a composite LaNi$_5$ with a glass (Fig.10.).

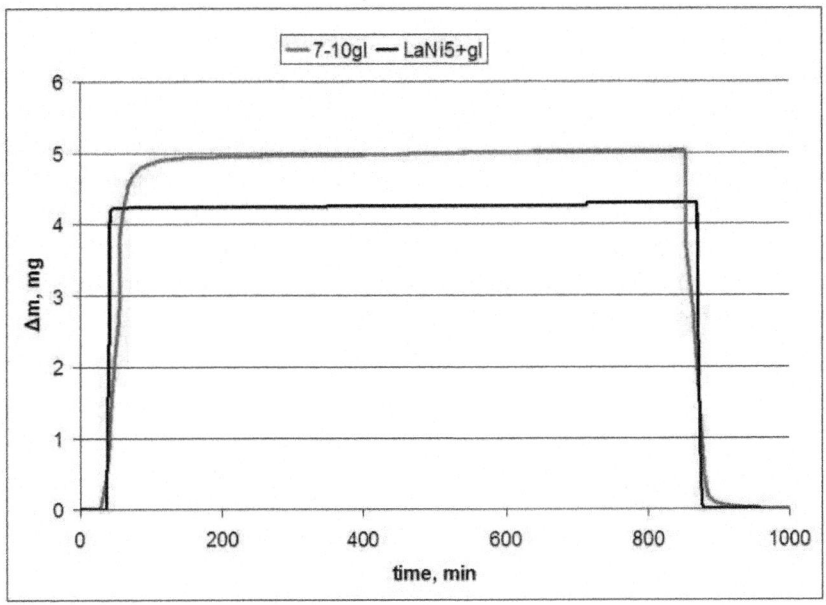

Figure. 10: HPB data of the weight change in time of the composite alloy 7-10 +glass (dashed line) and the composite LaNi$_5$ + glass (solid line)

Table 5: Calculations of HPB data for raw alloy (7-10) and composite (7-10+glass)

7-10	Δm, mg	w%	x
Δm$_{total}$	5,507E-03	1,582	6,966
Δm$_{av}$	5,138E-03	1,477	6,499
Δm(1atm)	4,883E-03	1,405	6,177
7-10+glass	**Δm, mg**	**w%**	**x**
Δm$_{total}$	5,048E-03	1,643	7,242
Δm$_{av}$	4,616E-03	1,504	6,621
Δm(1atm)	4,393E-03	1,433	6,302

The calculations of amount of the absorbed hydrogen in a raw alloy 7-10 and in the composite proved that the composite of 7-10 and glass have absorbed more than pure alloy 7-10 and are showed in the table 2, where Δm total – total change of the weight, Δmav – average change of the weight during cycling, Δm (1 atm) weight remaining at the 1 atm of pressure, w% - weight percents of hydrogen in alloy, x = a value from stoichiometric formula of hydride AB_5H_x The XRD analysis of hydrogenated samples showed a good agreement with calculations of HPB data (Fig. 11, Table 6). From the XRD plot is well observable that the diffraction peaks of the hydrogenated alloy 7-10 are largely shifted to the smaller angles than that of the starting alloy, indicating that the $\square$-phase of hydride is changed into the $^\lrcorner$-phase and the lattice parameters and cell volume of the hydride is larger than that of the starting alloy. An observed shift of XRD peaks after hydrogenation for the composite is even larger than that for the pure 7-10 alloy. Also corresponding lattice parameters and a cell volume for the hexagonal P6/mmm symmetry accordingly is larger of hydrogenated composite sample as for fully hydrogenated alloy 7-10 (Table 6).

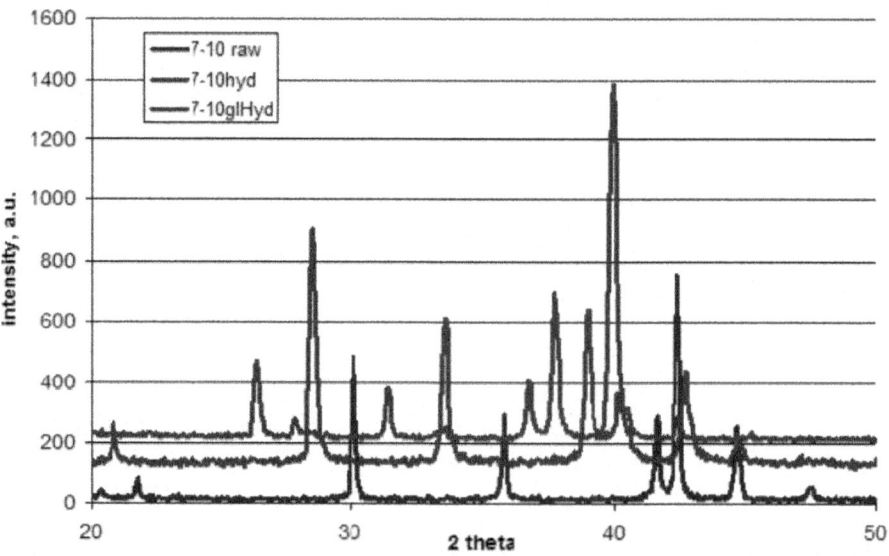

Figure. 11: XRD plot of raw (7-10raw), hydrogenated pure 7-10 (7-10H) and hydrogenated alloy 7-10 with glass phase (7-10gl-H)

Table 6: Structural parameters of raw and hydrogenated samples of alloy 7-10 and composite

Cell size / Sample	a, Å	c, Å	V, Å3
7-10	5,0083	4,0567	88,12
7-10hydrogenated	5,326	4,234	104,0
7-10+glass hydrogenated	5,369	4,2754	106,78

Conclusions

The HPB data treatment and calculations showed that hydrogen uptake in the composite alloy 7-10 with glass exceeds the pure alloy that can be explained as the spillover from the AB_5 catalyst. The following mechanism is deduced - the hydrogen chemisorbs at the surface sites found on the AB5 (mostly Ni sites). Bridges between the catalyst and glass particles allow the chemisorbed hydrogen to migrate onto the glass surface. Desorption occurs directly from the relatively lower energy glass sites without migration back to the catalyst. Hydrogen spillover depends upon the glass-catalyst contact. The contact changes with the quality of the mixing and milling, as well as the position of alloy 7-10 grains in the mixture. It was observed from the X-ray diffraction patterns, that the beta phase of the ball milled composite sample occurred faster than in the pure alloy sample and the peak shift to the smaller angles is noticeable larger. It is possible to assume, that the gamma hydride phase (γ) is forming when the alloy 7-10 is mixed in composite with Pyrex glass that isn't observable for pure alloy. Large lattice distortions in the γ phase are caused by hydrogen atom location sites close to Ni atoms in the elementary cell that produces inhomogeneous distribution of atoms but in the same time allows include more hydrogen atoms in the cell volume than the β phase.

REFERENCES

1. Ares J. R., Cuevas F., Percheron-Guegan A., (2004). Influence of thermal annealing on the hydrogenation properties of mechanically milled AB5-type alloys, Materials Science and Engineering B,vol. 108, No. 1-2, (Apr., 2004) 76-80, ISSN 0921-5107

2. Bittner H. F. & Badcock C. C., (1983). Electrochemical Utilization of Metal Hydrides, Journal of Electrochemical Society Vol. 130, No. 5 (May, 1983) pp. 193C-198C, ISSN 0013-4651

3. Das.D, & Veziroglu, T.N. (2001). Hydrogen production by biological processes: a survey of literature. Intern. J. Hydrogen Energy, Vol. 26, No.

1, (Jan., 2001) pp. 13–28, ISSN 0360-3199

4.   Das D., & Veziroglu N.T. (2008). Advances in biological hydrogen production processes. Intern. J. Hydrogen Energy, Vol. 33, No. 21, (Nov., 2008) pp. 6046-6057, ISSN 0360-3199

5.   Dilworth L.R. & W.J. Wunderlin (1968). Allis-Chalmers manufacturing Company (US) Patent US- 3,405,008.

6.   Donohue T.J., & Cogdell R.J. (2006). Microorganisms and clean energy. Nature Reviews Microbiology Vol. 4, 800. doi:10.1038/nrmicro1534 (Nov. 2006), ISSN : 1740-1526

7.   Flynn T., Ghirardi M.L., & Seibert M. (2002). Accumulation of O2-tolerant phenoltypes in H2-producing strains of Chlamydomonas reinhardtii by sequential applications of chemical mutagenesis and selection. Intern. J. Hydrogen Energy, Vol. 27, No. 11-12, (Nov.-Dec., 2002) pp. 1421-1430, ISSN 0360-3199

8.   Ghirardi M, Togasaki R.K., & Seibert M. (1997). Oxygen sensitivity of algal H2-production. Applied Biochem Biotechnol. Vol. 63-65, pp.141–151, ISSN: 0273-2289

9.   Hallenbeck P.C., & Benemann J.R. (2002). Biological hydrogen production; fundamentals and limiting processes. Intern. J. Hydrogen Energy, Vol. 27, No. 11-12, (Nov. –Dec.2002) pp. 1185-1193, ISSN 0360-3199

10.   Holladay J.D., Hu J., King D.L., & Wang Y. (2009). An overview of hydrogen production technologies. Catalysis Today, Vol. 139, No. 4, (Jan., 2009) pp 244-260, ISSN 0920-5861

11.   Kotay Meher S., & Das D. (2008). Biohydrogen as a renewable energy resource – prospects and potentials. Intern. J. Hydrogen Energy, Volume 33, Issue 1, January 2008, Pages 258-263 ISSN 0360-3199

12.   Levin D.B., Pitt L., & Love M. (2004). Biohydrogen production: prospects and limitations topractical application. Intern. J. Hydrogen Energy, Vol. 29, No. 2, (Feb. 2004) pp. 173-185, ISSN 0360-3199

13.   Lindholm I., (1966) Allmanna Svenska Elektriska Aktiebolaget (SE) Patent US- 3,262,816 Maeda T., Sanchez-Torres V. & Wood T.K. (2008). Enhanced hydrogen production from glucose by metabolically engineered Escherichia coli. Applied Microbiology and Biotechnology, Vol. 77, Nr. 4, pp. 879-890, DOI: 10.1007/s00253-007-1217-0, ISSN 1432-0614

14.   Midilli A., Ay M., Dincer I., Rosen M. A., On hydrogen and hydrogen energy strategies: I: current status and needs., Renewable and Sustainable Energy Reviews Vol. 9, No. 3,(Jun., 2005) pp. 255-271, ISSN 1364-0321

15. Nath K., & Das D. (2004). Improvement of fermentative hydrogen production: various approaches. Appl. Microbiol. Biotechnol., Vol. 65 (Oct., 2004) pp. 520–529, ISSN: 1432-0614

16. Navarro R.M., del Valle F., Villoria de la Mano J.A., Álvarez-Galván M.C., Fierro J.L.G. (2009) Photocatalytic Water Splitting Under Visible Light: Concept and Catalysts Development Advances in Chemical Engineering, Advances in Chemical Engineering - Photocatalytic Technologies, Vol.36, (Jul, 2009) pp. 111-143, ISSN: 0065-2377

17. Roland U., Braunschweig T., Roessner F., (1997) On the nature of spilt-over hydrogen, Journal of Molecular Catalysis A: Chemical, Vol. 127, No. 1-3, (Dec. 1997) pp 61-84, ISSN 0733-9372

18. Sandrock, G., A panoramic overview of hydrogen storage alloys from a gas reaction point of view. Journal of Alloys and Compounds, Vol. 293-295, (Dec., 1999) pp. 877-888, ISSN 0925-8388

19. Scarano D., Bordiga S., Lamberti C., Ricchiardi G., Bertarione S., Spoto D G., Applied Catalysis A: General,, Vol. 307, No. 1, (Jun., 2006) pp 3-12, ISSN 0926-860X

20. Schlapbach L., (1992) Surface properties and activation. In: Hydrogen in intermetallic compounds II, Vol. 67, ed. L. Schlapbach. Pp. 15-95, Springer: Berlin Heidelberg,ISBN 9780387546681, New York

21. Schlapbach L. and Zuttel A., (2001) Hydrogen-storage materials for mobile applications Nature, Vol. 414, (Nov., 2001), pp. 353-358, ISSN 0028-0836

22. Wang J., & Wan W. (2009). Factors influencing fermentative hydrogen production: A review. Intern. J. Hydrogen Energy, Vol. 34, No. 2, (Jan., 2009) pp. 799-81, ISSN 0360-3199

23. Wilkins J. R., Stoner G.E., & Boykin E.H. (1974). Microbial Detection Method Based on Sensing Molecular Hydrogen. Applied Microbiology, Vol. 5 (May 1974), pp. 949–952, ISSN 1365-2672

24. Zaluska A., Zaluski L., & Ström-Olsen J. O. (2001) Structure, catalysis and atomic reactions on the nano-scale: a systematic approach to metal hydrides for hydrogen storage, Applied Physics A: Materials Science & Processing, Vol. 72, No 2, pp. 157 – 165, ISSN 0947-8396

25. Züttel A., (2003) Materials for hydrogen storage, Materials Today, Vol. 6, No 9 (Sept., 2003) p. 24-33, ISSN 1369-7021

26. Züttel, A., (2004) Hydrogen storage methods. Naturwissenschaften, Vol. 91. No. 4, (Apr. 2004) p. 157-172, ISSN 0028-1042.

Chapter 2

GENOME ANNOTATION PROVIDES INSIGHT INTO CARBON MONOXIDE AND HYDROGEN METABOLISM IN RUBRIVIVAX GELATINOSUS

Karen Wawrousek[1] , Scott Noble[2] , Jonas Korlach[3] , Jin Chen[4] , Carrie Eckert[1] , Jianping Yu[1] , Pin-Ching Maness[1]

[1]. Department of Chemical and Petroleum Engineering, University of Wyoming, Laramie, Wyoming, United States of America

[2]. Biosciences Center, National Renewable Energy Laboratory, Golden, Colorado, United States of America

[3]. Pacific Biosciences, Menlo Park, California, United States of America

[4]. Department of Energy Plant Research Laboratory, Michigan State University, East Lansing, Michigan, United States of America

ABSTRACT

We report here the sequencing and analysis of the genome of the purple non-sulfur photosynthetic bacterium *Rubrivivax gelatinosus* CBS. This microbe is a model for studies of its carboxydotrophic life style under anaerobic condition, based on its ability to utilize carbon monoxide (CO) as the sole carbon substrate and water as the electron acceptor, yielding CO_2 and H_2 as the end products. The CO-oxidation reaction is known to be catalyzed by two enzyme complexes, the CO dehydrogenase and hydrogenase. As expected, analysis of the genome of *Rx. gelatinosus* CBS reveals the presence of genes encoding both enzyme complexes. The CO-oxidation reaction is CO-inducible, which is consistent with the presence of two putative CO-sensing transcription factors in its genome. Genome analysis also reveals the presence of two additional hydrogenases, an uptake hydrogenase that liberates the electrons in H_2 in support of cell growth, and a regulatory hydrogenase that senses H_2 and relays the signal to a two-component system that ultimately controls synthesis of the uptake hydrogenase. The genome also contains two sets of hydrogenase maturation genes which are known to assemble the catalytic metallocluster of the hydrogenase NiFe active site. Collectively, the genome sequence and analysis information reveals the blueprint of an intricate network of signal transduction pathways and its underlying regulation that enables *Rx. gelatinosus* CBS to thrive on CO or H_2 in support of cell growth.

INTRODUCTION

Rubrivivax gelatinosus CBS was originally isolated from soil in Denver, Colorado. It is a purple non-sulfur (PNS) photosynthetic bacterium belonging to the family of *Rhodospirillaceae* [1]. Similar to most PNS in this family, *Rx. gelatinosus* is versatile with various modes of growth and energy metabolism. It carries out anoxygenic photosynthesis using electrons derived from organic acids and energy from sunlight, aerobic respiration using organic acids, and N_2 fixation to ammonium in support of cell growth [2], [3]. Moreover, upon exposure to carbon monoxide (CO) in the culture gas phase, two *Rx. gelatinosus* strains, i.e., CBS and S1 can both carry out a water-gas shift reaction according to the equation $CO + H_2O \rightarrow CO_2 + H_2$ [4]-[7]. The CO_2 and H_2 products are then assimilated into new cell mass either in the light or in darkness, both under anaerobic condition. The latter growth mode in darkness can use CO as the sole carbon and energy source, hence coupling CO oxidation to energy conservation by a proton/sodium gradient generated by electron transfer via the CO-linked hydrogenase [8], [9].

The CO metabolic pathway in the PNS *Rhodospirillum rubrum* is well characterized and has served as the model system to unravel the genes and enzymes involved in CO metabolism. Upon exposure to CO, two *coo* (CO oxidation) operons are induced, with the *cooFSCTJ* operon encoding CO dehydrogenase (CODH) and related Ni-insertion proteins [10], [11], and the *cooMKLXUH* operon encoding a NiFe-hydrogenase [12], [13]. Transcription of both operons is under the control of the heme-containing CO-sensing transcription factor *cooA* [14]. The CooF protein contains iron-sulfur (FeS) clusters which likely mediates electron transfer between CODH and the CO-linked CooMKLXUH hydrogenase [15], [16]. Prior research in *Rx. gelatinosus* CBS has identified a *cooMKLXUH* operon encoding a CO-inducible hydrogenase[17]. This hydrogenase is likely a hexameric protein with high degree of amino acid identity to its homologs in *Rs. rubrum*, *Carboxydothermus hydrogenoformans* [18], *Methanosarcina barkeri* [19], and *Thermoanaerobacter tengcongensis* [20]. This type of hexameric NiFe-hydrogenases shares a common feature in that the hydrophilic subunits CooLXUH display sequence similarity to the energy-conserving NADH: quinone oxidoreductase (complex I). This class of hydrogenases is classified as the Group 4 energy-converting hydrogenase (Ech) [21], consistent with its role in proton-pumping reaction to yield energy from CO oxidation and H_2production [9], [22]. The CO-dependent H_2 production hence has important ramifications in microbial energy generation during C1 carbon metabolism.

Although the genomes of the *Rx. gelatinosus* strains ATCC17011 and IL144 have been sequenced, neither contains homologs of the *cooFSCTJ* or the *cooMKLXUH* genes [23]. When tested, the strain ATCC17011 failed to metabolize CO (NREL, unpublished work). As such the versatility of energy metabolism in *Rx. gelatinosus* CBS especially regarding CO metabolism and CO-linked H_2 production prompted us to sequence its genome [24], which is the first sequenced genome for a *Rx gelatinosus* strain capable of metabolizing CO.

Detailed genome annotation revealed a *cooFSC* and *cooMKLXUH* gene cluster likely encoding the protein machinery responsible for CO oxidation and H_2 production. We uncovered a set of hydrogenase maturation genes (*hypABFCDE*), which is clustered near the *coo* operon and presumably assembles the active site of the Ech hydrogenase [25], [26]. Moreover, we uncovered a H_2-uptake hydrogenase and a H_2-sensing hydrogenase along with a second set of *hyp* maturation genes, with a genome arrangement similar to that in *Ralstonia eutropha* [27]. The *Rx. gelatinosus* CBS genome hence provides the blueprint to an intricate signal transduction network governing H_2 and CO sensing, regulation, and metabolism.

MATERIALS AND METHODS

Genome Sequencing

SMRTbell template libraries were prepared as previously described [28]. Two different sized SMRTbell template libraries were employed. Genomic DNA samples were either sheared to an average size of ~800 base pairs via adaptive focused acoustics (Covaris; Woburn, MA, USA) or to a target size of approximately 8–10 kilobase pairs using Covaris g-TUBEs (Woburn, MA, USA). Fragmented DNA was then end repaired and ligated to hairpin adapters. Incompletely formed SMRTbell templates were digested with a combination of Exonuclease III (New England Biolabs; Ipswich, MA, USA) and Exonuclease VII (Affymetrix; Cleveland, OH, USA). SMRT Sequencing was carried out on the Pacific Biosciences *RS* (Menlo Park, CA, USA) using C2 chemistry with standard protocols for either small or large insert SMRTbell template libraries.

Genome Assembly

The *de novo* genome assembly was performed using hybrid assembly protocols, which is based on error correction of the long-insert library of SMRT sequencing reads with the short-insert library, circular consensus sequencing (CCS) SMRT sequencing reads [29], [30]. The algorithm is available in SMRT Analysis version 1.3.3. The initial assembly resulted in three contigs with sizes of 4,724,250, 362,142 and 241,393 bases. The 241 kb contig was determined to be a distinct genomic element without any sequence similarity to the other two contigs, and was concluded to be circularly closed due to overlapping ends from the assembler (Figure S1). The remaining two contigs were found to have a region of high sequence similarity near the end of the 4.7 Mb and the very beginning of the 362 kb contig. It overlaps with a coverage anomaly at the end of the 4.7 Mb contig, indicating the presence of a 47 kb high-copy phage/plasmid element that has sequence identity with this region in the chromosome over a stretch of ~43 kb (Figure S2). The 47 kb high-copy element was represented by a new contig, and the 4.7 Mb and 362 kb contigs were connected to yield the full-length, 5.1 Mb bacterial chromosome. The chromosome had overlapping, self-similar ends, indicating that it is circularly closed. Such end-overlaps are typical for *de novo* assemblies on circularly closed genomic elements, and were trimmed manually. Approximately 93% of the long-insert library reads mapped back to the *de novo* assembly which is typical for SMRT sequencing.

We also applied a more recent hierarchical genome assembly algorithm (HGAp) on the long-insert library data only, and obtained identical results for the final genome assembly. HGAp is available at www.pacbiodevnet.com/hgap. Dot plots were generated using Gepard [31]. For the quality control of remapping reads back to the assembly, reads were mapped using the BLASR mapper (http://www.pacbiodevnet.com/SMRT-Analysis/Algorithms/BLASR) and the Pacific Biosciences' SMRT Analysis pipeline (http://www.pacbiodevnet.com/SMRT-Analysis/Software/SMRT-Pipe) using the standard mapping protocol.

RESULTS AND DISCUSSION

Overview of Subsystem Category

Hybrid, *de novo* genome assembly from long-insert library SMRT sequencing reads and short-insert library CCS SMRT sequencing reads resulted in three contigs representing one circularly closed bacterial chromosome with a size of 5,075,070 bases and a GC content of 71.3%, a 235,5122 basepair plasmid with a GC content of 64.9%, and a 47,366 basepair plasmid with a 73.5% GC content (Figure 1). In order to annotate the *Rx. gelatinosus* CBS genome, we submitted all contigs to RAST (Rapid Annotation using Subsystem Technology), a fully-automated service for annotating bacterial and archaeal genomes [32]. With RAST, we discovered 4,910 genes in *Rx. gelatinosus* CBS. The subsystem distribution in the genome shown in Figure 2 reveals the majority of genes encode regulons and genes involved in metabolism of carbohydrates, amino acids and derivatives. Specifically, we found 41 predicted hydrogenase genes and 158 predicted dehydrogenase genes in the genome, including 6 new hydrogenases genes and 5 new dehydrogenases genes that have not been reported in [24].

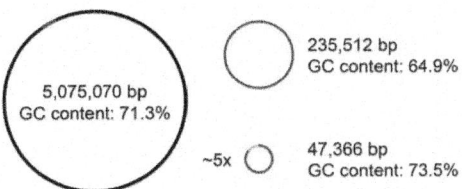

Figure 1: *De novo* genome assembly results for *Rubrivivax gelatinosus* CBS, resulting in one bacterial chromosome and two satellite DNA elements.

The common section of the small satellite and the chromosome are highlighted in red. Not drawn to scale.

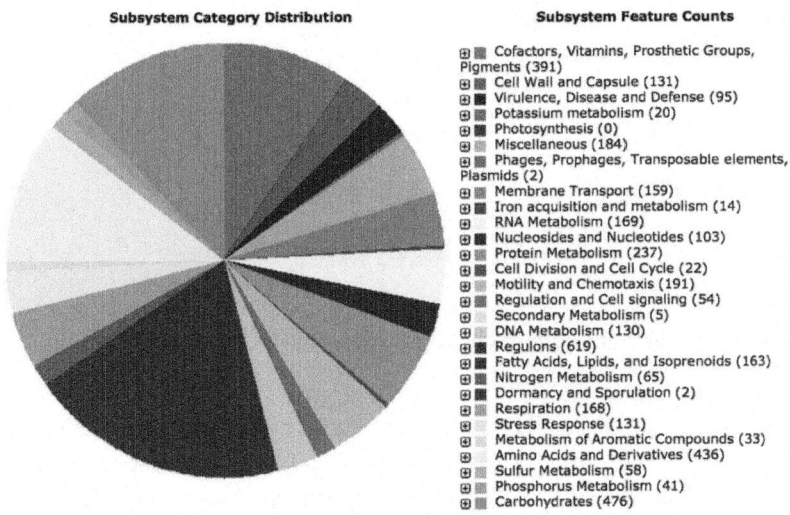

Figure 2: Organism Overview for *Rubrivivax gelatinosus* CBS.

There are 643 subsystems, 4852 coding sequences, and 58 RNAs.

CO Metabolism via CO Dehydrogenase

Genome annotation in *Rx. gelatinosus* CBS identified *cooFCS* genes (Figure 3), which display high levels of homology with their counterparts in *Rs. rubrum*, *C. hydrogenoformans*, and *D. vulgaris* str. Hildenborough (Table 1). These genes are presumably responsible in *Rx. gelatinosus* CBS for CO oxidation catalyzed by CODH (encoded by *cooS*), followed by electron transfer to CooF, an FeS protein. Critical residues for the Ni-Fe-S active site in CooS are highly conserved [33], and in CBS these conserved residues are H266, C293, G457, C458, C489, C540, T581, and K583. CooS is predicted to have one FeS cluster in the N-terminal half of the protein coordinated by conserved cysteine residues at C48, C51, C56, and C70. The CooF protein in *Rs. rubrum* contains conserved cysteine motifs to coordinate up to four FeS clusters[15], [16], while the CBS CooF protein sequence predicts only three FeS clusters. However, EPR data of *Rs. rubrum* CooF revealed that there are likely only two FeS clusters present per CooF monomer [16]. CooF likely mediates electron transfer from the CooS active site to the Coo hydrogenase (discussed below).

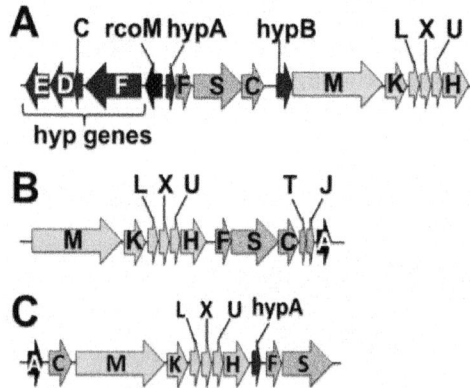

Figure 3: Organization of genes encoding anaerobic CO metabolism in multiple bacteria.

(A) Gene structure of CO metabolism genes in *Rubrivivax gelatinosus* CBS. The *hyp*genes putatively responsible for *coo* hydrogenase maturation are clustered among the*coo* CODH and hydrogenase genes. The gene encoding RcoM, the putative transcription factor for the *coo* CODH and hydrogenase genes is located among the *hyp*genes within this cluster of CO metabolism genes. (B) Gene structure of CO metabolism genes in *Rhodospirillum rubrum*.

The *cooA* gene, encoding the CooA CO-responsive transcription factor, directly follows the CODH genes. (C) Gene structure of CO metabolism genes in *Carboxydothermus hydrogenoformans*. *C. hydrogenoformans* has five CODH complexes, but the genes for only one CODH are co-located with the genes for an energy-conserving hydrogenase [36]. The *cooA* and *hypC* genes precede the*cooMKLXUH* hydrogenase genes, which are followed by *hypA* and the *cooFS* CODH genes.

Table 1: Similarity and identity of select CO dehydrogenase proteins and CO-responsive transcription factors between *Rubrivivax gelatinosus* CBS with other bacterial species, generated by EMBOSS 6.3.1: matcher search [62]

Protein and organism	Similarity %	Identity %
CooF		
Rs. rubrum	71	60
C. hydrogenoformans	63	49
D. vulgaris	65	48
CooS		
Rs. rubrum	76	61
C. hydrogenoformans	77	59
D. vulgaris	60	44
CooC		
Rs. rubrum	70	52
C. hydrogenoformans	65	46
D. vulgaris	58	46
RcoM		
B. xenovorans RcoM-1	52	32
B. xenovorans RcoM-2	58	35
Rs. rubrum	43	28
CooA		
Rx. gelatinosus IL144	100	98
Rx. benzoatilyticus JA2	95	90
Rs. rubrum	57	36
C. hydrogenoformans	61	37
CowN		
Rx. gelatinosus IL144	100	99
Rx. benzoatilyticus JA2	92	91
Rs. rubrum	50	31
CoxS		
Rx. gelatinosus IL144	100	100
O. carboxidovorans	74	58
CoxL		
Rx. gelatinosus IL144	99	99
O. carboxidovorans	54	38
CoxM		
Rx. gelatinosus IL144	100	100
O. carboxidovorans	58	43
CoxD		
Rx. gelatinosus IL144	100	99
O. carboxidovorans	64	48

Similar to other complex metalloproteins, the CODH enzyme requires maturation factors, or chaperones, for the correct insertion of metals to form the holoprotein. In *Rs. rubrum*, it was found that CooCTJ are involved in the insertion of Ni into the CooS protein [10]. CooC is a membrane-associated protein that is believed to couple ATP hydrolysis with Ni insertion into the CODH active site [34]. There is less Ni insertion into the active site of CooS in either *cooT* or *cooJ* mutants, but there was virtually no CODH activity in a *cooC* mutant [34], [35]. This implies that CooC is the most important of these proteins for Ni insertion. A CooC homolog is evident in *Rx. gelatinosus* CBS, but the genes encoding CooT and J are not present (Figure 3), similar to *C. hydrogenoformans*. Yet even without clear homologs of *cooT* and *cooJ*, both *Rx. gelatinosus* CBS and *C. hydrogenoformans* produces an active CODH, as evidenced by growth on CO as a sole carbon source [1], [36]. This suggests that either CooC alone may suffice for Ni insertion into the CooS active site in the latter microbes, or Ni insertion is assisted by some yet to be identified proteins.

In addition to the *coo* genes for anaerobic CO oxidation, *Rx. gelatinosus* CBS was found to also have the *coxSLMD* genes for aerobic CO oxidation. Aerobic CO oxidation has been best studied in *Oligotropha carboxidovorans*, and in this organism *coxSLM* encode the carbon monoxide dehydrogenase enzyme. CoxS is the small subunit and contains 2 FeS clusters. CoxM, the medium-sized subunit, contains an FAD binding site, and the enzymatic CoxL is the large, catalytic subunit and contains the Mo-Cu active site. The CoxD protein is involved in biosynthesis of the active site metallocluster [37]. While *Rx. benzoatilyticus* does not contain any of the *coxSLMD* genes, the *Rx. gelatinosus* IL144 genome does. The IL144 strain is similar to *Rx. gelatinosus* CBS in genome size (around 5 Mb) and GC content (around 71%), as well as numbers of predicted genes (close to 5000) [23]. As shown in Table 1, the *Rx. gelatinosus* IL144 CoxSLMD predicted proteins share 99–100% sequence identity with those predicted in *Rx. gelatinosus* CBS. While the genes encoding *coxSLMD* have been identified in both the *Rx. gelatinosus* CBS and *Rx. gelatinosus* IL144 genomes, no expression data exists for these genes in either organism and the functionality of an aerobic mode of CO metabolism has not been confirmed in either strain.

Transcription Regulation of CO Metabolism

Anaerobic carbon monoxide metabolism in both *Rs. rubrum* and *Rx. gelatinosus* CBS is regulated by CO. In *Rs. rubrum*, CO induces increased expression of the genes encoding CODH and hydrogenase [12], [38], [39]. CooA is the CO-responsive transcription factor that regulates expression of *coo* genes in *Rs. rubrum* [12], [14], [40], [41]. The CooA protein contains an N-terminal b-type heme for gas sensing and a C-terminal helix-turn-helix motif to bind DNA [42]. CO metabolism in *Rx. gelatinosus* CBS is also regulated by the presence of CO with CO-dependent growth and the appearance of Coo hydrogenase proteins CooH and CooL strictly depending on the presence of CO [1], [17]. Genome annotation revealed a *cooA* gene in*Rx. gelatinosus* CBS, yet it is not clustered with the *cooFSC* genes as in *Rs rubrum*. CooA in*Rx. gelatinosus* CBS is only 36% identical and 57% similar to the well-studied CooA in *Rs. rubrum* (Table 1). As expected based on protein alignment, the CooA in *Rx. gelatinosus* CBS is most similar to its counterpart in *Rx. gelatinosus* IL144 *and Rx. benzoatilyticus* (Table 1). However, neither of the latter two microbes metabolizes CO anaerobically and genes encoding the CODH and Coo hydrogenase necessary for anaerobic CO metabolism are not present [23].

Interestingly, the *Rx. gelatinosus* CBS *cooA* gene is adjacent to the *cowN* gene in the genome. In *Rs. rubrum*, the CowN protein is responsible for the protection of nitrogenase from inactivation by CO [43]. While it is known that *Rx. gelatinosus* CBS can fix nitrogen while consuming CO as the sole carbon source [1], the mechanism of the protection of nitrogenase is unknown. The *Rx. gelatinosus* CBS, *Rx. gelatinosus* IL144, and *Rx. benzoatilyticus* genomes do all contain the *cowN* gene, with protein sequence identities ranging from 91-99% with the predicted *Rx. gelatinosus* CBS CowN protein (Table 1). Since neither *Rx. gelatinosus* IL144 nor*Rx. benzoatilyticus* contain genes for the anaerobic metabolism of CO, the role of CooA and CowN proteins is unclear if these genes are indeed expressed. It is worth noting that the genome arrangement in *Rx. gelatinosus* CBS is the reverse of what is observed in *Rs. rubrum*, where *cooA* is clustered with coo genes and the gene encoding the RcoM transcription factor (described below) is adjacent to the *cowN* gene [43], suggesting that *cooA* may regulate *cowN*expression in *Rx. gelatinosus* CBS.

Rx. gelatinosus CBS harbors a single copy of the *rcoM* gene, which is clustered with the *coo*genes responsible for anaerobic CO oxidation (Figure 3). RcoM (Regulator of CO Metabolism) was identified in 2008 and is expected to regulate genes responsible for CO metabolism in*Geobacter* spp. and *Pelobacter carbinolicus* DSM 2380, since these microbes do not harbor a CooA homolog [44]. Like CooA, RcoM is a single-component transcription factor; with an N-terminal PAS sensor domain presumably used to sense CO and a C-terminal domain containing a LytTR DNA-binding domain [44], [45]. The *Rx. gelatinosus* CBS RcoM protein is predicted to contain these conserved motifs [46] and is similar to RcoM in other organisms (Table 1). Therefore, based on the location of the *rcoM* gene in the genome along with evidence that *rcoM* is absent in strains of *Rubrivivax* that do not oxidize CO, we predict that RcoM regulates anaerobic CO metabolism genes in *Rx. gelatinosus CBS*.

HYDROGEN METABOLISM AND MULTIPLE HYDROGENASES

CO-Linked Hydrogenase.

The *Rx. gelatinosus* CBS genome contains a *cooMKLXUH* operon (Figure 3), with genes displaying high levels of homology with their counterparts in *Rs. rubrum* and *C. hydrogenoformans* [17]. These genes encode a membrane anchored hexameric NiFe hydrogenase [47] that is responsible in *Rx. gelatinosus* CBS for CO-linked H_2 production [7]. This operon was initially identified via transposon mutagenesis; a *cooH* mutant strain completely lost CO-linked H_2 production [17]. The CooH subunit harbors the NiFe active site; CooL contains an FeS cluster predicted to serve as an electron relay to/from the active site in the CooH subunit; CooX also harbors FeS clusters for electron relay based on a study on its counterpart in *Methanosarcina barkeri* [17], [48]. CooU is predicted to be a soluble protein of 180 amino acids (20 kDa). Its amino acid sequence shows 39% identity and 56% similarity to CooU (annotated as NADH dehydrogenase subunit) in *Rs. rubrum*, and 34% identity and 57% similarity to CooU in *C. hydrogenoformans*. However, the specific role of CooU in the CooMKLXUH hydrogenase has yet to be determined. Gene expression from this operon is regulated by CO, as discussed above. Not only is hydrogenase activity in *Rx. gelatinosus* CBS induced by CO, so is gene transcription (data not shown) and accumulation of subunits CooL and CooH [17].

The CooM and CooK hydrogenase subunits are membrane-associated proteins. CooM is predicted to be a large membrane protein of 1255 amino acids (130 kDa) with 29–33 transmembrane segments, as predicted by DAS (http://www.sbc.su.se/~miklos/DAS/). The amino acid sequence shows 52% identity and 66% similarity to *Rs. rubrum* CooM (annotated as NADH dehydrogenase) and 46% identity and 62% similarity to *C. hydrogenoformans* CooM (annotated as carbon monoxide-induced hydrogenase, membrane anchor subunit). CooK is predicted to be a membrane protein of 319 amino acids (34 kDa), with 8 transmembrane segments predicted by DAS. Its amino acid sequence exhibits 57% identity and 74% similarity to CooK in *Rs. rubrum* (annotated as membrane bound hydrogenase subunit, MbhM), and 50% identity and 66% similarity to CooK in *C. hydrogenoformans*. Besides serving as a membrane anchor for the hydrogenase, CooM and CooK may be involved in energy generation from the

CO oxidation-H_2 production pathway catalyzed by Ech hydrogenase including that in *Rx. gelatinosus* CBS [9], [47].

The *cooMKLXUH* operon is absent from the genome of *Rx. gelatinosus* IL144 [23]. Comparison of the two *Rx. gelatinosus* genomes shows that this operon may have been present in a common ancestor and subsequently lost in the IL144 strain. This is evidenced by the observation that genes flanking the upstream and downstream of the operon are present in both microbes, including a 48-bp region identical to the 3′ of *hypE* in *Rx. gelatinosus* CBS (Figure S3).

H_2-Uptake and Sensor Hydrogenases.

Membrane-bound hydrogenases (MBH) are a class of NiFe hydrogenases that couple H_2 oxidation to the respiratory chain for energy generation, hence allowing organisms to utilize H_2 as an electron and energy source [21]. The extensively studied MBH of *R. eutropha* H16 serves as the model enzyme for comparison [49], [50]. The MBH operon of *R. eutropha* H16 contains 22 genes, including genes encoding the MBH structural subunits (*hoxKG*) and a regulatory or sensor hydrogenase (*hoxBC*), as well as associated signaling factors (*hoxJ* and *hoxA*). HoxBC (homologs of HupUV in other organisms) functions with a two-component system composed of a histidine kinase (HoxJ/HupT) and a DNA-binding response regulator (HoxA/HupR) which regulate MBH operon expression in the presence of H_2 [51]–[53]. Analysis of the *Rx. gelatinosus* CBS genomic sequence reveals a 17,891 basepair MBH operon with 20 genes. Comparison at the nucleotide level reveals that the MBH operon in *Rx. gelatinosus* CBS displays 98% identity with the MBH operon in *Rx. gelatinosus* IL144 spanning 99% query coverage, yet shows only 80% nucleotide identity with the MBH operon in *R. eutropha* spanning 20% query coverage. Therefore, we adopted the *Rx. gelatinosus* IL144 "*hup*" gene nomenclature for the genes in the *Rx. gelatinosus* CBS MBH operon. Nevertheless, high amino acid conservation is observed between *Rx. gelatinosus* CBS, *Rx. gelatinosus* IL144, and *R. eutropha* H16 as to the respective MBH small subunit (HupA/HoxK), large subunit (HupB/HoxG), and the sensor hydrogenase small subunit (HupU/HoxB) and large subunit (HupV/HoxC) (Table 2), with their expression likely under the influence of H_2 [51]–[53].

Table 2: Similarity, identity, and coverage of select hydrogenase proteins comparing the *Rubrivivax gelatinosus* CBS sensor and uptake hydrogenases to other bacterial species including *Ralstonia eutropha* H16

		Similarity %	Identity %	Coverage %
Sensor Hydrogenase				
HupU	*Rx. gelatinosus* IL144	99	98	100
	R. eutropha H16 (HoxB)	84	74	98
HupV	*Rx. gelatinosus* IL144	99	98	100
	R. eutropha H16 (HoxC)	75	66	100
Uptake Hydrogenase				
HupA	*Rx. gelatinosus* IL144	100	100	100
	R. eutropha H16 (HoxK)	96	89	87
HupB	*Rx. gelatinosus* IL144	99	99	100
	R. eutropha H16 (HoxG)	92	86	100

Examining the sequence of the *Rx. gelatinosus* CBS MBH small subunit HupA reveals that it is likely an O_2-tolerant hydrogenase. It contains two conserved supernumerary cysteines (Cys98 and Cys197) that align with the characterized Cys19 and Cys120 in the O_2-tolerant NiFe-hydrogenases of *R. eutropha* H16 MBH [54] and *E. coli* Hyd-1 [55], whereas the O_2-sensitive hydrogenase contain glycine residues instead. *Rx. gelatinosus* CBS contains HupI (HoxR homolog in *R. eutropha* H16), a rubredoxin-type FeS protein deemed essential in assembling the O_2-tolerant MBH in *R. eutropha* H16 when cultured in aerobic environment [54], [56]. Future studies may reveal whether the supernumerary cysteines contribute to the O_2-tolerance of the MBH in *Rx. gelatinosus* CBS [7].

NiFe Hydrogenase Maturation Factors.

Genome sequencing also reveals the presence of two copies of the pluripotent *hyp*hydrogenase maturation genes. One copy (hereafter *hyp1FCDEAB*) clusters near the *coo*operon (Figure 3) and another (hereafter *hyp2ABFCDE*) is located in the MBH operon. The respective proteins from *hyp1* and *hyp2* share identities ranging from 35% to 54% (Table 3). However, *Rx. gelatinosus* CBS Hyp2 proteins display much higher identity (ranging from 60 to 77%) with the respective counterparts in the *R. eutropha* H16 MBH operon (Table 3). It is therefore likely that the *hyp1* cluster is responsible for assembling the CO-linked, H_2-evolving hydrogenase while the *hyp2* cluster assembles the MBH in *Rx. gelatinosus* CBS. Genetic knockout of *hyp1* and/or *hyp2* will directly test this hypothesis. No *hyp* genes were found near the *coo* operon in the *Rs. rubrum* sequenced genome (Figure 3). Two sets of *hyp* maturation genes were found in the *Rs. rubrum* genome, with one partial set (*hypABC*) clustered near its MBH (HupAB) and a second set (*hypFCDEB*) clustered near a ferredoxin hydrogenase. The latter is likely a fermentative hydrogenase linking H_2 production to formate oxidation [57] since a nearby operon contains genes encoding formate dehydrogenase, a formate transporter, and a chaperon protein for the biosynthesis of molybdenum cofactor. A second *hypA* was found in a distant location in the genome. *Rs. rubrum* therefore might recruit these *hyp* genes for maturation of its CO-linked hydrogenase.

Table 3: Identity percentage of *Rubrivivax gelatinosus* CBS Hyp1, Hyp2 proteins and the Hyp proteins from *Ralstonia eutropha* H16, generated by NCBI P-BLAST search

	Rx. gelatinosus Hyp1 vs. Hyp2%	Rx. gelatinosus Hyp1 vs. R. eutropha Hyp %	Rx. gelatinosus Hyp2 vs. R. eutropha Hyp %
HypA	35	31	66
HypB	54	52	67
HypC	39	38	60
HypD	51	47	77
HypE	54	54	74
HypF	37	38	60

Physiological Relevance of CO and H_2 Metabolism.

Genetic elements and gene arrangements controlling the CO-dependent H_2 evolution system are very similar between *Rx. gelatinosus* CBS and *Rs. rubrum*. It is therefore logical to compare the physiological relevance of the two strains regarding CO and H_2 metabolism. CODH was first purified from *Rs. rubrum* without adding CO for its induction [38], although adding CO does increase its CODH activity significantly and CO is required for the CO-induced hydrogenase activity [58]. To the contrary, in *Rx. gelatinosus* CBS the presence of CO during growth is required to afford CODH activity linking CO oxidation to the reduction of methyl viologen [7]. The CO-inducible hydrogenase activity in *Rs. rubrum* is reported to be insensitive to CO, retaining 40% activity in 100% CO gas phase (880 µM) [39]. However, the chromatophore membranes used for the above hydrogenase assay also contained CODH activity; its high rate of CO oxidation could protect hydrogenase from CO inhibition. Yet once partially purified from extracts, the CO-inducible hydrogenase in *Rx. gelatinosus* CBS is extremely sensitive to CO, with 50% inhibition observed at 3.9 µM dissolved CO [7]. The latter is consistent with the 5.8 to 40 µM Ki values of CO reported for most hydrogenases [59], [60]. This dramatic difference in tolerance to CO is likely attributed to the removal of the bulk of CODH from the partially purified*Rx. gelatinosus* CBS hydrogenase preparation used for the assay. While *Rs. rubrum*, *Rx. gelatinosus* strain CBS and strain S1 can grow in CO in darkness, with CO serving as a carbon substrate and energy source [4], [9], [61], an equal comparison of growth rates is complicated by the fact that all strains were grown under different conditions and with differing media components. No matter the growth rate, it is now well recognized that CO-inducible hydrogenase couples H_2 evolution to proton translocation based on additional evidence including the effects of the electron transfer uncoupler carbonyl-cyanide *m*-chlorophenylhydrazone (CCCP) and the ATP synthesis inhibitor *N, N′*-dicyclohexylcarbodiimide (DCCD), with effects of the latter observed in both *Rs. rubrum* and *Rx. gelatinosus* CBS [9],[22].

Conclusion

The genomic information compiled in this report for *Rx. gelatinosus* CBS reveals features responsible for an autotrophic life style based on CO utilization. The *coo, hup* and *hyp* gene repertoire along with genes encoding the Calvin-Benson-Bassham (CBB) pathways (data not shown) enable growth using CO as the sole carbon substrate (Figure 4). This microbe contains a CO sensor, likely encoded by *rcoM*, which upon sensing CO initiates the transcription of the*coo* operon to enable CO oxidation and H_2 production. This microbe also contains the H_2-sensing hydrogenase HupUV, which is expected to work in concert with its cognate regulatory two-component system HupRT to regulate the expression of an H_2-uptake hydrogenase HupAB. H_2 oxidation ultimately provides electrons; that along with ATP generated from photosynthesis, participate in the CO_2 fixation reaction via the CBB pathway leading to cell growth. *Rx. gelatinosus* CBS is therefore a model photosynthetic bacterium to study a network of intricate signal transduction pathways and the underlying regulations controlling the assimilation of one-carbon compounds such as CO and CO_2.

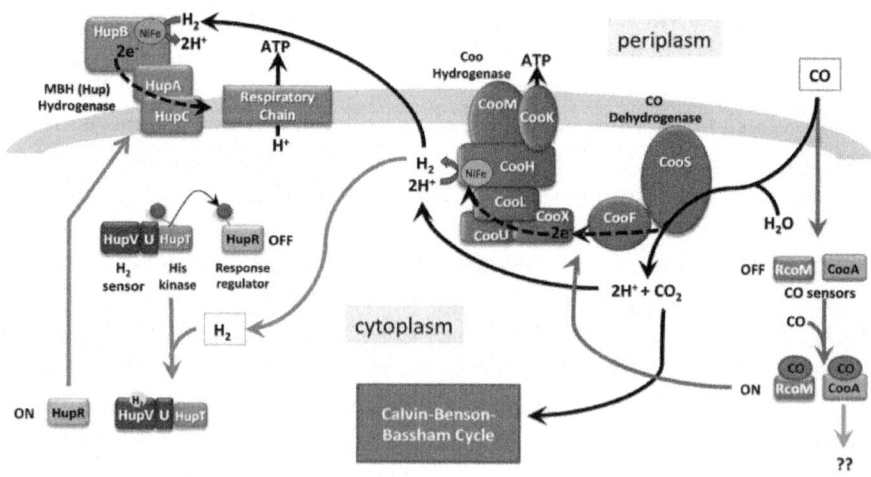

Figure 4: An overview of the CO and H_2 signal transduction pathways and metabolism in *Rubrivivax gelatinosus* CBS.

ACKNOWLEDGMENTS

We acknowledge Tyson A. Clark and Khai Luong at Pacific Biosciences for assistance with the sequencing and data analysis.

AUTHOR CONTRIBUTIONS

Conceived and designed the experiments: PCM JY. Performed the experiments: JK JC SN KW CE Analyzed the data: JK JC SN KW CE Contributed reagents/ materials/analysis tools: SN KW JK JC CE JY PCM Wrote the paper: JK JC SN KW CE JY PCM

REFERENCES

1. Maness P-C, Weaver PF (1994) Production of poly-3-hydroxyalkanoates from CO and H2 by a novel photosynthetic bacterium. Applied Biochemistry and Biotechnology 45–46:395–406 doi:10.1007/BF02941814.

2. Cogdell RJ, Isaacs NW, Howard TD, McLuskey K, Fraser NJ, et al. (1999) How Photosynthetic Bacteria Harvest Solar Energy. Journal of Bacteriology 181:3869–3879.

3. Tang K-H, Tang YJ, Blankenship RE (2011) Carbon Metabolic Pathways in Phototrophic Bacteria and Their Broader Evolutionary Implications. Front Microbio 2. doi:10.3389/fmicb.2011.00165.

4. Uffen RL (1983) Metabolism of carbon monoxide by Rhodopseudomonas gelatinosa: cell growth and properties of the oxidation system. Journal of Bacteriology 155:956–965.

5. Champine JE, Uffen RL (1987) Membrane topography of anaerobic carbon monoxide oxidation in Rhodocyclus gelatinosus. Journal of Bacteriology 169:4784–4789. doi: 10.1111/j.1574-6968.1987.tb02306.x

6. Maness P-C, Weaver PF (2002) Hydrogen production from a carbon-monoxide oxidation pathway in Rubrivivax gelatinosus. International Journal of Hydrogen Energy 27:1407–1411 doi:10.1016/S0360-3199(02)00107-6.

7. Maness PC, Smolinski S, Dillon AC, Heben MJ, Weaver PF (2002) Characterization of the Oxygen Tolerance of a Hydrogenase Linked to a Carbon Monoxide Oxidation Pathway in Rubrivivax gelatinosus. Appl Environ Microbiol 68:2633–2636 doi:10.1128/AEM.68.6.2633-2636.2002.

8. Uffen RL (1976) Anaerobic growth of a Rhodopseudomonas species in

the dark with carbon monoxide as sole carbon and energy substrate. Proc Natl Acad Sci USA 73:3298. doi: 10.1073/pnas.73.9.3298

9.   Maness P-C, Huang J, Smolinski S, Tek V, Vanzin G (2005) Energy Generation from the CO Oxidation-Hydrogen Production Pathway in Rubrivivax gelatinosus. Appl Environ Microbiol 71:2870–2874. doi: 10.1128/aem.71.6.2870-2874.2005

10.   Kerby RL, Hong SS, Ensign SA, Coppoc LJ, Ludden PW, et al. (1992) Genetic and physiological characterization of the Rhodospirillum rubrum carbon monoxide dehydrogenase system. Journal of Bacteriology 174:5284–5294.

11.   Kerby RL, Ludden PW, Roberts GP (1997) In vivo nickel insertion into the carbon monoxide dehydrogenase of Rhodospirillum rubrum: molecular and physiological characterization of cooCTJ. Journal of Bacteriology 179:2259–2266.

12.   Fox JD, He Y, Shelver D, Roberts GP, Ludden PW (1996) Characterization of the region encoding the CO-induced hydrogenase of Rhodospirillum rubrum. Journal of Bacteriology 178:6200–6208.

13.   Drennan CL, Heo J, Sintchak MD, Schreiter E, Ludden PW (2001) Life on carbon monoxide: X-ray structure of Rhodospirillum rubrum Ni-Fe-S carbon monoxide dehydrogenase. Proceedings of the National Academy of Sciences 98:11973–11978 doi:10.1073/pnas.141230698.

14.   Shelver D, Kerby RL, He Y, Roberts GP (1995) Carbon monoxide-induced activation of gene expression in Rhodospirillum rubrum requires the product of cooA, a member of the cyclic AMP receptor protein family of transcriptional regulators. Journal of Bacteriology 177:2157–2163.

15.   Ensign SA, Ludden PW (1991) Characterization of the CO oxidation/H2 evolution system of Rhodospirillum rubrum. Role of a 22-kDa iron-sulfur protein in mediating electron transfer between carbon monoxide dehydrogenase and hydrogenase. Journal of Biological Chemistry 266:18395–18403.

16.   Singer SW, Hirst MB, Ludden PW (2006) CO-dependent H2 evolution by Rhodospirillum rubrum: Role of CODH: CooF complex. Biochimica et Biophysica Acta (BBA) - Bioenergetics VL - 1757:1582–1591 doi:10.1016/j.bbabio.2006.10.003.

17.   Vanzin G, Yu J, Smolinski S, Tek V, Pennington G, et al. (2010) Characterization of Genes Responsible for the CO-Linked Hydrogen Production Pathway in Rubrivivax gelatinosus. Appl Environ Microbiol 76:3715–3722. doi: 10.1128/aem.02753-09

18. Soboh B, Linder D, Hedderich R (2002) Purification and catalytic properties of a CO-oxidizing: H2-evolving enzyme complex from Carboxydothermus hydrogenoformans. European Journal of Biochemistry 269:5712–5721 doi:10.1046/j.1432-1033.2002.03282.x.

19. Kunkel A, Vorholt JA, Thauer RK, Hedderich R (1998) An Escherichia coli hydrogenase-3-type hydrogenase in methanogenic archaea. European Journal of Biochemistry 252:467–476 doi:10.1046/j.1432-1327.1998.2520467.x.

20. Soboh B, Linder D, Hedderich R (2004) A multisubunit membrane-bound [NiFe] hydrogenase and an NADH-dependent Fe-only hydrogenase in the fermenting bacterium Thermoanaerobacter tengcongensis. Microbiology 150:2451–2463 doi:10.1099/mic.0.27159-0.

21. Vignais PM, Billoud B, Meyer J (2001) Classification and phylogeny of hydrogenases. FEMS Microbiology Reviews 25:455–501 doi:10.1111/j.1574-6976.2001.tb00587.x.

22. Fox JD, Kerby RL, Roberts GP, Ludden PW (1996) Characterization of the CO-induced, CO-tolerant hydrogenase from Rhodospirillum rubrum and the gene encoding the large subunit of the enzyme. Journal of Bacteriology 178:1515–1524.

23. Nagashima S, Kamimura A, Shimizu T, Nakamura-Isaki S, Aono E, et al. (2012) Complete Genome Sequence of Phototrophic Betaproteobacterium Rubrivivax gelatinosus IL144. Journal of Bacteriology 194:3541–3542 doi:10.1128/JB.00511-12.

24. Hu P, Juan L, Wawrousek K, Yu J, Maness P-C, et al. (2012) Draft genome sequence of Rubrivivax gelatinosus CBS. Journal of Bacteriology 194:3262 doi:10.1128/JB.00515-12.

25. Lutz S, Jacobi A, Schlensog V, Böhm R, Sawers G, et al. (1991) Molecular characterization of an operon (hyp) necessary for the activity of the three hydrogenase isoenzymes in Escherichia coli. Mol Microbiol 5:123–135 doi:10.1111/j.1365-2958.1991.tb01833.x.

26. Hube M, Blokesch M, Böck A (2002) Network of Hydrogenase Maturation in Escherichia coli: Role of Accessory Proteins HypA and HybF. Journal of Bacteriology 184:3879–3885 doi:10.1128/JB.184.14.3879-3885.2002.

27. Pohlmann A, Fricke WF, Reinecke F, Kusian B, Liesegang H, et al. (2006) Genome sequence of the bioplastic-producing |[ldquo]|Knallgas|[rdquo]| bacterium Ralstonia eutropha H16. Nat Biotechnol 24:1257–1262 doi:10.1038/nbt1244.

28. Travers KJ, Chin CS, Rank DR, Eid JS, Turner SW (2010) A flexible and efficient template format for circular consensus sequencing and

SNP detection. Nucleic Acids Research 38:e159–e159 doi:10.1093/nar/gkq543.

29. Bashir A, Klammer AA, Robins WP, Chin C-S, Webster D, et al. (2012) A hybrid approach for the automated finishing of bacterial genomes. Nat Biotechnol 30:701–707 doi:10.1038/nbt.2288.

30. Koren S, Schatz MC, Walenz BP, Martin J, Howard JT, et al. (2012) Hybrid error correction and de novo assembly of single-molecule sequencing reads. Nat Biotechnol 30:693–700 doi:10.1038/nbt.2280.

31. Krumsiek J, Arnold R, Rattei T (2007) Gepard: a rapid and sensitive tool for creating dotplots on genome scale. Bioinformatics 23:1026–1028. doi: 10.1093/bioinformatics/btm039

32. Aziz RK, Bartels D, Best AA, DeJongh M, Disz T, et al. (2008) The RAST Server: Rapid Annotations using Subsystems Technology. BMC Genomics 9:75 doi:10.1186/1471-2164-9-75.

33. Doukov TI, Iverson TM, Seravalli J, Ragsdale SW, Drennan CL (2002) A Ni-Fe-Cu Center in a Bifunctional Carbon Monoxide Dehydrogenase/Acetyl-CoA Synthase. Science 298:567–572 doi:10.1126/science.1075843.

34. Jeon WB, Cheng J, Ludden PW (2001) Purification and Characterization of Membrane-associated CooC Protein and Its Functional Role in the Insertion of Nickel into Carbon Monoxide Dehydrogenase from Rhodospirillum rubrum. Journal of Biological Chemistry 276:38602–38609. doi: 10.1074/jbc.m104945200

35. Watt RK, Ludden PW (1999) Ni2+ Transport and Accumulation inRhodospirillum rubrum. Journal of Bacteriology 181:4554–4560.

36. Wu M, Ren Q, Durkin AS, Daugherty SC, Brinkac LM, et al. (2005) Life in Hot Carbon Monoxide: The Complete Genome Sequence of Carboxydothermus hydrogenoformans Z-2901. PLoS Genet 1:e65 doi:10.1371/journal.pgen.0010065.

37. Pelzmann A, Ferner M, Gnida M, Meyer-Klaucke W, Maisel T, et al. (2009) The CoxD Protein of Oligotropha carboxidovorans Is a Predicted AAA+ ATPase Chaperone Involved in the Biogenesis of the CO Dehydrogenase [CuSMoO2] Cluster. Journal of Biological Chemistry 284:9578–9586 doi:10.1074/jbc.M805354200.

38. Bonam D, Murrell SA, Ludden PW (1984) Carbon monoxide dehydrogenase from Rhodospirillum rubrum. Journal of Bacteriology 159:693–699.

39. Bonam D, Lehman L, Roberts GP, Ludden PW (1989) Regulation of carbon monoxide dehydrogenase and hydrogenase in Rhodospirillum

rubrum: effects of CO and oxygen on synthesis and activity. Journal of Bacteriology 171:3102–2107.

40. He Y, Shelver D, Kerby RL, Roberts GP (1996) Characterization of a CO-responsive Transcriptional Activator from Rhodospirillum rubrum. Journal of Biological Chemistry 271:120–123 doi:10.1074/jbc.271.1.120.

41. Aono S, Nakajima H, Saito K, Okada M (1996) A Novel Heme Protein That Acts as a Carbon Monoxide-Dependent Transcriptional Activator inRhodospirillum rubrum. Biochemical and Biophysical Research Communications 228:752–756 doi:10.1006/bbrc.1996.1727.

42. Lanzilotta WN, Schuller DJ, Thorsteinsson MV, Kerby RL, Roberts GP, et al. (2000) Structure of the CO sensing transcription activator CooA. Nat Struct Biol 7:876–880 doi:10.1038/82820.

43. Kerby RL, Roberts GP (2011) Sustaining N2-Dependent Growth in the Presence of CO. Journal of Bacteriology 193:774–777. doi: 10.1128/jb.00794-10

44. Kerby RL, Youn H, Roberts GP (2008) RcoM: A New Single-Component Transcriptional Regulator of CO Metabolism in Bacteria. Journal of Bacteriology 190:3336–3343. doi: 10.1128/jb.00033-08

45. Nikolskaya AN, Galperin MY (2002) A novel type of conserved DNA-binding domain in the transcriptional regulators of the AlgR/AgrA/LytR family. Nucleic Acids Research 30:2453–2459. doi: 10.1093/nar/30.11.2453

46. Pellequer J-L, Wager-Smith KA, Kay SA, Getzoff ED (1998) Photoactive yellow protein: A structural prototype for the three-dimensional fold of the PAS domain superfamily. Proc Natl Acad Sci USA 95:5884–5890. doi: 10.1073/pnas.95.11.5884

47. Hedderich R, Forzi L (2005) Energy-Converting [NiFe] Hydrogenases: More than Just H_2 Activation. J Mol Microbiol Biotechnol 10:92–104 doi:10.1159/000091557.

48. Forzi L, Koch J, Guss AM, Radosevich CG, Metcalf WW, et al. (2005) Assignment of the [4Fe-4S] clusters of Ech hydrogenase from Methanosarcina barkeri to individual subunits via the characterization of site-directed mutants. FEBS Journal 272:4741–4753 doi:10.1111/j.1742-4658.2005.04889.x.

49. Burgdorf T, Lenz O, Buhrke T, van der Linden E, Jones AK, et al. (2005) [NiFe]-Hydrogenases of *Ralstonia eutropha* H16: Modular Enzymes for Oxygen-Tolerant Biological Hydrogen Oxidation. J Mol Microbiol Biotechnol 10:181–196 doi:10.1159/000091564.

50. Ludwig M, Cracknell JA, Vincent KA, Armstrong FA, Lenz O (2008) Oxygen-tolerant H2 Oxidation by Membrane-bound [NiFe] Hydrogenases of Ralstonia Species: Coping with Low Level H2 in Air. Journal of Biological Chemistry 284:465–477 doi:10.1074/jbc.M803676200.

51. Lenz O, Friedrich B (1998) A novel multicomponent regulatory system mediates H2 sensing in Alcaligenes eutrophus. Proceedings of the National Academy of Sciences 95:12474–12479 doi:10.1073/pnas.95.21.12474.

52. Vignais PM, Elsen S, Colbeau A (2005) Transcriptional regulation of the uptake [NiFe]hydrogenase genes in Rhodobacter capsulatus. Biochemical Society Transactions: 28–32.

53. Elsen S, Duché O, Colbeau A (2003) Interaction between the H2 Sensor HupUV and the Histidine Kinase HupT Controls HupSL Hydrogenase Synthesis in Rhodobacter capsulatus. Journal of Bacteriology 185:7111–7119 doi:10.1128/JB.185.24.7111-7119.2003.

54. Fritsch J, Scheerer P, Frielingsdorf S, Kroschinsky S, Friedrich B, et al. (2011) The crystal structure of an oxygen-tolerant hydrogenase uncovers a novel iron-sulphur centre. Nature 479:249–252 doi:10.1038/nature10505.

55. Volbeda A, Amara P, Darnault C, Mouesca J-M, Parkin A, et al. (2012) X-ray crystallographic and computational studies of the O2-tolerant [NiFe]-hydrogenase 1 from Escherichia coli. Proceedings of the National Academy of Sciences 109:5305–5310. doi: 10.1073/pnas.1119806109

56. Fritsch J, Siebert E, Priebe J, Zebger I, Lendzian F, et al. (2014) Rubredoxin-related Maturation Factor Guarantees Metal Cofactor Integrity during Aerobic Biosynthesis of Membrane-bound [NiFe] Hydrogenase. Journal of Biological Chemistry 289:7982–7993. doi: 10.1074/jbc.m113.544668

57. Maness P-C, Weaver PF (2001) Evidence for three distince hydrogenase activities in Rhodospirillum rubrum. Appl Microbiol Biotechnol: 751–756. Available: http://f. Accessed 15 May 2013.

58. Bonam D, Ludden PW (1987) Purification and characterization of carbon monoxide dehydrogenase, a nickel, zinc, iron-sulfur protein, from Rhodospirillum rubrum. J Biol Chem 262:2980–2987.

59. Adams MW, Mortenson LE, Chen JS (1980) Hydrogenase. Biochimica et Biophysica Acta (BBA) - Bioenergetics VL - 594:105–176. doi: 10.1016/0304-4173(80)90007-5

60. Peters JW (1998) X-ray Crystal Structure of the Fe-Only Hydrogenase (CpI) from Clostridium pasteurianum to 1.8 Angstrom Resolution. Science 282:1853–1858 doi:10.1126/science.282.5395.1853.

61. Kerby RL, Ludden PW, Roberts GP (1995) Carbon monoxide-dependent growth of Rhodospirillum rubrum. Journal of Bacteriology 177:2241–2244.

62. Rice P, Longden I, Bleasby A (2000) EMBOSS: The European Molecular Biology Open Software Suite. Trends in Genetics 16:276–277 doi:10.1016/S0168-9525(00)02024-2.

Chapter 3

POSTSYNTHESIS TREATMENT INFLUENCE ON HYDROGEN SORPTION PROPERTIES OF CARBON NANOTUBES

Andrew Basteev[1], Leonid Bazyma[1], Michail Obolensky[2], Andrew Kravchenko[2], Vladimir Beletsky[2], Yuri Petrusenko[3], Valeriy Borysenko[3], Sergey Lavrynenko[3], Oleg Kravchenko[4], Irina Suvorova[4] and Vladimir Golovanevskiy[5]

[1]National Aerospace University "Kharkov Aviation Institute", Ukraine

[2]V.N. Karazin Kharkiv National University, Ukraine

[3]National Science Center - Kharkov Institute of Physics and Technology, Ukraine

[4]A.N. Podgorny Institute for Mechanical Engineering Problems, Ukraine

[5]Western Australian School of Mines, Curtin University, Australia

INTRODUCTION

It has been shown in the previous investigations (Pradhan et al., 2002; Obolensky et al., 2011a, 2011b) that the postsynthesis treatment (e.g. chemical, heat treatment, milling, irradiation etc.) of single-walled carbon nanotubes (SWCNT) can essentially change their sorption properties. Methods of SWCNT synthesis and features of the postsynthesis treatment of the SWCNT show in numerous experiments a wide range of gas (e.g. hydrogen etc.) mass contents in SWCNT ranging from extremely low (less than 0.1 wt%) to significantly high (more than 6 wt%) values. Unfortunately, none of the claims of hydrogen storage exceeding the DOE limit have been confirmed. For recent review of effect of synthesis method and post-treatment of CNTs, see Ref. (Chang & Hui-Ming, 2005).

So, it is well established that hydrogen storage capacity of CNTs depends on many parameters including their pretreatment, types and structural modifications etc. One of the ways seems to be highly influential and it is the generation of appreciable structural defects in the tube walls. However, the increase in defects in graphitized layers of CNTs should be limited otherwise their structure would be destroyed which might lead to lowering of interaction or potential energy between the hydrogen molecules and carbon atoms. For instance, non-optimized electron irradiations severely destroy the graphitic network of CNTs (Banhart et al., 2002) which is not desirable for application purposes.

It has been reported that postsynthesis treatment can easily induce defects in the wall of SWCNT (Pradhan et al., 2002; Banhart et al., 2002; Hashimoto et al., 2004). The possibility of molecular hydrogen adsorption by the defect sites can also be considered. The experimental investigations of above mentioned factors as well as the result of irradiation on to SWCNT sorption capability are reported in this study.

EXPERIMENTAL

Materials

The arc-derived as-prepared (AP) SWNT material used in the present studies was obtained from Carbon Solutions, Inc.. All data reported here are collected using materials from the same batch of SWNT material.

The impurities in AP-SWNTs may be divided into two categories: carbonaceous (amorphous carbon and graphitic nanoparticles) and metallic (typically transition-metal catalysts). Thermogravimetric analysis (TGA) can be used to analyze the amount of metallic components quantitatively whereas Raman spectroscopy can be used to estimate qualitatively the carbonaceous impurities in AP and purified-SWNTs.

The AP SWNT (Carbon Solutions, Inc.) material has 21 to 31 percent of impurities which is consistent with the previously reported (Itkis et al., 2003) 34.5 ±1.8 percent impurities range. For purification of SWNTs the following technique has been used (Pradhan et al., 2002; Obolensky et al., 2011b). The SWNT (Carbon Solutions, approximately 1g amount) was oxidized in open air at 350^0C temperature for 30 minutes and then heated in 2.6 M HNO_3 for 20 hours. After above procedure the treated CNTs were flushed in methanol and then in water until the neutral reaction was achieved, dried in air and then were annealed in deep vacuum (10^{-8} bar) for 20 hours at 800°C temperature. Identification of the gases allocated in an act of warming of a sample at 850°C was carried by mass-spectrometry (MX-7203). Working pressure in the chamber of the mass-spectrometer was 10^{-4} Па. Approbation of nanotubes purify technology was carried out and 700 milligrams of the purified nanotubes were obtained (metal 6-9 at. %).

After chemical treatment, CNTs were milled at liquid nitrogen temperature in a stainless steel ball mill. Milling duration was 60 minutes with interruption after each 5 minutes for CNT samples' microscopic control. After the milling procedure finishing the "ball mill" was heated up to the room temperature and CNT samples were sieved through a 63 μm sieve (Obolensky et al., 2011b).

Experimental Facility

The method for SWNT electron bombardment irradiation was elaborated. In contrast to (Obolensky et al., 2011a) where the exposure of γ – quantum 10^5 Rad was used (this dose is equal to regime of electron bombardment ~$3*10^{12}$ e⁻/cm²) the fluence in the work reported here was 10^{13}-$2*10^{15}$e⁻/cm². The electron bombardment procedure was carried out at room temperature on the ELIAS linear accelerator (National Science Center "Kharkov Institute of Physics and Technology"). The electron energy was 2.3 MeV and beam intensity was 0.2 μA/cm². The main parameters of ELIAS accelerator are specified in Table 1.

Study of sorption/desorption process was carried out using the standard volumetric method (Obolensky et al., 2011a, 2011b). Hydrogen desorption out from SWNT was also studied on the mass spectrometer MX7203. The schematics of the experimental facility used with both measurement methods is shown inFigure 1.

Table 1: Parameters of ELIAS accelerator

Parameters	Values
Energy of accelerated electrons	0.5- 3.0 MeV
Beam current (without scanning)	0.5-150 μA
Max beam current (with scanning):	up to 500 μA
Electron beam disperse	10^{-4} radian
Diameter of electron beam without focusing	1,0 cm
Diameter of electron beam with focusing, for 90% capacity	<1 mm
Vacuum in electron beam line	1×10^{-7} mbar
Power consumption	20 кW

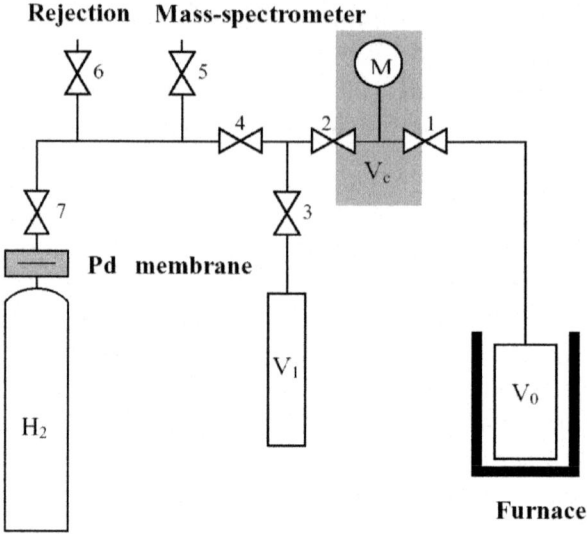

Figure 1: Experimental facility schematics (Obolensky et al., 2011b). Vc – total accurate including manometer and supply pipelines; V_1 – vessel with pure gas; V_0 – ampoule with carbon nanotubes sample; H_2 – non-purified hydrogen source

Raman spectra studies at ambient conditions with 514 nm laser line were carried out for some samples at the Physics Institute of the Penn State University.

RESULTS

Adsorption Studies of Non-Irradiated Samples

Study of sorption/desorption process was carried out using the standard volumetric method on a custom designed and manufactured manufactured vacuum stainless steel facility which included the known good unit volume V_E and measuring vessel with volume V_M. Part of measuring vessel V_{78} was cooled down to liquid nitrogen vaporisation temperature of $\sim$ 78 K. Before measurements the facility was calibrated. The good unit with volume V_E was inflated by gaseous helium or hydrogen with pressure $P_I \sim$ 10 bar at room temperature which has been measured with $\pm$ 0.5 K precision. The pressure was controlled by electronic manometer GE Druck 104 with 1 mbar resolution. After the total facility volume V = VE + VM was filled up, the pressure dropped to the value of PF = VEPI(VE + VM - VA). This circumstance allowed exact determination of the volumes relation, taking into account the ampoule volume V_A without sample. Above relation for hydrogen and helium was found equal with approximately ~ 0.01 percent accuracy and therefore the hydrogen sorption by vacuum system elements could be excluded from consideration. After this procedure, the container with ampoule was cooled to temperature 78 K and relation P_W/P_F was determined, where P_W is the pressure in the system after cooling.

The measurement of physically absorbed hydrogen was conducted in accordance with the following procedure. Firstly, after cooling down the container to temperature 78 K the system was evacuated to pressure ~ 0.1 mbar for gaseous hydrogen elimination. The next step was pressure increase registration in the system during container heating, caused by hydrogen desorption process.

The samples of SWNT with mass of 80 – 300 mg (estimated bulk density ~ 1000 kg/m³) were used for the experiments on the volumetric facility. The pressure drop character dynamics during container with non-irradiated samples cooling are shown in Figure 2.

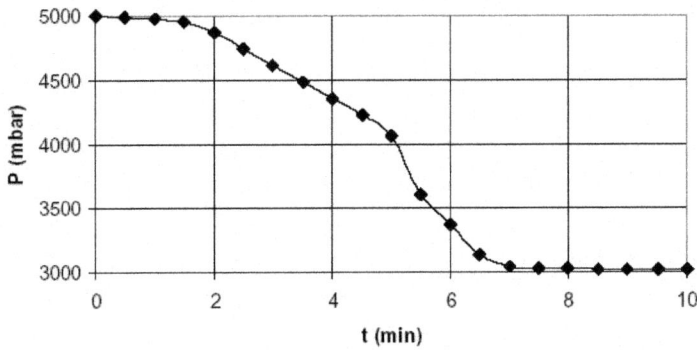

Figure 2: The dependence of pressure drop in the container with non-irradiated CNT samples upon time during container with samples cooling (P_F = 5035 mbar) (Obolensky et al., 2011b)

In Figure 2, one can see three sections with different sorption dynamics. On the base of comparison of this pressure drop dynamics with pressure drop dynamics for calibration stroke the following can be noted. The pressure drop relative to the empty container was observed already at room temperature during the letting-to-hydrogen and this pressure was noted at a ~ 20 mbar level. The additional pressure decrease was observed during further container cooling to 78 K temperature and at different cooling cycles this value was 10 – 15 mbar.

The pressure in the system during whole system heating was less than the pressure before the beginning of cooling by 10 – 12 mbar. Apparently this fact could be explained by any amount of hydrogen staying in binding state at room temperature after desorption. We didn't register this phenomenon on repeat of the heating-cooling cycles.

Figure 3 illustrates the correlation of pressure and temperature in the system for two evacuating regimes: fast (3 min) and gradual (15 min). Apparently already at 78 K temperature the increase of the duration of evacuation causes significant desorption and consequently elimination of part of hydrogen out from the system even before the beginning of heating. The duration of container heating for both regimes was approximately 25 – 30 min.

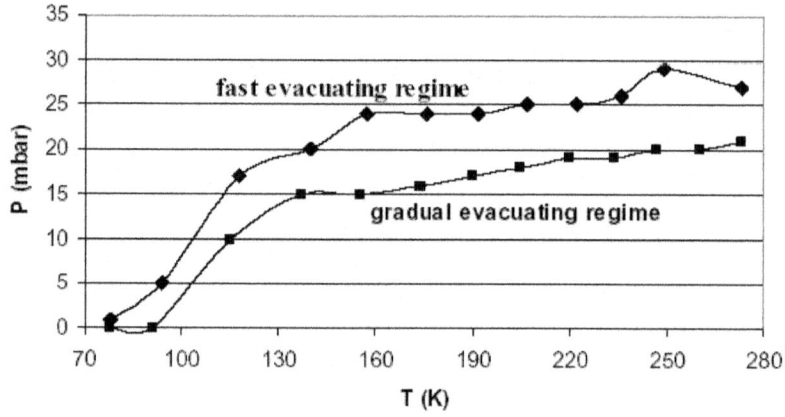

Figure 3: The dependence of pressure in the container with non-irradiated CNT samples upon temperature during container heating for different evacuation rates (Obolensky et al., 2011b)

The temperature dependences of hydrogen density in volume situated at heating regime (V_{78}) and in the rest facility volume at ambient temperature are shown in Figure 4. It could be seen that the main hydrogen mass is exuded at temperature T< 160 K that is caused by relatively low value of physical absorption activation energy. Pressure difference occurring at container heating without evacuating and initial pressure P_F as noted earlier is caused by other mechanism with higher values of desorption energies and it can be related to chemosorption regime with higher characteristic temperatures.

Our estimations show that at different sorption/desorption cycles the amount of hydrogen located in non-irradiated SWNT was equal to 0.12 ±0.2 mass percent.

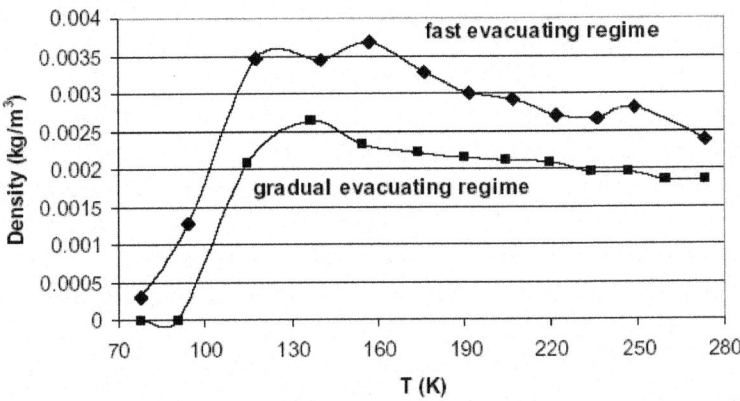

Figure 4: The dependence of hydrogen density in container with non-irradiated CNT samples upon temperature during container heating for different evacuation rates (Obolensky et al., 2011b)

Adsorption Studies of Irradiated Samples

Hydrogen storage was studied with 2.5-90 mg samples. Saturation was carried out at liquid nitrogen temperatures (~78 K), and 300 K and at pressures between 3 and 10 bar. The measurement of physically absorbed hydrogen was conducted in accordance with the following procedure.

First, after cooling the container down to 78 K temperature the system was evacuated to ~ 0.1 mbar of pressure for gaseous hydrogen elimination. Next, pressure in the system during container heating up to room temperature caused by hydrogen desorption process was registered. Hydrogen desorption out from SWNT was also studied in the 0-900 C temperature range by mass-spectrometry method on the MX 7203 mass spectrometer.

Our estimations (Obolensky et al., 2011b) show that at different sorption/ desorption cycles the amount of hydrogen located in non-irradiated SWNT was equal to 0.12 ±0.2 mass percent. Hydrogen desorption from treated and exposed to physical sorption/desorption procedure material at pressures ~3000 mbar as it is described above has been studied with the use of mass-spectrometry method within the 0 – 900 °C temperature interval on the MX 7203 mass spectrometer. The dependencies of the amount of hydrogen extracted from non-irradiated (a) and irradiated up to the fluence of 10^{14} e-/cm^2 (b) SWNT samples upon temperature are shown in arbitrary units in Figure 5.

As can be seen from Figure 5 the amount of hydrogen desorbed from irradiated material is approximately 2.5 times higher compared with that desorbed from the non-irradiated material. We have to draw attention to the fact that all procedures of the sorption/desorption processes were carried out at the 78 – 300 K temperature interval but at the same time significant amount of hydrogen is desorbed at temperatures higher than 300 K. There are additional peaks on the desorption curve which apparently correspond with different sorption sites appearing as the result of irradiation and can be characterized by various activation energies. It should be noted that over a long period of time of samples staying in air between irradiation procedures and hydrogen saturation, complete saturation generated by irradiation sites filling by the molecules of other gases has not been detected. The data in Figure 5 can be presented in another form (Figure 6) i.e. as additional amount of hydrogen desorbed from irradiated sample (Δ - the difference between the irradiated and not irradiated samples).

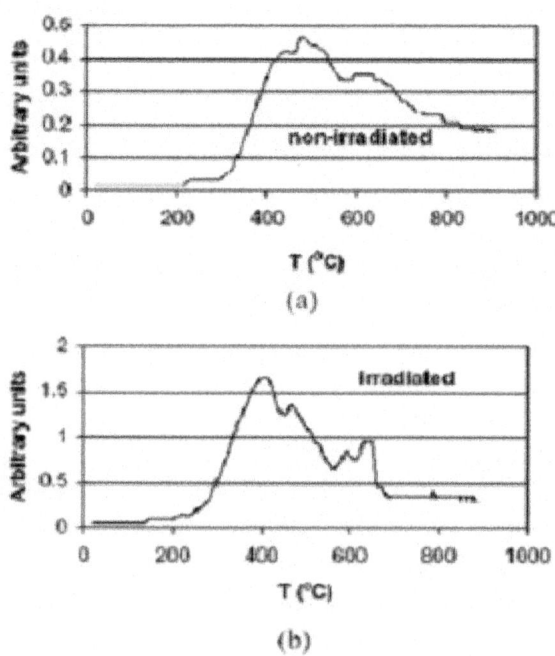

Figure 5: The dependencies of amount of hydrogen (arbitrary units) exuded from non-irradiated (a) and irradiated up to fluence 10^{14} e⁻/cm² (b) samples upon temperature (Obolensky et al., 2011b)

In order to control the structural transformations of CNTs, it is important to clearly understand the defect generation mechanism and to realize the extent of defects and their influence on sorption properties of CNTs. Acid or alkali treatment does not routinely improve the hydrogen storage capacity of CNT samples. A suitable ball-milling treatment and activation process can open the caps of CNTs and produce more structural defects, and therefore may be beneficial in improving their hydrogen storage properties (Chang & Hui-Ming, 2005). In this connection we investigated the effect of electron irradiation on the hydrogen adsorption property of SWNTs without mechanical milling.

Five AP samples (cf., Table 2) have been annealed in vacuum during 10 hours at 800°C temperature. The first sample (#1) was retained as the control sample. Sample #2 has remained for control and three samples (#3-5) were irradiated with various fluence ($5*10^{14}$, 10^{15} and $2*10^{15}$e⁻/cm²). After that, four samples (#2-5) were saturated at 10 bar pressure for 3 hours at 78 K temperature..

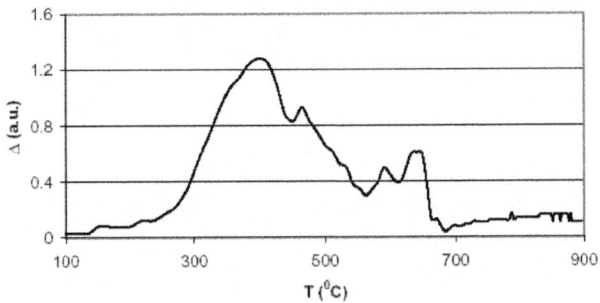

Figure 6: The dependence of additional amount of hydrogen (arbitrary units) desorbed from irradiated up to fluence 10^{14} e⁻/cm² sample upon temperature (Obolensky et al., 2011b)

Table 2: Sample history for adsorption studies

Sample	Vacuum anneal T(°C)/time (h)	Electron irradiation fluence (e⁻/cm²)/time (sec)	Hydrogen sorption T (K)/P (bar)	ΔH_2 (wt%)
1	800/10	-	-	-
2	800/10	-	~78/10	0.13
3	800/10	$5*10^{14}$/327	~78/10	0.12
4	800/10	10^{15}/706	~78/10	0.16
5	800/10	$2*10^{15}$/1340	~78/10	0.22

The dependencies of the amount of hydrogen extracted from the samples upon temperature are shown in arbitrary units in Figure 7. As can be seen from this Figure, the desorption curves were displaced in comparison with those for the samples exposed to all stages of treatment (i.e. oxidation, chemical treatment, annealing and milling shown in Figure 5.

For not irradiated samples the curves have remained similar, with some displacement of the maximum peak in the area of lower temperatures for the non-chemically treated and non-milled sample (2, Table 2). At the same time, the character of the curves for the irradiated samples (3-5, Table 2) essentially differs from the Ref. (Obolensky et al., 2011b). Characteristic peaks in the vicinity of 400 K and 650 K temperatures for the Ref. (Obolensky et al., 2011b) practically do not become visible for samples 3-5 (Table 2). With increase in fluence, the amount of hydrogen desorbed from irradiated samples is reduced for the given range of temperatures (Figure 8 and Figure 9).

As the temperature of samples at irradiation did not exceed 40°C (Figure 10), most likely the temperature factor could not affect decrease in the amount of hydrogen desorbed from the irradiated samples in comparison with the non-irradiated samples. This feature needs additional study.

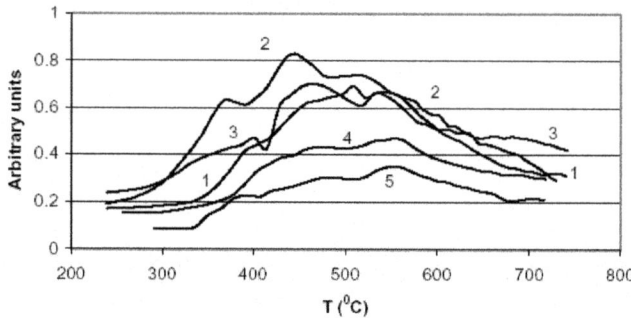

Figure 7: The dependencies of the amount of hydrogen (arbitrary units) exuded from samples (Table 1) upon temperature

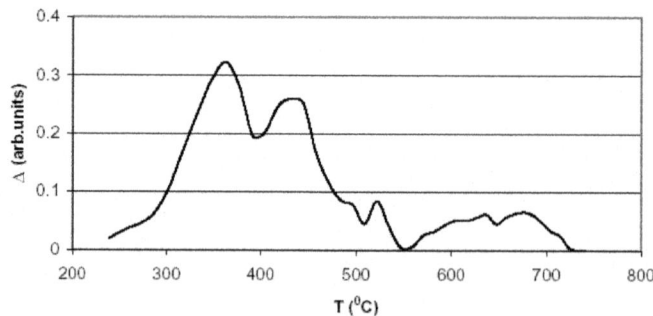

Figure 8: The dependence of the additional amount of hydrogen (the difference between saturated and non-saturated samples; arbitrary units) exuded from non-irradiated samples (Table 2) upon temperature

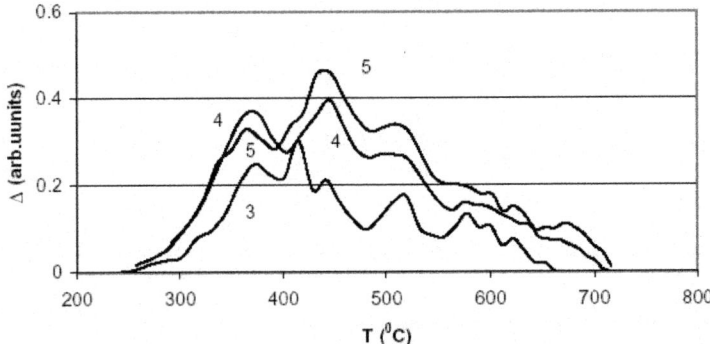

Figure 9: The dependencies of the additional amount of hydrogen (the difference between saturated non-irradiated sample and saturated irradiated samples; arbitrary units) for irradiated samples (Table 2)upon temperature

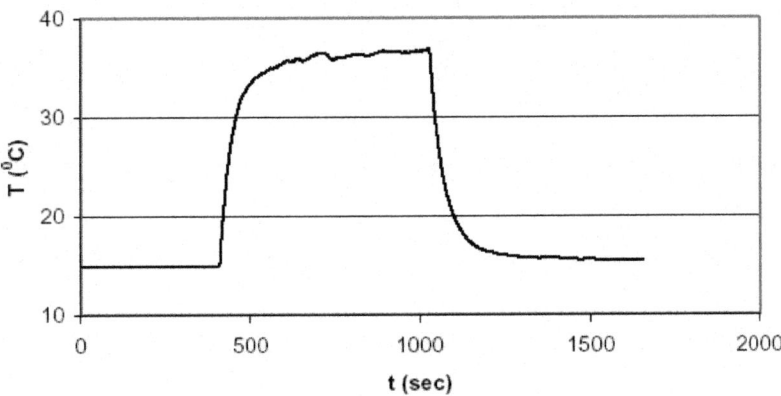

Figure 10: The dependence of samples temperature on the irradiation time (fluence $10^{15}e^-/cm^2$)

Raman Scattering Studies

Raman scattering is a sensitive probing of the structure and bonding in carbon materials, particularly carbon nanotubes. Raman scattering spectra for seven sets of samples (cf., Table 2) taken at room temperature in the range 100 to 3200 cm^{-1} are given in Figures 11-15..

The dominant spectral features include the low-frequency radial breathing modes (RBM) in the range approximately 150 to 200 cm^{-1} and the higher frequency modes in the range 1300 to 2700 cm^{-1} (Figure 11).

The SWNT tangential displacement modes observed near approximately 1600 cm^{-1} (G band) are related to the high frequency vibrational modes of a flat graphene sheet. The band (at 1300 cm^{-1}), commonly called the D band, has been observed in many sp^2-bonded carbon materials and is associated with disorder in the hexagonal carbon network.

Table 3: ample history for adsorption studies

Sample	Sample history	Vacuum anneal T(°C)/time (h)	Metal (at. %)
1	AP SWNTs [a], irradiation [b]	-	21-31
2	AP SWNTs [a] irradiation [b], sorption [c]	-	21-31
3	AP SWNTs [a], sorption [c]	1000/20	21-31
4	Treated [d], milling, sorption [c]	1000/20	6-9
5	Treated [d], milling, irradiation [b]	-	6-9
6	Treated [d], milling, irradiation [b], sorption [c]	-	6-9
7	Treated [d], milling, irradiation [e], sorption [c]	1000/20	6-9

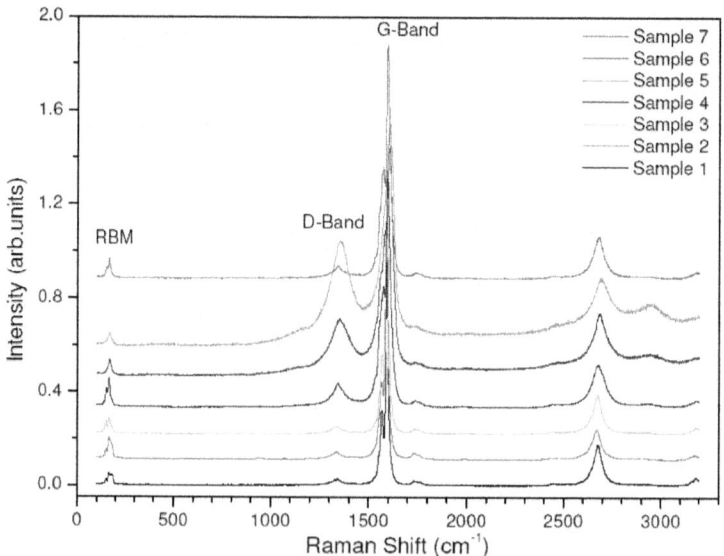

Figure 11: Room-temperature Raman spectra of bundles of arc-derived carbon nanotubes at various stages of post synthesis (Table 3). The Raman spectra were taken using 514-nm excitation

From the analysis of Raman spectra, it follows that the RBM frequency, ω_R, is inversely proportional to the tube diameter, d, while its value is up-shifted owing to the intertube interaction within a SWNT bundle (Pradhan et al., 2002). One of the empirical relations between d and ω_R, applicable for bundled SWNT is (Pradhan et al., 2002) is d (nm) = [224 cm^{-1} nm]/ [ω_R (cm^{-1}) - 12 cm^{-1}]. According to this equation, the main RBM peaks (Figure 12) at 163-166 cm^{-1} correspond to SWNT with a diameter of ~1.46 nm, whereas the shoulder at ~150 cm^{-1} is related to SWNT with a diameter of ~1.62 nm.

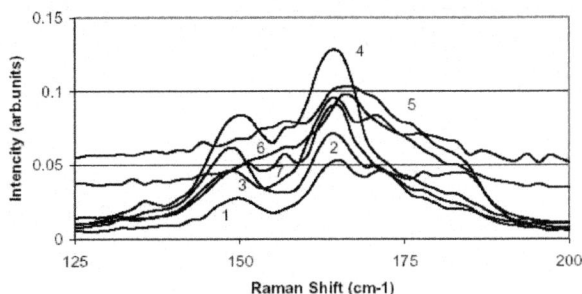

Figure 12: Raman spectra (RBM) at 514nm laser excitation energy (Table 3)

The intense Raman G-band at a higher energy corresponds to the C–C stretching vibrations in tangential and axial directions of the SWNT that splits to G$^-$ (tangential) and G$^+$ (axial) bands located at 1566-1571 cm^{-1} and 1588-1602 cm^{-1}, respectively. The shape of the G$^-$-band is sensitive to the electronic properties (strongly related to chirality) of SWNT.

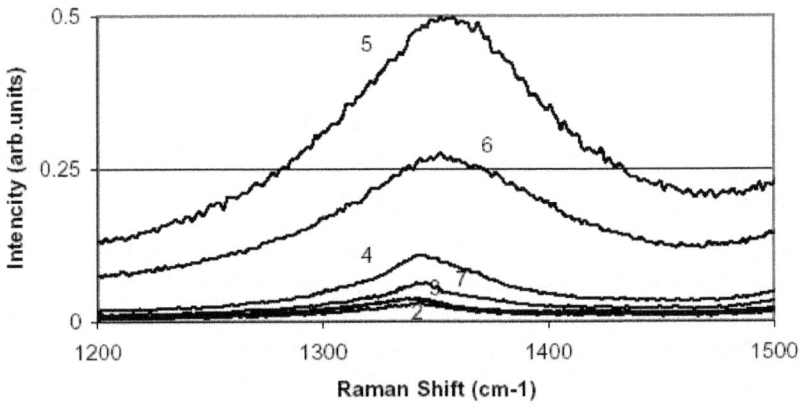

Figure 13: Raman spectra (D band) at 514nm laser excitation energy (Table 3)

D band (1341-1352 sm^{-1}, Figure 13) testifies to presence in samples of amorphous carbon and defects in structure of SWCNT. The ratio of peak intensity

D band to peak intensity G band allows to judge as a level of crystallization and quantity of defects. It is possible to see that for samples which were not exposed to chemical treating and cryogenic milling (#1-3), intensity of peak D is minimal (0.03-0.036; I(D)/I(G)= 0.333-0.793, fig.14) and quantity of defects in structure of SWCNT was small.

Intensity of peak D band culminates for sample #5 (0.498; I(D)/I(G)= 2.129, Figure 14) which was exposed to chemical treating and cryogenic milling, but was not exposed to high-temperature heating in vacuum, to an irradiation and to hydrogen saturation.

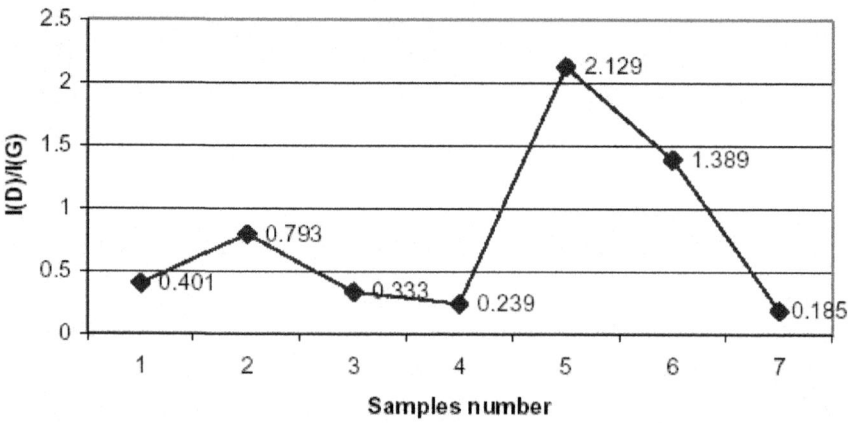

Figure 14: Distribution of the intensity ratio I(D)/I(G) for samples (Table 3)

Intensity of peak D for sample #6 which differed from sample #5 only on stages of hydrogen sorption is a little bit lower (0.275; I(D)/I(G)= 1.389, fig.14), but considerably exceeds intensity of peak D for samples #1-3. Apparently, during hydrogen sorption there was "treatment" of defects due to introduction of hydrogen in vacant sites of SWCNT.

Addition to the above-stated procedures of high-temperature heating in vacuum (samples #7 and #4) leads to essential decrease in intensity of peak D (0.062 and 0.107; I(D)/I(G)= 0.185 and 0.239, Figure 14). Most likely, at once two factors here dominate: treatment due to heating and treatment due to formation of C-H bond.

Range G' (2672-2686 cm^{-1}, Figure 15) which characterizes presence of "positive" or "negative" defects in SWCNT is submitted in spectra and has the obvious tendency: hydrogen sorption lowers intensity of this peak and leads to displacement of peak "to the left" in short-wave area of a spectrum.

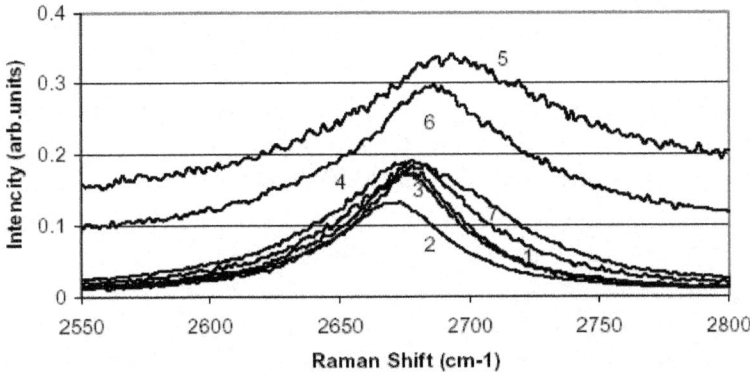

Figure 15: Raman spectra (G' band) at 514nm laser excitation energy (Table 3)

CONCLUSION

The amount of hydrogen desorbed from non-irradiated samples was 0.12 ±0.2 mass percent. The significant increase (more than two times) of hydrogen mass absorbed in irradiated SWCNT samples (10^{14} e$^-$/cm^2) relative to non-irradiated ones has been proved. For samples irradiated with fluences more than 10^{14} e-/cm^2 the decrease of hydrogen desorption was observed.

ACKNOWLEDGEMENT

This research was supported by Science and Technological Center in Ukraine, project #4957. The Raman measurements were performed at the Physics Institute of the Penn State University by Dr. Humberto R. Gutierrez and Dr. Xiaoming Liu.

REFERENCES

1.   F. Banhart, J. X. Li, A. V. Krasheninnikov, 2005Carbon nanotubes under electron irradiation: stability of the tubes and their action as pipes for atom transport. Phys. Rev. B 7124140841098-0121

2.   Solutions. Carbon, Inc, web, site: http://www.carbonsolution.com

3.   Chang Liu, & Hui-Ming Cheng.2005Carbon nanotubes for clean energy applications. J. Phys. D: Appl. Phys. 38: R231R252, 0022-3727

4.   M. E. Itkis, D. E. Perea, S. Niyogi, S. M. Rickard, M. A. Hamon, H. Hu, B. Zhao, R. C. Haddon, 2003Purity Evaluation of As-Prepared Single-Walled Carbon Nanotube Soot by Use of Solution-Phase Near-IR Spectroscopy. Nano Lett., 33309314c.

5.   A. Hashimoto, K. Suenaga, A. Gloter, K. Urita, S. Iijima, 2004Direct evidence for atomic defects in graphenelayers. Nature 4308708730028-0836

6.   M. A. Obolensky, A. V. Basteev, L. A. Bazyma, 2011Hydrogen storage in irradiated low dimensional structures. Fullerenes, Nanotubes, and Carbon Nanostructures, 1911331360153-6383X

7.   M. Obolensky, A. Kravchenko, V. Beletsky, Yu. Petrusenko, V. Borysenko, S. Lavrynenko, Basteev. A. Andrew, Bazyma. L. Leonid, 2011Thermal, Chemical and Radiation Treatment Influence on Hydrogen Adsorption Capability in Single Wall Carbon Nanotubes. Fundamentals of Low-Dimensional Carbon Nanomaterials. Edited by John J. Boeck. Mater. Res. Soc. Symp. Proc. 1284125130978-1-60511-261-9

8.   0884-291491722092016Pradhan, B.K.; Harutyunyan, A.R.; Stojkovic, D.; Grossman, J.C.; Zhang, P.; Cole, M.W.; Crespi, V.; Goto, H.; Fujiwara, J. & Eklund, P.C., (2002) Large cryogenic storage of hydrogen in carbon nanotubes at low pressures. J. Mater. Res. 17(9): 2209-2016, ISSN 0884-2914

Chapter 4

HIGH CAPACITY HYDROGEN STORAGE IN NI DECORATED CARBON NANOCONE: A FIRST-PRINCIPLES STUDY

S. Abdel Aal, A. S. Shalabi, K. A. Soliman

Department of Chemistry, Faculty of Science, Benha University, Benha, Egypt

ABSTRACT

Hydrogen adsorption and storage on Ni-decorated CNC has been investigated by using DFT. A single Ni atom decorated CNC adsorbs up to six H_2 with a binding energy of 0.316 eV/H_2. The interaction of $3H_2$ with Ni-CNC is irreversible at 603 K. In contrast, the interaction of $4H_2$ with Ni-CNC is reversible at 456 K. Further characterizations of the two reactions are considered in terms of the projected densities of states, electrophilicity, and statistical thermodynamic stability. The free energy of the reaction between $4H_2$ and Ni-CNC, surface coverage and rate constants ratio meet the ultimate targets of DOE at 11.843 atm, 0.925 and 1.041 respectively. The Ni-CNC complexes can serve as high-capacity hydrogen storage materials with capacities of up to 11.323 wt.%. It is illustrated that unless the access of oxygen to the surface is restricted, its strong bond to the decorated systems will preclude any practical use for hydrogen storage.

INTRODUCTION

Hydrogen is considered to be an ideal fuel for many energy converters because of its low mass density and nonpolluting nature. Hydrogen can also be directly used in fuel cells in transportation applications. However, hydrogen storage, which is safe, effective and stable, remains a notable challenge to be overcome before hydrogen's use in any automotive applications [1] [2] . Since the U.S. Department of Energy (DOE) has revised the targets for an on-board hydrogen storage medium (HSM), carbon-based materials have been considered candidates for a hydrogen storage media [3] . Accordingly, the medium should have a storage capacity of 4.50 wt.% by 2010 and 5.50 wt.% by 2015 [4] [5] .

Carbonaceous nanostructures have attracted considerable interest due to the search for new materials with specific applications. Zero-dimensional C_{60}, one-dimensional nanotubes, and two-dimensional graphene sheets have been in focus due to their mechanical and electric transport properties [6] -[8] . Modification of the carbon nanostructure surface can be achieved by doping with B and Al [9] [10] or decorating with alkali metal [11] , alkaline earth metal [12] , transition metals (Ti, Ni, Pd and Pt) [13] -[17] , and metal hydrides [18] . These modifications can be used to increase the interaction binding energy and therefore the hydrogen storage capacity. Lee et al. [19] have reported that functionalized SWCNT with Ni atoms can yield a storage capacity of 10 wt.%.

Various types of non-planar graphitic structures, such as carbon nanocones (CNCs), have been generated by using carbon arc and other related techniques [20] . CNCs are defined as hollow structures that are composed of carbon with a conical shape. The CNCs with different disclination angles have been observed in the pyrolysis of hydrocarbons [21] . Mechanical stability and sharp tip structures at the interface of CNCs usually have a lower density than carbon nanotubes, which makes them appropriate for field emissions due to the screening effect [22] .

Carbon nanohorns (CNHs) are a subclass of the carbon nanocones CNCs family. CNHs are the fifth allotropic form of carbon. CNHs have been selected and investigated for the use in hydrogen storage capacity because a significant amount of hydrogen is evolved at ambient temperatures [23] . Although the theoretical hydrogen storage on CNCs has not been heavily researched, nickel-doped CNCs are not reported. Ming-Liang Liao [24] investigated hydrogen adsorption behaviors of single walled carbon nanocones (SWCNCs) by molecular dynamics simulations. A. Gotzias et al. [25] examined hydrogen adsorption on CNHs and CNCs by using the grand canonical Monte Carlo method. Q. Wang et al. [26] tested the ability to store hydrogen by using the gradient corrected density functional theory. A. S. Shalabi et al. [27] investigated the reactions for hydrogen storage on Ti decorated carbon nanocones (CNC) by using density functional theory (DFT) calculations. Finally, J. Yang, et al. [28] presented an overview of experimental and computational techniques employed in the field of hydrogen storage materials research.

Theoretical studies have predicted that carbon-based materials decorated with transition metal (TM) atoms, such as Ti, Ni, Sc, and V, should be capable of binding up to five hydrogen molecules per metal atom with a binding energy between 7 and 12 kcal·mol^{-1} and a gravimetric density higher than 7 wt.% [11] [29] -[31] . Lee et al. [19] suggested that a single Ni atom deposited on a carbon nanotube could store up to five hydrogen molecules with a binding energy of 0.26 eV/H$_2$ at a desorption temperature of 328 K [19] . Nickel decoration on carbonaceous nanostructures can be used to improve the capacity of these compounds toward hydrogen storage [19] [28] [32] [33] . To date, the focus of previous studies has been on hydrogen adsorption on single metal atom [14] [30] . It was suspected that metal clustering might have reduced the area of hydrogen holding and thus reduced the hydrogen uptake [34] . It has been observed that the low diffusion barrier of the Ni atom on the SWCNC leads to high mobility of metal atoms and therefore leads to large formation energy of the Ni clusters that drives atomic aggregation. For the aggregation of metals on the SWCNCs, the major driving force is metal-to-metal attraction.

The aim of this work is to examine hydrogen storage capacity and the possibility of hydride formation upon hydrogen storage operation and to determine hydrogen storage capacity in the presence of oxygen molecules at the Ni decorated CNC. Finally, aggregation of the metal atoms on the adsorption media may occur (e.g., at ambient and elevated operational temperature) and should be carefully considered before one can assess the potential of the material for hydrogen storage.

COMPUTATIONAL METHODOLOGY

Electronic calculations to determine structural optimization and total energy were performed using DFT. The DFT calculations were performed by simultaneously using Becke's three parameter exchange function (B3) and the Lee Young Parr (LYP) correlation function [35] -[38] . B3LYP correctly reproduced the thermochemistry of many compounds, including transition metal atoms [39] -[42] . The analysis of electronic and thermochemical properties of CNC molecules was performed using the Gaussian 09 program [43] . The optimal geometries were visualized using the corresponding Gauss View software.

Full geometry optimizations, without symmetry constraints, were performed for CNC with a disclination angle of 120°, a height of 7 Å and with 72 carbon atoms. The optimal geometries of CNC, Ni-CNC, nH_2-Ni-CNC, nH_2-O_2-Ni-CNC and nH_2-Ni_2-CNC (n = 1 - 6) were determined at the B3LYP level of theory using 6 - 31 G (d, p) as the basis. This set uses Gaussian type functions (GTOs), adds d-type polarization functions to carbon, f-type polarization functions to nickel, and p-type polarization functions to hydrogen. The adsorption mechanisms were determined with natural bond orbital (NBO) analysis and partial density of states (PDOS) plots, which are capable of providing a definitive description for charge redistribution. The projected densities of states (PDOS) and Fermi levels were performed using the Gauss Sum 2.2.5 program [44] .

RESULTS AND DISCUSSION

Structure of Pure and Ni Doped CNC

It is well-known that the decoration of carbonaceous nanostructures with TM atoms may be an attractive alternative to improve hydrogen storage capacity. Hence, the first SWCNC, decorated with single Ni atom, has been investigated.

The adsorption energy ($E_{ads.}$) of Ni atom on the surface was calculated as

$$E_{ads.} = E(\text{Ni-CNC}) - E(\text{CNC}) - E(\text{Ni}) \tag{1}$$

where E(Ni-CNC) and E(CNC) are the total energy of the fully relaxed Ni-CNC and CNC, respectively. E(Ni) is the energy of the isolated Ni atom. The negative binding energy corresponds to an exothermic process. The adsorption energy of Ni atom is −4.369 eV, and the average distance between Ni atom and the nearest C atoms is 1.843 Å. This strong binding interaction may originate from the hybridization between the nickel atom and CNC as the atomic projected density of states (PDOS). The corresponding interactions to the pure Ni atom and doped Ni atom have been shown inFigure 1 and Figure 2. Due to low ionization potentials of the Ni atom, the s electrons are easily donated to the CNC. These donated electrons partially fill the unoccupied states of CNC as is indicated by the PDOS near the Fermi level (shown in Figure 1). The occupied s orbital of Ni, just below the Fermi energy at −6 eV Figure 2(a), reduced in Figure 2(b) due to the charge transfer from the Ni atom to the CNC. Conversely, there is a finite probability for the CNC to donate back part of its received electrons to the low-lying d orbitals of the Ni atoms, resulting in strong hybridizations between the Ni and CNC.

The NBO analysis shows that a charge of 0.590 e is transferred from Ni to the CNC, as indicated by the decreased peak of Ni if it is supported on the CNC (Figure 2(b)). Consequently, the increased dipole moment of Ni-CNC (19.097 D), compared to the case of pristine CNC (at 12.774 D), is sufficient to enhance the van der Waal's interaction between H_2 molecules and the Ni-CNC. A fraction of the charge, between −0.223 and −0.225, is transferred from Ni and distributed to the nearest neighbor carbon atoms. This finding confirmed that Ni donates electrons to the neighboring C atoms of CNC, thereby resulting in the d-orbitals of the Ni atom overlapping with the sp^2 orbitals of the C atoms. Such strong hybridizations can be observed from the resonated peaks in PDOS near −7 eV, as shown in Figure 2(b).

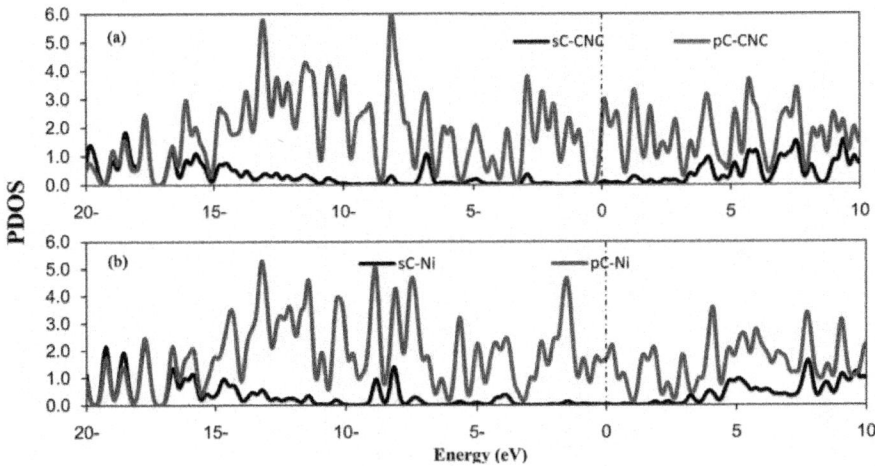

Figure 1: The partial density of states (PDOS) of (a) CNC (b) Ni doped CNC. The Fermi level is set to zero and indicated by a dotted line.

H$_2$ Adsorption in Ni Doped CNC

Next, the interaction between the Ni/CNC complexes with H$_2$ molecules has been investigated. The average adsorption energy of a H$_2$ molecule at a Ni decorated CNC surface is obtained from the expression

$$E_b\left(H_2\right) = \left[\frac{E\left(nH_2\text{-Ni-CNC}\right) - E\left(\text{Ni-CNC}\right) - E\left(nH_2\right)}{n}\right] \tag{2}$$

In this equation, "E" is the total energy of the optimized structure, and "n" represents the number of adsorbed H$_2$ molecules. Negative energy indicates a stable exothermic process.

Figure 3 shows optimized geometries corresponding to a supported Ni atom that is surrounded by one to six H$_2$ molecules. The average adsorption binding energies and their corresponding distance per hydrogen molecule are presented in Table 1. From the viewpoint of adsorption geometry, it was noticed that the first 3 H$_2$ molecules were adsorbed on Ni by chemisorption. Further adsorption occurred as physisorption. This finding suggests that the first three hydrogen molecules interacted with the supported Ni-CNC via chemical bonds instead

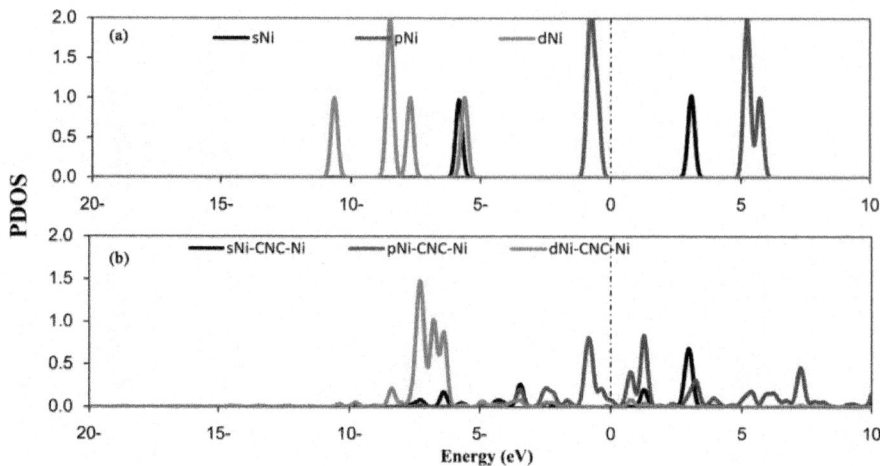

Figure 2: PDOS of (a) pure nickel atom (b) nickel doped CNC. The Fermi level is set to zero and indicated by a dotted line.

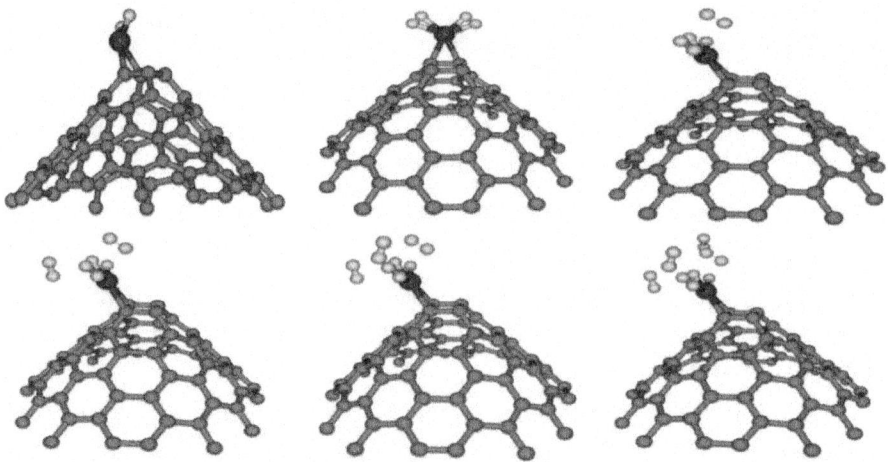

Figure 3. Optimized structure of one to six H_2 adsorbed at nickel doped CNC (nH_2-Ni-CNC) where n (1 - 6). Circles in grey, blue and white denote carbon, nickel and hydrogen atoms, respectively. For interpretation of the references to color in this figure legend, the reader is referred to the web version of this article.

Table 1: Structural and energy properties of the optimized systems. Average adsorption energies of H_2 on Ni/CNC complex $E_b(H_2)$ in eV, average distances between Ni and H_2 (d_{Ni-H}), average H_2 bond length (d_{H-H}) in (Å), charges (Q) in a.u.

	$E_b(H_2)$	d_{Ni-CNC}	d_{Ni-R}		d_{H-H}		Q_{Ni}	Q_C	Q_{H-H}
			$n=1-2$	$n=3-6$	$n=1-2$	$n=3-6$			
1H$_2$-Ni-CNC	−1.068	1.89	1.558	-	0.84	-	0.61	−0.45	−0.042
2H$_2$-Ni-CNC	−0.926	1.93	1.599	-	0.82	-	0.45	−0.39	0.099
3H$_2$-Ni-CNC	−0.622	1.93	1.598	3.081	0.82	0.74	0.45	−0.39	0.099
4H$_2$-Ni-CNC	−0.468	1.93	1.599	3.095 4.089	0.82	0.74	0.45	−0.40	0.091
5H$_2$-Ni-CNC	−0.387	1.93	1.611	3.737 4.180	0.82	0.74	0.44	−0.40	0.099
6H$_2$-Ni-CNC	−0.316	1.93	1.612	4.061 4.298 4.322 4.406	0.82	0.74	0.43	−0.40	0.100

of the Kubas interaction. Under this condition, the H-H bonds were slightly elongated from 0.743 to 0.826 Å compared to the bond distance of 0.74 Å of an isolated H_2 molecule. This result confirmed that the bonds between H_2 and the supported Ni atom had both physical and chemical bond characteristics. The s electrons of hydrogen were slightly hybridized with the d orbital of Ni, which weakened the interaction between Ni and C atoms of CNC. Therefore, the average adsorption distances between the CNC and the decorated Ni atom were enhanced due to the subsequent addition of H_2 molecules. The corresponding Ni-C distance also elongated to between 1.896 and 1.926 Å. The average H_2-Ni adsorption distances were also found to have increased when more H_2 molecules were adsorbed. These results are reported in Table 1.

It is observed that there are three different mechanisms for hydrogen adsorption on metal decorated Carbonaceous compounds: 1) polarization under the effect of the electrical field induced by the supported metal atoms; 2) the hybridization between H_2 molecules and metal atoms that is modulated by the electrostatic potential induced by metal atoms; and 3) the formation of hydrogen super-molecules. The substantial charge redistribution (due to the Ni atoms donating their s electrons to the C-sp^2 in the CNC) can lead to a high electric field near the Ni atoms. Consequently, the electric field causes the polarization of H_2 molecules. In the interactions of H_2 molecules with a decorated Ni atom, the positive charge on the Ni atom and the Coulomb's repulsive energy also reduces. Theoretically, the interaction of H_2 with TM basically arises due to the hybridization of the d levels of TM with H_2 levels. To elucidate the hybridization between the Ni atoms and H_2 molecules, the PDOS of three and four H_2 per Ni atom on CNC have been plotted in Figure 4 and Figure 5. In Figure 4, the peaks are centered at approximately −14 - 16 eV, which corresponds to the hybridization of the Ni 3d with the H_2 σ orbitals. The peaks that are centered at approximately 1 - 3 eV correspond to the hybridization of the Ni 3d with the H_2σ* orbitals. Conversely, the hydrogen

PDOS are shown in the range between −17 eV and −12 eV and are split into several peaks. This finding confirms the formation of bonding and antibonding states between the H_2 molecules, which results in the formation of the hydrogen super-molecule.

INTERACTIONS OF 3H$_2$ AND 4H$_2$ WITH NI-CNC

Two types of interactions between nH_2 and Ni-CNC can be identified from Table 1: 1) irreversible interactions between nH_2 (n = 1 - 3) and Ni-CNC 2) reversible interactions between nH_2 (n = 4 - 6) and Ni-CNC. The irreversible interactions are outside the range of the desirable energy (−0.2 to −0.6 eV), as recommended by the DOE for practical applications. In contrast, the reversible interactions fall within this range. To characterize the nature of these two types of interactions, we considered the following theoretical descriptors.

Electronic Properties

The electronic descriptors, formation energy (ε), ionization potential (IP), electron affinity (EA), chemical potential (l), electronegativity (χ), chemical hardness(η) and electrophilicity index (ω) of the complexes nH_2-Ni- CNC (where n = 3, 4) were considered in Table 2.

The IP and EA can be calculated from the highest occupied (HOMO) and the lowest unoccupied (LUMO)

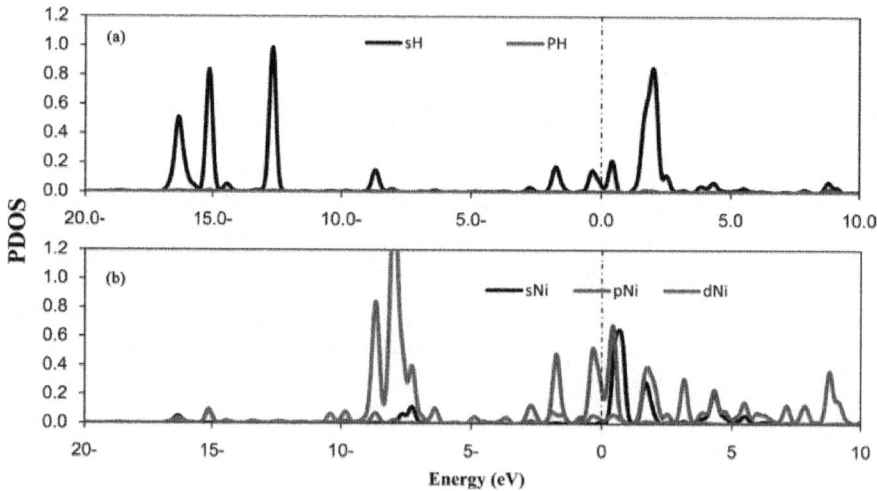

Figure 4: PDOS of (a) three H_2 molecules, (b) decorated Niatomsin the (3H$_2$-Ni-CNC) system. The Fermi level is set to zero and indicated by a dotted line.

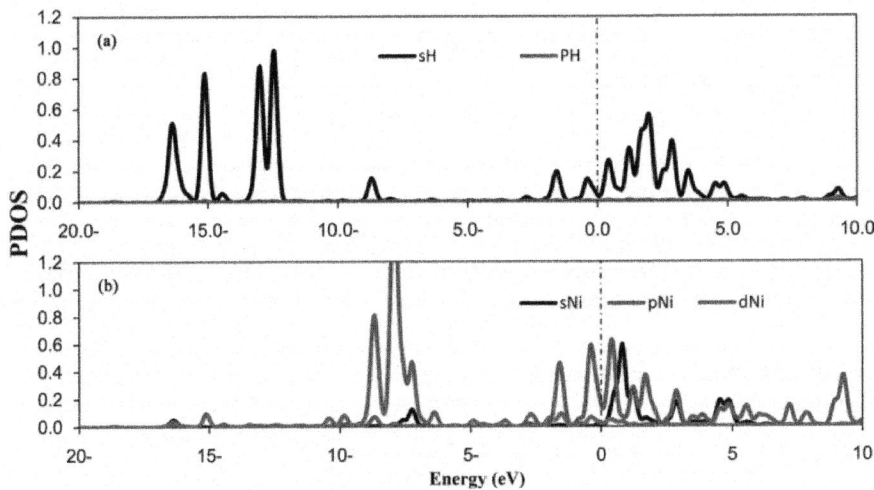

Figure 5: PDOS of (a) four H_2 molecules (b) decorated Ni atoms in the $(4H_2$-Ni-CNC) system. The Fermi level is set to zero and indicated by a dotted line.

Table 2: The total energy (E_{total}/Eh), formation energy (ε/eV), energy gap (Eg), ionization potential (I/eV), electron affinity (A/eV), chemical hardness (η/eV), electronegativity (χ/eV), and electrophilicity (ω/eV) of the complexes nH_2-Ni-CNC (for n = 3, 4)

System	E_{total}	ε	E_g	I	A	η	χ	S	ω
$3H_2$-Ni-CNC	−4253.369	−1.868	0.498	5.558	5.061	0.249	5.309	2.009	56.642
$4H_2$-Ni-CNC	−4254.548	−1.871	0.489	5.555	5.066	0.244	5.310	2.046	57.704

molecular orbital energies using Koopmans' approximation [45] where, IP— HOMO and EA—LUMO. The chemical potential μ and χ are defined as [45]

$$\mu = -\chi = -\frac{IP + EA}{2}.$$

Pearson [46] introduced two parameters 'chemical hardness (η)' and 'chemical softness (S)' to account for the stability of a molecule. Hardness (η) can also be expressed in terms of HOMO and LUMO, implying a finite

difference approach [45] , as follows: $\eta \approx \dfrac{IP + EA}{2} \approx \dfrac{E_{LUMO} - E_{HOMO}}{2}$. The softness can

be defined as $S = \dfrac{1}{2\eta}$ [47] . Parr [48] defined electrophilicity index (ω) as: $\dfrac{\mu^2}{2\eta} = \dfrac{\chi^2}{2\eta}$

which measures the energy stabilization when the molecule accepts an additional electrical charge from the environment. It is noted that a lower energy gap (E_g) between the LUMO and HOMO of a compound implies a greater and easier possibility of the electron transition between these energy levels. Additionally, a small value for E_g for the compound is an indicator of lower chemical stability. In other words, the respective chemical hardness (η) should be low and electrophilicity (ω), which is a parameter indicating reactivity, should be high.

The results presented in Table 2 show that the $4H_2$-Ni-CNC shows a notably lower value of E_g and η and a maximum value compared to the $3H_2$-Ni-CNC. The η value for a compound essentially represents how chemically hard a compound is. Because $4H_2$-Ni-CNC is chemically softer, it implies that it is a better candidate in electronic transport. The electrophilicity index () is one of the DFT based parameters which quantifies how reactive a molecule or compound is.It is clear from the present analysis that the lower E_g and η values and higher value for the $4H_2$-Ni-CNC (compared to the $3H_2$-Ni-CNC compound) implies its favorable nature for the possible electronic transport and conductivity. Polarizability is closely associated with the softness (S) of a system, where the more polarizable a chemical system is, the softer the compound, and vice versa.

Thermodynamic Properties

The thermodynamics of the hydrogen storage reaction is one of the most fundamental properties of the hydrogen storage material. Thermodynamic properties indicate that the pressure of desorbed hydrogen and operating temperature are required for a fuel cell. The heat requirements for desorption and the potential for on-board recharge (or "reversibility") are also associated with the thermodynamic properties of the storage reaction.

The thermodynamic properties of the $3H_2$-Ni-CNC and $4H_2$-Ni-CNC complexes can be calculated from standard statistical mechanical equations to include the finite-temperature translational, rotational and vibrational energies. The enthalpy (H_r) can be calculated as follows:

$$H_r = E_{elec.}(T=0) + E_{vib.}(T=0) + E_{vib.}(T) + E_{rot.}(T) + E_{tra.}(T) = PV \tag{4}$$

where $E_{elec.}(T=0\,K)$ is the total electronic energy, $E_{vib.}(T=0\,K)$ is the zero point vibrational energy (ZPVE), which is a linear sum of the fundamental harmonic frequencies, and $E_{vib.}(T)$, $E_{rot.}(T)$ and $E_{tra.}(T)$ are vibrational, rotational, and translational contributions, respectively.

Similarly, the total entropy (S) can be expressed as

$$S = S_{elec.} + S_{vib.} + S_{rot.} + S_{tran.} \tag{5}$$

where $S_{elec.}$, $S_{vib.}$, $S_{rot.}$, and $S_{tran.}$ are the electronic, vibrational, rotational, and translational terms, respectively. The change in the standard Gibbs free energy is given by

$$\Delta G = \Delta H_r - T\Delta S \tag{6}$$

where $\Delta H_r = H_{rP} - H_{rR}$ and $\Delta S = S_P - S_R$, or simply

$$\Delta G = G_P - G_R \tag{7}$$

where P = product, and R = reactants.

The results of thermochemical properties of entropy, enthalpy, Gibbs free energy changes, thermal energies (E_t), and heat capacities at constant volume (C_v), for the reactions 8 and 9 processed from 100 K to 700 K are presented in Table 3.

$$3H_2 + \text{Ni-CNC} \rightarrow 3H_2\text{-Ni-CNC} \tag{8}$$

and

$$4H_2 + \text{Ni-CNC} \rightarrow 4H_2\text{-Ni-CNC} \tag{9}$$

It can be clearly deduced from this table that as temperature (T) increases, the values of enthalpy (H) and entropy (S) increase, while the Gibbs free energy (G) decreases. The negative Gibbs free energy (G) also indicates an exothermic process, where the system releases energy to its surroundings during the adsorption process. The system then gradually reaches a stable condition of equilibrium. Thus, the reaction with G = −31.276 kcal/mol at 100 K has a higher probability of occurring than that of G = 6.036 kcal/mol at 700 K, (shown in Figure 6).

The polynomial regression

$$y = a_5 x^5 + a_4 x^4 + a_3 x^3 + a_2 x^2 + a_1 x + a_0 \tag{10}$$

was subsequently applied to reactions (8) and (9) from 100 to 700 K after replacing (y) by (T) and (x) by (ΔG). The residual sum of squares (rss = 0) value at $\Delta G = 0$ occurs at T = 603 K for reaction (8), and at T = 454 for reaction (9). Therefore, reactions (8) and (9) reverse above 603 K and 454 K, respectively. This implies that the two complexes, $3H_2$-Ni-CNC and $4H_2$-Ni-CNC, tend to release all hydrogen molecules above 603 and 454 K, respectively. In other words, the higher the amount of hydrogen molecules at the Ni-CNC interface, the lower temperature of hydrogenation. This can be explained by the relatively lower stability of the higher hydrogenated Ni-CNC. The other statistical thermodynamic parameters also characterized the two types of interactions, where ΔS, E_t, and C_v values of the reversible interaction were always greater than those of irrreversible interaction at the same temperature range.

Optimal Reaction Enthalpy

The temperature and pressure range at which a hydrogen storage system should operate is dictated by the environment and the requirements of the fuel cell. This approach translates vehicle operating constraints into thermodynamic constraints, which can be used to guide material development. Enthalpy is considered as the quantity of heat that must be added to (or subtracted from) the system during hydrogen release (or uptake). It is demonstrated that materials that have large enthalpies of desorption are undesirable because they require high temperatures for hydrogen release. In principle, a system with a small desorption enthalpy is capable of liberating

Table 3: Temperature (T/K), Gibbs free energy ΔG (kcal/mol), enthalpy change ΔH (kcal/mol), entropy change ΔS (cal/mol.K), thermal energy E_t(K.cal/mol), and heat capacity at constant volume Cv (cal/mol.K) of the complexes nH_2-Ni-CNC (for n = 3, 4)

T	ΔG		ΔH		ΔS		E_t		Cv	
	n = 3	n = 4	n = 3	n = 4	n = 3	n = 4	n = 3	n = 4	n = 3	n = 4
100	−31.276	−28.289	−36.862	−35.838	−55.866	−75.496	28.793	37.201	20.597	26.973
200	−25.372	−20.447	−37.587	−36.431	−61.072	−79.908	31.19	40.427	26.555	36.169
300	−19.171	−12.430	−37.975	−36.487	−62.679	−80.186	34.01	44.274	29.592	40.342
400	−12.872	−4.441	−38.145	−36.25	−63.186	−80.527	37.078	48.444	31.679	42.907
500	−6.549	3.466	−38.149	−36.513	−63.198	−80.915	40.33	52.832	33.307	44.774
600	−0.241	11.268	−38.313	−36.820	−63.455	−81.479	43.729	57.385	34.628	46.235
700	6.036	15.789	−38.764	−37.251	−63.574	−81.896	47.249	60.123	35.731	48.987

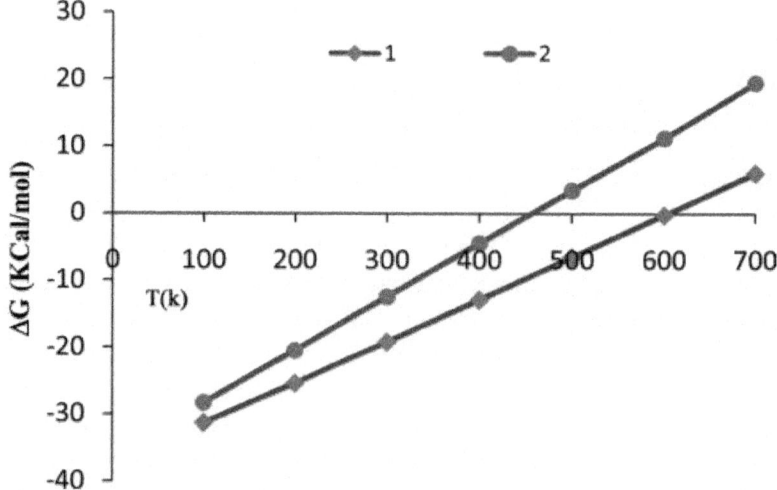

Figure 6: Variations of Gibbs free energy (ΔG) with temperature (T) of the complexes 1: $3H_2$-Ni-CNC and 2: $4H_2$-Ni-CNC.

hydrogen at low temperatures but will require notably high pressures to recharge. Consequently, the enthalpy is an important engineering design parameter [28] . It is possible to identify a range of reaction enthalpies that satisfy the ultimate DOE targets of temperatures and pressures by using the van't Hoff equation. These targeted temperatures and pressures are presented in Table 4. The van't Hoff equation can describe the equilibrium between gas phase hydrogen and one or more condensed phases [49] :

$$P_{H_2} = P_o \exp\left(\frac{-\Delta H}{RT} + \frac{\Delta S}{R}\right)$$

(11)

where P_{H_2} and P_o are referred to the equilibrium pressure and a reference pressure (typically atmospheric pressure), respectively, T is the absolute temperature, R is the gas constant, S and H respectively represent the change in entropy and enthalpy, which accompanies the hydrogen storage reaction to describe H_2 uptake in sorbent materials [28] .

The thermodynamics of $4H_2$-Ni-CNC complex have been calibrated with the ultimate targets of DOE at (−0.2 to −0.6 eV) for physisorption, (−40°C to 105°C) for (min./max.) temperature, (0.3/1.2MPa) for (min./max.) pressure, and (7.5%). The results shown in Table 4 illustrate that the equilibrium pressure of 1.2 MPa is equal to the ultimate target for maximum H_2 pressure that is required in the fuel cell. The desorption reaction requires a desorption enthalpy of −25.45 kJ/mol H_2, which is in agreement with the previous theoretical calculations of Yang et al. [28] , who reported efficiency considerations that target the lower third of this range (H = 20 - 30 kJ/mol H_2), are optimal.

In addition, the Langmuir isotherm is based on the monolayer adsorption on the active sites of the adsorbent. Langmuir suggested that the mechanism of the adsorption process is defined as $A_g + S = AS$, where A is a gas molecule and S is an adsorption site. The direct and inverse rate constants are k and k_{-1}. Surface coverage, which is defined as the fraction of the number of adsorption sites occupied in the equilibrium, is shown in Equation (12).

$$K = \frac{k}{k_{-1}} = \frac{\theta}{(1-\theta)P}$$

(12)

where P is the partial pressure of the gas. Substituting the values of (0.925) and P (11.843 atm.) of the reversible reaction from Equation (9) gives the value (k = 1.041).

Effect of Clustering Ni on CNC

For the next step, we examined the interactions and clustering effects of Ni ((e.g. Ni dimer at CNC) on the nature of hydrogen uptake. Full geometry optimizations at the B3LYP/6-31g (d, p) level of theory were carried out for the complexes nH_2-Ni_2-CNC (n = 2, 4, 6, 8, 10, 12). The corresponding geometry is shown in Figure 7. The average adsorption energy, of n = 2, 4, 6, 8, 10, 12 hydrogen molecules at Ni dimer decorated CNC, is obtained from the expression,

$$E_b\left(H_2\right) = \left[\frac{E\left(nH_2\text{-}Ni_2\text{-CNC}\right) - E\left(Ni_2\text{-CNC}\right) - E\left(nH_2\right)}{n}\right]$$

The average adsorption energies of with their geometric parameters are listed in Table 5. The results indicate that the bond length of Ni-Ni is 2.43 Å and it is elongated compared to that of the free Ni dimer at 2.01 Å. The bond lengths of Ni-C are higher than those in Figure 3, due to the internal interactions between the two Ni atoms that weaken the Ni?C interaction. Conversely, the interaction between CNC and Ni dimer weakens the Ni-Ni bond. The average adsorption energies of nH_2-Ni-CNC (n = 1 - 6) complexes are approximately −1.068, −0.926, −0.622, −0.468, −0.387 and −0.316 eV per H_2, which is stronger than that on Ni dimer at −1.057, −0.820, −0.546, −0.416, −0.333 and −0.278 eV, respectively. These values indicate that the formation of the Ni dimer also weakens the interaction between the H_2 and the doped CNC. Figure 7 shows the saturated hydrogen uptake on a Ni_2 cluster. Note that 2 of the 6 adsorbed hydrogen molecules are completely dissociated and bonded in the atomic form (irreversible reaction), while the other four H_2 remain in molecular form (reversible reaction). This suggests that clustering of Ni not only alters the binding energy of the H_2 molecules but also the nature of H bonding. This adsorption behavior may be explained based on the charge-transfer mechanism discussed earlier. In fact, the average charge transfer from the Ni dimer of nH_2-Ni_2-CNC (n = 2 - 12) complexes to the CNC is only 0.48 and 0.39 eV per Ni atom, which is less than that of a single Ni atom of nH_2-Ni-CNC (n = 1 - 6)

Table 4: The (min./max.) temperature (T), (min./max.) pressure (P), free energy change (ΔG), enthalpy change (ΔH), and surface coverage (θ) of the highest hydrogen storage capacity reaction $4H_2$ + Ni-CNC = $4H_2$-Ni-CNC

T(°C/K°)	P(MPa/atm.)	ΔG(kcal/mol)	ΔH(kJ/mol H_2)	θ
105/378.15	1.2/11.8430792	−13.61	−25.33	0.925
105/378.15	0.3/2.9607698	−9.44	−25.33	0.744
−40/233.15	1.2/11.8430792	−22.38	−25.45	0.925
−40/233.15	0.3/2.9607698	−19.81	−25.45	0.744

Table 5: Average adsorption energies of H2 $E_b(H_2)$ on Ni dimer at CNC complex, average distance between Ni and CNC d(Ni-CNC), average distances between Ni and H_2 (d_{Ni-H}), average H_2 bond length (d_{H-H}), charges (Q) in a.u. and the expected hydrogen storage capacity (wt.%) of the complexes nH_2-Ni_2-CNC (n = 2, 4, 6, 8, 10, 12). Complete surface coverage affords up to 24 Ni and 144 H_2. Energies are given in eV and lengths in Å

System	$E_b(H_2)$	d_{Ni-CNC}	d_{Ni-Ni}	Q_{Ni}	Q_C	Q_{H-H}	d_{H-H} n = 1 - 2	d_{H-H} n = 3 - 6	Capacity/wt.%
2H$_2$-Ni$_2$-CNC	−1.057	1.873 1.891	2.477	0.475 0.475	−0.712	−0.066	0.842		2.084
4H$_2$-Ni$_2$-CNC	−0.820	1.913 1.912	2.786	0.378 0.375	−0.639	0.217	0.815		4.082
6H$_2$-Ni$_2$-CNC	−0.546	1.913 1.913	2.784	0.375 0.380	−0.641	0.230	0.816	0.743	6.001
8H$_2$-Ni$_2$-CNC	−0.416	1.913	2.785	0.391 0.394	−0.648	0.226	0.815	0.744	7.845
10H$_2$-Ni$_2$-CNC	−0.333	1.913	2.779	0.395 0.391	−0.651	0.225	0.814	0.743	9.617
12H$_2$-Ni$_2$-CNC	−0.278	1.913	2.778	0.395 0.395	−0.652	0.226	0.815	0.743	11.323

Optimized Ni—Ni free without CNC = 2.0085 Å.

Optimized Ni—Ni free without CNC = 2.0085 Å.

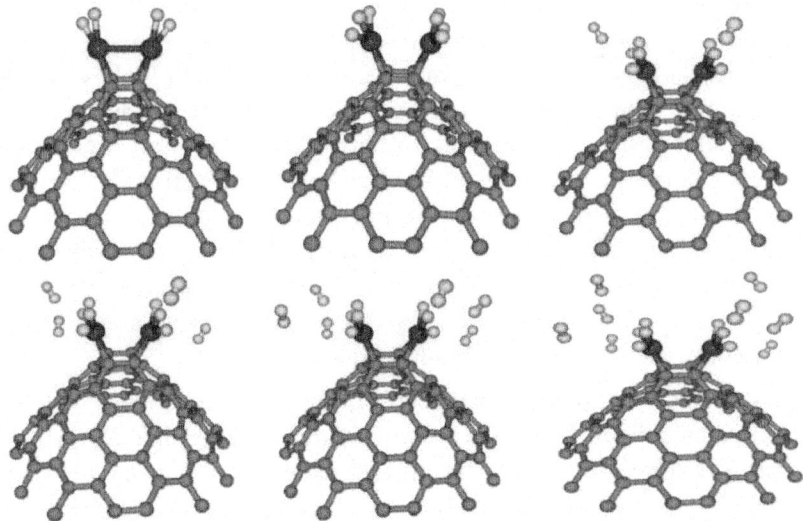

Figure 7: Optimized structure of one to six H_2 adsorbed at nickel dimer doped CNC (nH_2-Ni_2-CNC) where n [1 - 6]. The circles in grey, blue and white denote carbon, nickel and hydrogen atoms, respectively. For interpretation of the references to color in this figure legend, the reader is referred to the web version of this article.

complexes to the CNC at 0.61 and 0.43 eV, respectively.

The expected gravimetric hydrogen storage capacities for nH_2-Ni_2-CNC (n = 2 - 12) complexes are listed in Table 5. For nH_2-Ni_2-CNC (n = 2, 4) complexes, the average adsorption energies of H_2 are -1.057 and -0.820 eV, respectively, which is beyond the highest adsorption requirement (-0.60 eV). The hydrogen storage capacities of these two configurations are calculated to be 2.084% and 4.082%, respectively. The average adsorption energies of nH_2-Ni_2-CNC (n = 6, 8, 10, 12) complexes are -0.55, -0.50, -0.42 and -0.35 eV, respectively, which meet the DOE energy domain (-0.20 to -0.60 eV). The hydrogen storage capacities of these four configurations are expected to be 6.001%, 7.845%, 9.617%, and 11.323%, respectively. Consequently, these results indicate that the Ni-decorated C_{72} stories capable of storing 144 hydrogen molecules attached to 24 Ni atoms. Thus, the hydrogen storage capacity is up to 11.323 wt.%, which exceeds the DOE target for 2015 [5] .

Interaction of Oxygen with Ni Decoration

Decoration of different carbonaceous materials with metals has been investigated as an alternative to improve their capacity for hydrogen storage [9] [14] [19] [30] [32] [50] [51] . To follow experimental conditions, the effect of oxygen contamination from any residual atmosphere or pre-existing oxygen impurities on the nanotube surface, was examined. Rojas et al. [52] found that low quantities of oxygen present in the gas phase should yield the oxidation of the titanium atoms, even when hydrogen is stored in the system. They concluded that if the experimental system is exposed to air, titanium atoms on these surfaces are expected to oxidize to titanium dioxide. Felten et al. [53] studied the role of oxygen at the interface between titanium and carbon nanotubes. They observed that the presence of oxygen significantly weakened the Ti-CNT interaction and the Ti atoms at the surface preferentially bonded to oxygenated sites. Among these, oxygen molecules were found to be a strong inhibitor competitor for hydrogen adsorption.

To study the competition between O_2 and H_2 molecules at the Ni-CNC surface, we calculated the oxygen displacement energy that corresponded to the energy required to replace the adsorbed oxygen atom by nH_2 adsorbed hydrogen molecules. This displacement energy was calculated according to

$$E_{dis} = E\left(O_2\text{-Ni-CNC}\right) + n_{H_2}E_{H_2} - E_{O_2} - E\left(nH_2\text{-Ni-CNC}\right)$$

(13)

where $E(nH_2\text{-Ni-CNC})$ and $E(O_2\text{-Ni-CNC})$ denote the total energy of the substrate (Ni-CNC system) in the presence of nH_2 hydrogen molecules and O_2 oxygen molecules, respectively, n is the number of hydrogen molecules, and EO_2 and EH_2 are the energies of the isolated oxygen and hydrogen molecules respectively. In all cases, the values for E_{dis} are shown in Table 6, which indicate that the replacement of oxygen by hydrogen is energetically impeded.

The results show that the two O atoms bind strongly to a single Ni atom (−3.799 eV) and Ni-O bonds of 1.76 Å, which significantly weakens the Ni-C interaction (bonds dilated to 2.02 Å). Consequently, the quantity of oxygen present in the initial surface layer crucially controls the Ni-CNC interaction. This is because initial Ni deposition preferentially forms Ni-O bonds, strongly reducing the interaction between the hydrogen molecules and Ni-CNC.

The average adsorption energy of H_2 molecule at the surface of O_2-Ni-CNC was calculated according to

$$E_b\left(H_2\right) = \frac{E\left(nH_2\text{-}O_2\text{-Ni-CNC}\right) - E\left(O_2\text{-Ni-CNC}\right) - n_{H_2} E_{H_2}}{nH_2}$$

(14)

where $E(nH_2\text{-}O_2\text{-Ni-CNC})$ and $E(O_2\text{-Ni-CNC})$ are defined earlier. The geometries obtained are shown in Figure 8. The results are reported in Table 6 and show that even at low oxygen concentrations, the interaction of the O_2 molecule with the Ni decoration on CNC leads to the irreversible formation of nickel dioxide, and displaces the hydrogen molecule. The resulting nickel dioxide produces a lower storage capability than nickel at CNC. The formation of nickel dioxide was confirmed by further theoretical calculations [54].

The destabilization of the O_2 molecule by the adsorbed hydrogen molecule may be expressed as

$$E_b\left(O_2\right) = E\left(nH_2\text{-}O_2\text{-Ni-CNC}\right) - E\left(nH_2\text{-Ni-CNC}\right) - E_{O_2}$$

(15)

$E_b(O_2)$ is shown in the third column of Table 6, where it is found that the addition of hydrogen molecules exhibited a decrease in the binding energy of the oxygen molecules. However, O_2 molecules remained strongly adsorbed on the surface, even in the presence six hydrogen molecules, with a binding energy of −2.633 eV.

Table 6: E_{dis} Oxygen displacement energy corresponding to the energy required to replace the adsorbed oxygen atom by nH_2 adsorbed hydrogen molecules. $E_b(H_2)$ adsorption energy (per hydrogen atom) of hydrogen species on a surface where one oxygen molecule is already adsorbed on the Ni atom. $E_b(O_2)$ binding energy of the oxygen molecule on a surface where nH_2 hydrogen molecules are already adsorbed on the Ni atom

System	$E_b(H_2)$	ΔE_{dis}	$E_b(O_2)$	d_{Ni-CNC}	d_{Ni-H}	d_{H-H}	Q_{Ni}	Q_C	Q_{H-H}	QO_2
1H₂-O₂-Ni-CNC	−0.694	−2.781	−3.475	2.018	1.658	0.789	0.729	−0.326	0.099	−0.623
2H₂-O₂-Ni-CNC	−0.376	−0.973	−2.698	2.019	1.659 3.941	0.789 0.745	0.734	−0.324	0.090	−0.629
3H₂-O₂-Ni-CNC	−0.230	−0.647	−2.694	2.019	1.659 4.177 4.257	0.789 0.745 0.744	0.738	−0.325	0.091	−0.630
4H₂-O₂-Ni-CNC	−0.189	−0.482	−2.686	2.020	1.656 4.068 4.470 4.480	0.745 0.745 0.745 0.789	0.739	−0.323	0.074	−0.649
5H₂-O₂-Ni-CNC	−0.166	−0.373	−2.633	2.019	1.659 3.461 3.779 4.490 5.117	0.743 0.745 0.745 0.745 0.789	0.746	−0.324	0.077	−0.645
6H₂-O₂-Ni-CNC	−0.146	−0.317	−2.778	2.021	1.657 3.588 3.828 4.503 5.161 5.314	0.789 0.745 0.745 0.745 0.745 0.744	0.755	−0.324	0.071	−0.655

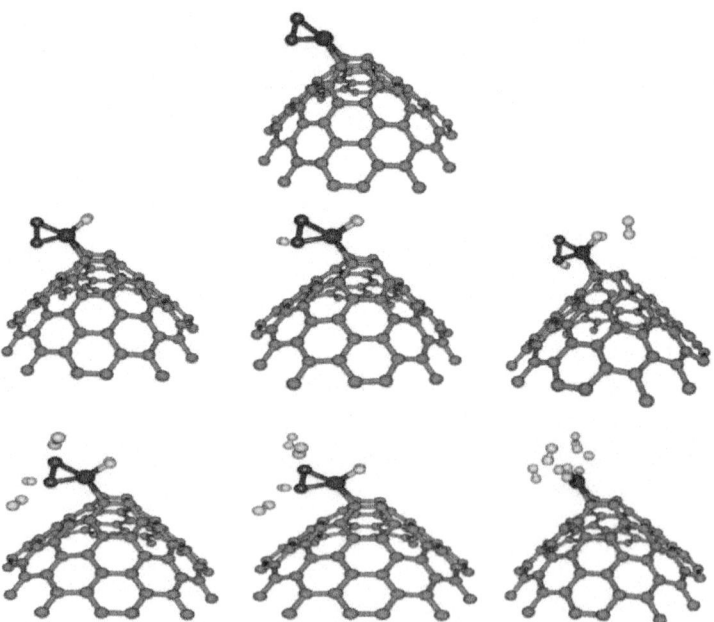

Figure 8: Optimized structure of competition between hydrogen and oxygen adsorption on the nickel doped CNC of one to six H₂. Balls in grey, blue, red and white

denote carbon, nickel, oxygen and hydrogen atoms respectively. For interpretation of the references to color in this figure legend, the reader is referred to the web version of this article.

Consequently, under normal conditions, oxygen interferes strongly with hydrogen adsorption.

The calculations show strong qualitative variations in the charge transfer behavior due to the adsorption of O_2 molecule at Ni-CNC surface. From Table 1, the interaction of Ni atom on CNC shows a strong net charge transfer from the Ni to the CNC resulting in a Ni charge of +0.589 e, where the H_2 molecules transfer charge to the decorated Ni atom. In contrast, the interaction of oxygen molecules with Ni-CNC causes the charge at the Ni atom to increase to +0.971 e. This is due to the O_2 molecules being more electronegative and withdrawing electrons from the metal charge on the oxygen atoms (−0.655 e). Therefore, the dipole moment of O_2-Ni-CNC and Ni-CNC is calculated to be 10.831 and 19.097, respectively. This is sufficient to reduce the van der Waal's interaction between hydrogen and O_2-Ni-CNC. Figure 9 confirms the results in Table 6 and the adsorption properties that were discussed earlier. Ao et al. [55] [56] showed that the application of a perpendicular electric field may act as a catalyst for the dissociative adsorption of hydrogen on pristine and nitrogen-doped graphene. Therefore, the application of an external field could be an alternative to favor the selective hydrogen adsorption. This subject will be a topic of future research.

CONCLUSIONS

Using DFT calculations, Ni-decorated CNC has been investigated for hydrogen storage applications. PDOS and NBO analysis have been performed to understand the H_2 adsorption mechanism. The adsorption mechanism for H_2 on Ni decorated CNC was primarily attributed to the polarization induced by electrostatic field of metal atoms on CNC and the hybridization between the Ni atom and hydrogen molecules. Two types of reactions (reversible and irresversible) were characterized in terms of (PDOS), electrophilicity, and statistical thermodynamic

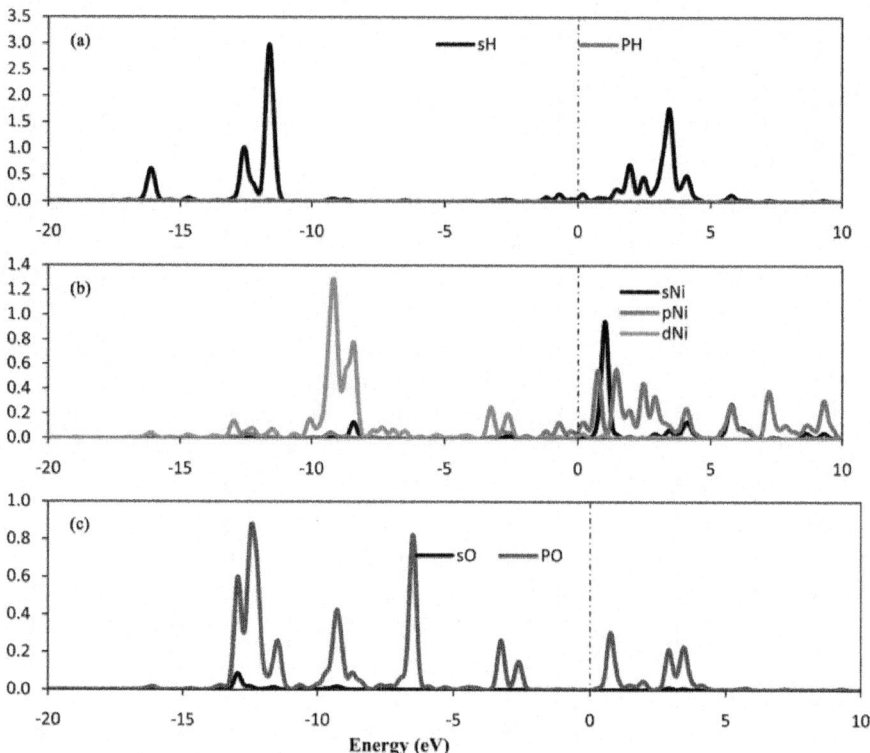

Figure 9: PDOS of (a) oxygen molecule (b) decorated Ni atoms in the (4H$_2$-O$_2$-Ni-CNC) system. The Fermi level is set to zero and indicated by a dotted line.

stability descriptors. The thermodynamics of the nH$_2$-Ni-CNC (n = 3, 4) reactions with reference to the ultimate targets of the DOE for physisorption, gravimetric hydrogen storage capacity, minimum and maximum temperatures and pressures, and optimal reaction enthalpy, were analyzed in considerable detail.

The adsorption binding energy meets the ultimate values for achieving adsorption and desorption at near ambient conditions. However, due to lower migration barriers and strong metal-to-metal attraction, aggregation of metal atoms in the form of a metallic layer or cluster on typical experimental CNC seems inevitable. It was observed that the structure where Ni atoms cluster was −0.738 eV lower in energy than when they remained isolated. Ni$_2$-CNC can bind six H$_2$ molecules and their corresponding storage capacity is 11.323 wt.%, which is higher than the revised 2015 target of U.S. DOE. Consequently, Ni-CNC could be used as a high capacity hydrogen storage medium in onboard automobile applications. On the other hand, the present results show that the competitive adsorption between H_2 and O_2 molecules strongly favors oxygen. Therefore, O_2 strongly reduces the interaction of hydrogen molecules with the Ni-CNC. Consequently, this interaction will inhibit the practical use for hydrogen storage.

REFERENCES

1. Xu, W.C., Takahashi, K., Matsuo, Y., Hattori, Y., Kumagai, M., Ishiyama, S., et al. (2007) Investigation of Hydrogen Storage Capacity of Various Carbon Materials. International Journal of Hydrogen Energy, 32, 2504-2512. http://dx.doi.org/10.1016/j.ijhydene.2006.11.012

2. Sakintuna, B., Lamari-Darkrim, F. and Hirscher, M. (2007) Metal Hydride Materials for Solid Hydrogen Storage: A Review. International Journal of Hydrogen Energy, 32, 1121-1140. Zhang, Z.W., Li, J.C. and Jiang, Q. (2010) Hydrogen Adsorption on Eu/SWCNT Systems: A DFT Study. The Journal of Physical Chemistry C, 114, 7733-7737. http://dx.doi.org/10.1021/jp100017y

3. Schlapbach, L. (2009) Technology: Hydrogen-Fuelled Vehicles. Nature, 460, 809-811. http://dx.doi.org/10.1038/460809a

4. U.S. Department of Energy. http://www.eere.energy.gov/hydrogen and fuel cells/mypp/

5. Kuzmin, S. and Duley, W.W. (2010) Ab Initio Calculations of a New Type of Tubular Carbon Molecule Based on Multi-Layered Cyclic C6 Structures. Physics Letters A, 374, 1374-1379.

6. Baei, M.T., Peyghan, A.A., Bagheri, Z. and Tabar, M.B. (2012) B-Doping Makes the Carbon Nanocones Sensitive towards NO Molecules. Physics Letters A, 377, 107-111. http://dx.doi.org/10.1016/j.physleta.2012.11.006

7. Iijima, S. (1991) Helical Microtubules of Graphitic Carbon. Nature, 354, 56-58. http://dx.doi.org/10.1038/354056a0

8. Surya, V.J., Iyakutti, K., Venkataramanan, N., Mizuseki, H. and Kawazoe, Y. (2010) The Role of Li and Ni Metals in the Adsorbate Complex and Their Effect on the Hydrogen Storage Capacity of Single Walled Carbon Nanotubes Coated with Metal Hydrides, LiH and NiH2. International Journal of Hydrogen Energy, 35, 2368-2375. http://dx.doi.org/10.1016/j.ijhydene.2010.01.001

9. Cui, S.S., Zhao, N.Q., Shi, C.S., Feng, C., He, C.N., Li, J.J. and Liu, E.Z. (2014) Effect of Hydrogen Molecule Dissociation on Hydrogen Storage Capacity of Graphene with Metal Atom Decorated. The Journal of Physical Chemistry C, 118, 839-844.http://dx.doi.org/10.1021/jp409594r

10.   Wang, F.D., Zhang, N.N., Li, Y.H., Tang, S.W. and Sun, H. (2013) High-Capacity Hydrogen Storage of Na-Decorated Grapheme with Boron Substitution: First-Principles Calculations. Chemical Physics Letters, 555, 212-216.http://dx.doi.org/10.1016/j.cplett.2012.11.015

11.   Hussain, T., Pathak, B., Ramzan, M., Maark, T.A., Ahuja, R. and De Sarkar, A. (2012) Calcium Doped Graphane as a Hydrogen Storage Material. Applied Physics Letters, 100, Article ID: 183902. http://dx.doi.org/10.1063/1.4710526

12.   Reyhani, A., Mortazavi, S.Z., Mirershadi, S., Moshfegh, A.Z., Parvin, P. and Golikand, A.N. (2011) Hydrogen Storage in Decorated Multiwalled Carbon Nanotubes by Ca, Co, Fe, Ni, and Pd Nanoparticles under Ambient Conditions. The Journal of Physical Chemistry C, 115, 6994-7001. http://dx.doi.org/10.1021/jp108797p

13.   Yildirim, T. and Ciraci, S. (2005) Titanium-Decorated Carbon Nanotubes as a Potential High-Capacity Hydrogen Storage Medium. Physical Review Letters, 94, Article ID: 175501. http://dx.doi.org/10.1103/PhysRevLett.94.175501

14.   Lopez-Corral, I., German, E., Juan, A., Volpe, M.A. and Brizuela, G.P. (2012) Hydrogen Adsorption on Palladium Dimer Decorated Graphene: A Bonding Study. International Journal of Hydrogen Energy, 37, 6653-6665.http://dx.doi.org/10.1016/j.ijhydene.2012.01.039

15.   Liu, Y., Brown, C.M., Neumann, D.A., Geohegan, D.B., Puretzky, A.A. and Rouleau, C.M. (2012) Metal-Assisted Hydrogen Storage on Pt-Decorated Single-Walled Carbon Nanohorns. Carbon, 50, 4953-4964. http://dx.doi.org/10.1016/j.carbon.2012.06.028

16.   Vinayan, B.P., Sethupathi, K. and Ramaprabhu, S. (2013) Facile Synthesis of Triangular Shaped Palladium Nanoparticles Decorated Nitrogen Doped Graphene and Their Catalytic Study for Renewable Energy Applications. International Journal of Hydrogen Energy, 38, 2240-2250. http://dx.doi.org/10.1016/j.ijhydene.2012.11.091

17.   Seenithurai, S., Pandyan, R.K., Kumar, S.V. and Mahendran, M. (2013) H2 Adsorption in Ni and Passivated Ni Doped 4 Å Single Walled Carbon Nanotube. International Journal of Hydrogen Energy, 38, 7376-7381. http://dx.doi.org/10.1016/j.ijhydene.2013.04.085

18.   Lee, J.W., Kim, H.S., Lee, J.Y. and Kanga, J.K. (2006) Hydrogen Storage and Desorption Properties of Ni-Dispersed Carbon Nanotubes. Applied Physics Letters, 88, Article ID: 143126. http://dx.doi.org/10.1063/1.2189587

19.   Iijima, S., Ichihashi, T. and Ando, Y. (1992) Pentagons, Heptagons and Negative Curvature in Graphite Microtubule Growth. Nature, 356, 776-778. http://dx.doi.org/10.1038/356776a0

20.   Ge, M. and Sattler, K. (1994) Observation of Fullerene Cones. Chemical Physics Letters, 220, 192-196.http://dx.doi.org/10.1016/0009-2614(94)00167-7

21.   Huang, C.J., Chih, Y.K., Hwang, J., Lee, A.P. and Kou, C.S. (2003) Field Emission from Amorphous-Carbon Nanotips on Copper. Journal of Applied Physics, 94, 6796. http://dx.doi.org/10.1063/1.1620681

22.   Matelloni, P., Grant, D.M. and Walker, G.S. (2009) Supporting Metal Catalysts on Modified Carbon Nanocones to Optimize Dispersion and Particle Size. MRS Proceedings, 1216. http://dx.doi.org/10.1557/PROC-1216-W02-02

23.   Liao, M.-L. (2012) A Study on Hydrogen Adsorption Behaviors of Open-Tip Carbon Nanocones. Journal of Nanoparticle Research, 14, 837. http://dx.doi.org/10.1007/s11051-012-0837-1

24.   Gotzias, A., Heiberg-Andersen, H., Kainourgiakis, M. and Steriotis, T. (2011) A Grand Canonical Monte Carlo Study of Hydrogen Adsorption in Carbon Nanohorns and Nanocones at 77K. Carbon, 49, 2715-2724. http://dx.doi.org/10.1016/j.carbon.2011.02.062

25.   Wang, Q., Sun, Q., Jena, P. and Kawazoe, Y. (2009) Potential of AlN Nanostructures as Hydrogen Storage Materials. ACS Nano, 3, 621-626. http://dx.doi.org/10.1021/nn800815e

26.   Shalabi, A.S., Taha, H.O., Soliman, K.A. and Abeld Aal, S. (2014) Hydrogen Storage Reactions on Titanium Decorated Carbon Nanocones Theoretical Study. Journal of Power Sources, 271, 32-41.http://dx.doi.org/10.1016/j.jpowsour.2014.07.158

27.   Yang, J., Sudik, A., Wolverton, C. and Siegelw, D.J. (2010) High Capacity Hydrogenstorage Materials: Attributes for Automotive Applications and Techniques for Materials Discovery. Chemical Society Reviews, 39, 656-675.http://dx.doi.org/10.1039/B802882F

28.   Kubas, G.J. and Organo, J. (2001) Metal-Dihydrogen and σ-Bond Coordination: The Consummate Extension of the Dewar-Chatt-Duncanson Model for Metal-Olefin π Bonding. Journal of Organometallic Chemistry, 635, 37-68.http://dx.doi.org/10.1016/S0022-328X(01)01066-X

29.   Durgun, E., Ciraci, S. and Yildirim, T. (2008) Functionalization of Carbon-Based Nanostructures with Light Transition-Metal Atoms for Hydrogen Storage. Physical Review B, 77, Article ID: 085405. http://dx.doi.org/10.1103/PhysRevB.77.085405

30.   Samolia, M. and Kumar, T.J.D. (2014) Hydrogen Sorption Efficiency of Titanium-Functionalized Mg-BN Framework. The Journal of Physical Chemistry C, 118, 10859-10866. http://dx.doi.org/10.1021/jp501722z

31.   Zielinski, M., Wojcieszak, R., Monteverdi, S., Mercy, M. and Bettahar, M.M. (2007) Hydrogen Storage in Nickel Catalysts Supported on Activated Carbon. International Journal of Hydrogen Energy, 32, 1024-1032.http://dx.doi.org/10.1016/j.ijhydene.2006.07.004

32.   Ayala, P., Freire, F.L., Gu, L., Smith, D.J., Solo'rzano, I.G. and Macedo, D.W. (2006) Decorating Carbon Nanotubes with Nanostructured Nickel Particles via Chemical Methods. Chemical Physics Letters, 431, 104-109.http://dx.doi.org/10.1016/j.cplett.2006.09.039

33.   Sun, Q., Wang, Q., Jena, P. and Kawazoe, Y. (2005) Clustering of Ti on a C60 Surface and Its Effect on Hydrogen Storage. Journal of the American Chemical Society, 127, 14582-14583. http://dx.doi.org/10.1021/ja0550125

34.   Becke, A.D. (1993) Density-Functional Thermochemistry. III. The Role of Exact Exchange. The Journal of Chemical Physics, 98, 5648. http://dx.doi.org/10.1063/1.464913

35.   Vosko, S.H., Wilk, L. and Nusair, M. (1980) Accurate Spin-Dependent Electron Liquid Correlation Energies for Local Spin Density Calculations: A Critical Analysis. Canadian Journal of Physics, 58, 1200-1211.http://dx.doi.org/10.1139/p80-159

36.   Becke, A.D. (1988) Density-Functional Exchange-Energy Approximation with Correct Asymptotic Behavior. Physical Review A, 38, 3098-3100. http://dx.doi.org/10.1103/PhysRevA.38.3098

37.   Lee, C., Yang, W. and Parr, R.G. (1988) Development of the Colle-Salvetti Correlation-Energy Formula into a Functional of the Electron Density. Physical Review B, 37, 785-789.http://dx.doi.org/10.1103/PhysRevB.37.785

38.   Ricca, A. and Bauschlicher, C.W. (1994) Successive Binding Energies of Fe(CO)5+. The Journal of Physical Chemistry, 98, 12899-12903. http://dx.doi.org/10.1021/j100100a015

39.   Russo, T.V., Martin, R.I. and Hay, P.J. (1995) Application of Gradient-Corrected Density Functional Theory to the Structures and Thermochemistries of ScF3, TiF4, VF5, and CrF6. The Journal of Chemical Physics, 102, 8023.http://dx.doi.org/10.1063/1.469000

40.   Siegbahn, P.E. and Crabtree, R.H. (1997) Mechanism of C-H Activation by Diiron Methane Monooxygenases: Quantum Chemical

Studies. Journal of the American Chemical Society, 119, 3103-3113. http://dx.doi.org/10.1021/ja963939m

41.   Wolverton, C., Siegel, D.J., Akbarzadeh, A.R. and Ozolin, V. (2008) Discovery of Novel Hydrogen Storage Materials: An Atomic Scale Computational Approach. Journal of Physics: Condensed Matter, 20, Article ID: 064228.http://dx.doi.org/10.1088/0953-8984/20/6/064228

42.   Frisch, M.J., et al. (2009) Gaussian 09. Gaussian Inc., Pittsburgh.

43.   O'Boyle, N.M., Tenderholt, A.L. and Langner, K.M. (2008) Cclib: A Library for Package-Independent Computational Chemistry Algorithms. Journal of Computational Chemistry, 29, 839-845. http://dx.doi. org/10.1002/jcc.20823

44.   Parr, R.G. and Yang, W. (1989) Density Functional Theory of Atoms and Molecules. Oxford University Press, New York.

45.   Pearson, R.G. (1973) Hard and Soft Acids and Bases. Dowden, Hutchinson and Ross, Inc., Stroudsburg.

46.   Yang, W. and Parr, R.G. (1985) Hardness, Softness, and the Fukui Function in the Electronic Theory of Metals and Catalysis. Proceedings of the National Academy of Sciences of the United States of America, 82, 6723-6726.http://dx.doi.org/10.1073/pnas.82.20.6723

47.   Parr, R.G., Szentpály, L.V. and Liu, S.B. (1999) Electrophilicity Index. Journal of the American Chemical Society, 121, 1922-1924. http://dx.doi. org/10.1021/ja983494x

48.   Griessen, R. and Riesterer, T. (1988) Heat of Formation Models. In: Schlapbach, L., Ed., Topics in Applied Physics: Hydrogen in Intermetallic Compounds I, Springer, Berlin, Heidelberg, 219-284.

49.   Kim, H.S., Lee, H., Han, K.S., Kim, J.H., Song, M.S., Park, M.S., et al. (2005) Hydrogen Storage in Ni Nanoparticle-Dispersed Multiwalled Carbon Nanotubes. The Journal of Physical Chemistry B, 109, 8983-8986.http://dx.doi.org/10.1021/jp044727b

50.   Ni, M., Huang, L., Guo, L. and Zeng, Z. (2010) Hydrogen Storage in Li-Doped Charged Single-Walled Carbon Nanotubes. International Journal of Hydrogen Energy, 35, 3546-3549. Rojas, M.I. and Leiva, E.P.M. (2007) Density Functional Theory Study of a Graphene Sheet Modified with Titanium in Contact with Different Adsorbates. Physical Review B, 76, Article ID: 155415.http://dx.doi.org/10.1103/PhysRevB.76.155415

51.   Felten, A., Suarez-Martinez, I., Ke, X., Van Tendeloo, G., Ghijsen, J. and Pireaux, J.J. (2009) The Role of Oxygen at the Interface between Titanium and Carbon Nanotubes. ChemPhysChem, 10, 1799-1804.http://dx.doi.org/10.1002/cphc.200900193

52. Sigal, A., Rojas, M.I. and Leiva, E.P.M. (2011) Interferents for Hydrogen Storage on a Graphene Sheet Decorated with Nickel: A DFT Study. International Journal of Hydrogen Energy, 36, 3537-3546.http://dx.doi.org/10.1016/j.ijhydene.2010.12.024

53. Ao, Z.M. and Peeters, F.M. (2010) Electric Field Activated Hydrogen Dissociative Adsorption to Nitrogen-Doped Graphene. The Journal of Physical Chemistry C, 114, 14503-14509. http://dx.doi.org/10.1021/jp103835k

54. Ao, Z.M. and Peeters, F.M. (2010) Electric Field: A Catalyst for Hydrogenation of Grapheme. Applied Physics Letters, 96, Article ID: 253106.

Chapter 5

EFFECTS OF GAMMA IRRADIATION ON THE KINETICS OF THE ADSORPTION AND DESORPTION OF HYDROGEN IN CARBON MICROFIBRES

Cesar Mota[1], Mario Culebras[2], Andrés Cantarero[2], Antonio Madroñero[1], Clara Maria Gómez[2], Jose María Amo[1], Jose Ignacio Robla[1]

[1]Centro Nacional de Investigaciones Metalúrgicas (CSIC), Madrid, Spain

[2]Materials Science Institute, University of Valencia, Valencia, Spain

ABSTRACT

In this study, three types of carbon fibres were used, they were ex-polyacrylonitrile carbon fibres with high bulk modulus, ex-polyacrylonitrile fibres with high strength, and vapour grown carbon fibres. All the samples were subjected to a hydrogen adsorption process at room temperature in an over-pressured atmosphere of 25 bars. The adsorption process was monitored through electrical resistivity measurements. As conditioning of the fibres, a chemical activation by acid etching followed by γ-ray irradiation with ^{60}Co radioisotopes was performed. The surface energy was determined by means of the sessile drop test. Both conditioning treatments are supplementary; the chemical activation works on the outer surface and the γ-irradiation works in the bulk material as well. Apparently, the most significant parameter for hydrogen storage is the crystallite size. From this point of view, the most convenient materials are those with small grain size because hydrogen is accumulated mainly in the grain boundaries.

INTRODUCTION

The increase of the pollution level at the atmosphere and the oceans produced by the residues of fossil fuels and atomic residues is disquieting. The change of fossil fuels first and atomic fission later by "green energy" represents a drastic change in the World Economy, since the cost of the energy will necessarily increase during the next decades. In particular, the progress towards the hydrogen economy [1] is a demand and that must be accepted and implemented.

From this perspective, energy alternatives shall partially replace conventional technologies for energy production from fossil fuels that are increasingly scarce.

But the great handicap of alternative technologies is the storage of hydrogen, especially for hydrogen that will be used in fuel cars.

Different possibilities in this field are difficult and expensive, hydrogen as over-pressurized gas, liquid hydrogen, hydrogen stored in compounds is able to be released by chemical reactions, etc. Therefore, one of the most attractive possibilities is the technique of hydrogen absorbed in porous solids [2].

Among the solids potentially usable for this purpose, carbon materials have received special attention, perhaps because there was a wide prior experience in gases separation by pressure swing absorption, and the use of activated carbons in problems of waste water recovery.

The activated carbon fibre capacity is up to 4.5% - 5.3% w/w. The same order or capacity shows carbon nanotubes with some exceptional cases with even better capacity. With Li and K doped carbon nanotubes the storage capacity reaches 10% w/w. Graphene stores about 1% H_2 w/w at room temperature [3]. The minimal capacity for hydrogen storage should be at least 6% w/w according to the Department of Energy of USA Standards. In order to improve the storage capability, the usual practice should be the activation of surface carbon fibres with an etching; the mechanism is the increase of the surface porosity [4] because micro-pores work as active sites to promote hydrogen adsorption [5].

The present work deals with the possibility of improving the hydrogen storage capacity using an activation process consisting of γ rays irradiation [6]. The efficiency of the method is based upon the fact that irradiation produces defects in carbon materials [7].

The present study was done using two types of samples, micro-fibres obtained from hydrocarbon gas on behalf a chemical vapour process, and commercial ex-PAN fibres, commercial carbon fibres obtained from a polyacrylonitrile precursor.

The micro-fibres are similar to activated carbon fibres with a hydrogen capacity storage of about 5% w/w [8] and with proven improvement with activation process [9]. The ex-PAN carbon fibres were chosen because it is known that their good response to the activation with γ rays [10]. In order to monitor the variation of the adsorbed hydrogen, the electrical resistivity was monitored [11].

MATERIALS

In the present work, three types of carbon fibres were studied, vapour grown carbon fibres (VGCF) and commercial ex-PAN fibres, in the high-strength and highmodulus versions.

The VGCF were obtained from hydrocarbon gases involving a chemical vapour process. They were prepared in our laboratory using a quartz tube placed horizontally. The dimensions of the tube were 80 millimetres of internal diameter and 1 metre length. The quartz tube was heated by electrical heaters embraced to the tube. The temperature inside the furnace was controlled and programmed by means of a temperature controller CN616 TC1 and a solid state relay SSR330DC25, both manufactured by Omega™.

Hydrogen Premier Plus X50S, 99.999% purity, and Methane X50S, 99.995% purity, both supplied by Carburos Metálicos™, were the hydrocarbon gases utilized as reactants. A mass flow controller AWS-Digital, supplied by Witt™, was employed in order to work with the proper mixture and flow rate of gases at any time. Neither preheating nor de-oxygenation pre-treatments were performed before the hydrocarbon gases entered the furnace.

A solid iron catalyst was used to carry out the chemical production process that led to the fibre production. This catalyst also acted as a solid support from which the fibre was produced and accumulated during the whole process. The basic solid support structure was made of Grafoil®, acquired from Le Carbone Lorraine™. A 0.025-molar solution of nonahydrated ferric nitrate $Fe(NO_3)_3 \cdot 9H_2O$ in ethanol was nebulised over this support. Then, the support was introduced in an oven at 373 K for 10 minutes for organic solvent evaporation. As a result, the substrate surface was covered by many ferric nitrate microscopic grains. Finally, the substrate was placed right in the centre of the mullite furnace.

In a normal furnace operation, the mixture of hydrocarbon gases enters the furnace and is heated up. At the operation temperature, the gas mixture contacts the catalyst and there they suffer a decomposition that results in a production of a solid carbon deposit with filament shape. A vapour-liquid-solid (VLS) process [12,13] describes the gemination and growth mechanisms. A liquid drop of coronene ($C_{24}H_{12}$) is formed and it absorbs hydrogen from the atmosphere of hydrocarbon gas and hydrogen. When the solidification of this drop takes place, a VGCF with a considerable amount of hydrogen is obtained [14].

Every operation batch was divided into three stages. In the first one, pure hydrogen circulated and the furnace was heated up from room temperature to 1173 K at a heating rate of 20 K/min. The operating temperature was then stabilised for 5 minutes. In the next stage, a mixture of 70% of hydrogen and 30% of methane with a total flow rate of 950 cm^3/min was used. The heating rate in this stage was 15 K/min until 1425 K was achieved. This temperature was kept constant during 40 minutes. Finally, a flow of pure argon was employed to cool down the furnace and finish with the batch.

Commercial ex-PAN carbon fibres are polyacrylonitrile filaments after a carbonization treatment. In this study, we used ex-PAN fibre Fortafil 3, supplied by Great Lakes Carbon Corporation®, now SGL Carbon®, as a high strength carbon fibre. As high modulus carbon fibre, we used the ex-PAN fibre Fortafil 5, also supplied by Great Lakes Carbon®.

The properties of these ex-PAN fibres are known [15]. For Fortafil 3, the axial tensile strength is 3.8 GN/m^2 and its axial modulus is 227 GN/m^2. For Fortafil 5, its Young's modulus is 345 GN/m^2 and its axial tensile strength is 2.76 GN/m^2.

All these three types of carbon fibres were tested in three states: pristine, with surface treatment and after γ-rays irradiation without previous chemical treatment. Surface treatment means that the carbon fibres were submitted to an acid etching in HNO_3, 53% purity, supplied from Panreac®, as described in [16,17]. The etching time was 24 hours, with a finishing consisting of rinsing three times with distilled water. Finally, the humidity remaining in the sample was removed by means of a moderate stove heating at 373 K for 4 hours.

All the studied carbon fibres were also exposed to γ-rays produced by ^{60}Co isotopes in a radiation facility until an irradiation dose of 504 kGy was reached. The irradiation treatment was performed in the Nayade Facility of the Centre of Energy, Environment and Technology Research (CIEMAT), in Madrid, Spain [18].

The fibres have been labelled as "X" to point out that they are in pristine state, without any treatment at all. They are labelled with the term "HNO_3" if they are carbon fibres with chemical activation, and finally, they are labelled with "Z" when have been irradiated.

In the case of the ex-PAN fibres, the label "PANhs" points out that it is a Fortafil 3 fibre (high strength) and the label "PANhm" means that the fibre is Fortafil 5 (high bulk modulus).Therefore, the complete list of the used samples is: PANhs-X, PANhs-HNO_3, PANhs-Z, PANhm-X, PANhm-HNO_3, PANhm-Z, VGCF-X, VGCFHNO$_3$ and VGCF-Z.

METHODS

To characterise the physical parameters of our samples, three different techniques have been used: electrical resistivity measurements, Raman spectroscopy and sessile drop test for the evaluation of the surface energy. The choice of the electrical resistivity for tracking the hydrogen adsorption is because the resistivity and the thermoelectric power have been used in the past to study the nature of the adsorption of hydrogen in single-walled carbon nanotubes. Both parameters show a linear behaviour as a function of gas coverage, consistent with a physisorption process [19].

The samples studied in the present work were cleaned with trichloroethylene for one hour and afterwards washed in ethanol. Finally, they were dried at 90°C. After that, the samples were mounted in a frame and placed within an autoclave with a hydrogen atmosphere, over pressurised at 25 bar up to 40 hours. For the hydrogen desorption, the previously hydrogenated samples were located in an atmosphere of 1 bar of hydrogen.

In order to get representative data, each measurement was repeated with five different fibre samples; the measured data were averaged and plotted. The electrical resistivity of the fibre samples was measured using the four probe method [20]. A single fibre was place on an electrically insulating substrate, where four contacts were fixed to the fibre by means of the conductive epoxy CW2400®, supplied by Chemtronics®. In the four probe methods, there are two inner contacts and two outer contacts. Through the outer contacts, the electrical current is supplied, and the voltage measurements are taken from the inner contacts. The electrical resistance was measured with an Ethernet® multimetre and data acquisition system Keithley 2701®. The electrical resistivity has been obtained through the expression:

$$\rho = R \times \frac{A}{L}$$

(1)

where R is the electrical resistance of the specimen, measured in ohms (Ω), L is the length of the material in meters and A is its cross-sectional area in square meters, so the electrical resistivity ρ is expressed in $\Omega \cdot m$. The length L corresponds to the distance between the two inner contacts, having in this study the constant value of 5.0×10^{-4} meters.

For the measurement of the cross-sectional area A, a precise determination of the diameter of the fibres was needed. The diameters of the fibre samples were determined using the Fraunhofer diffraction phenomenon, as described in [21]. For this purpose, a red monochromatic laser, particularly a helium-neon gas laser with a wavelength value of 632.8 nm and a power of 0.95 mW, supplied by Uniphase®, was used. The fibre was placed on a support, which had a hole in the middle in order to permit the laser to go through after the interaction with the fibre. This support was attached to a simple structure that can be moved smoothly in both horizontal and vertical directions so that the fibre could be properly located in the laser beam trajectory.

The surface energy of the fibres was determined with the sessile drop method. It involves the measurement of the contact angle between the fibre and a liquid drop of any of the available standard components widely used. With the value of these contact angles, the Young's equation [22] can be applied:

$$\gamma_{lv} \cdot \cos \theta = \gamma_{sv} - \gamma_{sl} - \pi_e \tag{2}$$

The liquid-vapour surface energy is denoted as γ_{lv}, the solid-vapour surface energy as γ_{sv} and the term related to solid-liquid interfaces as γ_{sl}. Surface energy values are given in joules per square metre units. The term π_e is the equilibrium pressure of the adsorbed vapour on the solid surface, being the most of the times a very small value, so it is usually considered negligible.

As it is established in [23], the solid-liquid term can be separated into polar (ionic forces and hydrogen bonds) and non-polar interactions (Keesom, Debye and London forces):

$$\gamma = \gamma^h + \gamma^d \tag{3}$$

where the superscript h represents the polar forces and d the non-polar forces. Notice that these superscripts are not exponents. Given the separation of polar and non-polar forces, the solution of the Equation (2) can be found in literature as [23]:

$$1 + \cos \theta = F + Q \tag{4}$$

where

$$F = \frac{2 \cdot \sqrt{\gamma_s^d} \cdot \sqrt{\gamma_l^d}}{\gamma_l} \tag{5}$$

and

$$Q = \frac{2 \cdot \sqrt{\gamma_s^h} \cdot \sqrt{\gamma_l^h}}{\gamma_l} \tag{6}$$

Thus, in the case of a polar liquid, it would have two components, according to Equation (3). However, if the liquid is non-polar, it would have only the non-polar component γ_l^d, whereas the polar component $\gamma_l^h = 0$, resulting in $Q = 0$. Given this, the energy surface of the solid γ_s, and therefore of the fibre, can be calculated by measuring the contact angle θ of two liquids, one polar and one non-polar. Then, the Equations (4) to (6) can be applied to both liquids and the resulting system of equations can be solved easily thanks to the simplification for the non-polar liquid. Firstly, with these three equations for the non-polar liquid, the non-polar component of the energy surface of the solid γ_s^d is obtained. In the following step, the same three equations for the polar liquid are used in order to obtain the polar component γ_s^h. Finally, γ_s is calculated as the sum of the non-polar and polar components, as in Equation (3).

In order to be able to perform these calculations, the data of the surface energy of the liquids should be known. For this study, coconut oil as non-polar compound and bidistilled glycerol as polar compound, both supplied by CUVE®, were used.

The data found in literature for the coconut oil is [24]:

$$\gamma_l^d = 26.11 \times 10^{-3} \text{ J} \cdot \text{m}^{-2}$$

(7)

The values for the bidistilled glycerol are [23]:

$$\gamma_l^d = 37 \times 10^{-3} \text{ J} \cdot \text{m}^{-2}$$

(8)

$$\gamma_l^h = 26.4 \times 10^{-3} \text{ J} \cdot \text{m}^{-2}$$

(9)

The measurement of the contact angles was carried out by placing one fibre of each sample in horizontal position. The fibre could be fixed to the frame in the two fibre tips, keeping the fibre horizontally. The drops of coconut oil and bidistilled glycerol were spread using a nebuliser. In this way, some drops remain attached to the fibre. For this reason, it is possible to do the calculation of the three equations for several drops of different sizes set up in the same fibre.

To get the outline of each drop, a high magnification optical microscope Nikon® SMZ1500 that includes digital recording was used. In this way, the boundaries of both drop and fibre were registered. Thus, it was possible to adjust these outlines to a curve equation using the data analysis software Origin 8.0. Therefore, in the drop-fibre contact point, a value of the contact angle could be measured. A description of the method performed in this work for determining contact angles on fibres was found in literature [25] and it was also used in a previous research [26].

On other side, the determinations of the crystallite size and the H_2 content of the fibre samples were carried out by means of Raman spectrometry analysis. In previous works, the ratio of the intensities of the Raman bands D and G has been proven to be an appropriate tool for the evaluation of the crystallite size, L_a [27]. Similarly, the ratio between the areas of the peaks D and G are a valid parameter to evaluate the hydrogen content [28].

The Raman spectroscopy analysis was carried out in a Jobin Yvon T64000 spectrometer with a micro entrance. The excitation source was a laser Spectrum 70 with several excitation lines along the visible. The excitation wavelength was 647.1 nanometres (red) and the signal was recorded in a range of 1000 cm^{-1} to 3000 cm^{-1} with acquisition time of 100 seconds for two scans to remove the peaks produced by cosmic rays.

RESULTS AND DISCUSSION

The comparison between the evolution of the electrical resistivity of high strength and high modulus ex-PAN carbon fibres in a 25-bar hydrogen atmosphere is very interesting. Figure 1 shows the kinetics of the process of hydrogen intake in high strength carbon fibres, in the three presentations: as pristine (X), treated with acid etching (HNO$_3$) and γ-irradiated (Z). In Figure 2 we also show the kinetics of the process, but for high modulus ex-PAN carbon fibres. Finally, Figure 3 shows the results for VGCFs.

If we compare the data of all these pictures, we can see that in pristine state, the high strength fibres are two times more resistive than that of high modulus. The effect of irradiation is more remarkable in the high strength fibres (at the beginning of the adsorption time there is a

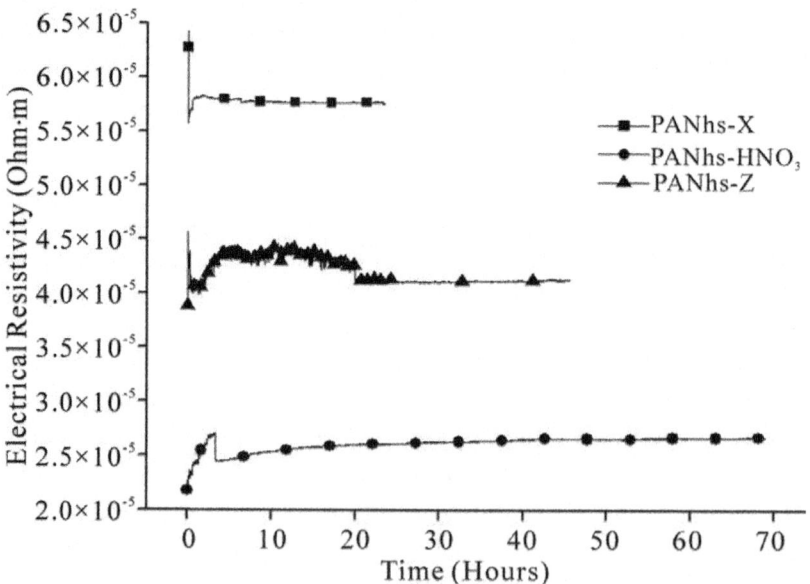

Figure 1: Adsorption of hydrogen in high strength ex-PAN fibres.

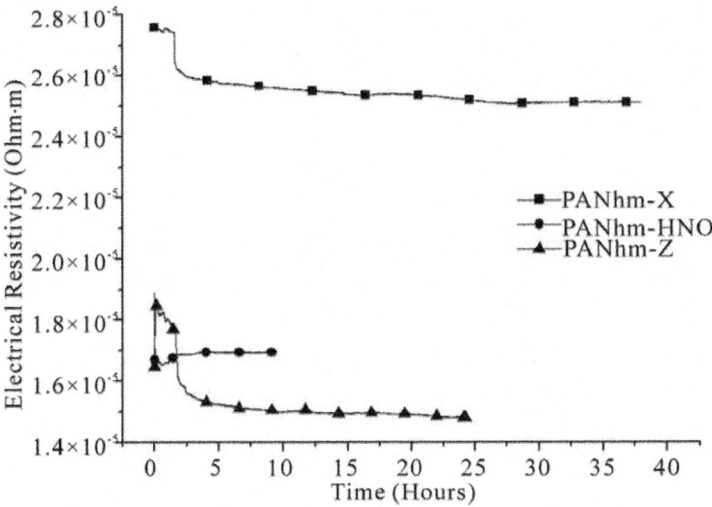

Figure 2: Adsorption of hydrogen in high modulus ex-PAN fibres.

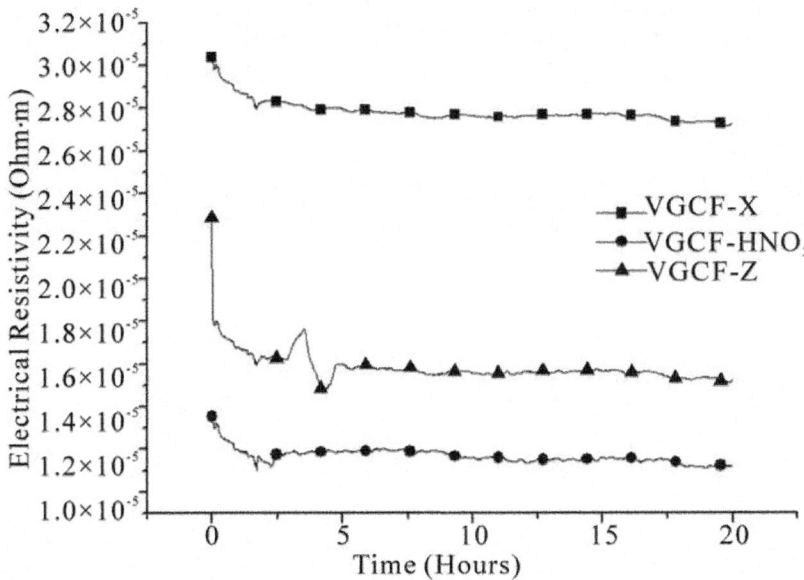

Figure 3: Adsorption of hydrogen in vapour grown carbon fibres.

transient regime) than in the high modulus fibres.

On the other side, there is a similar ratio between the stabilised resistivity of the irradiated and pristine samples for both high strength and high modulus fibres (70% and 60%, respectively).

This means that the γ-irradiation effect in the crystallinity of ex-PAN fibres is very relevant. This is due to Compton scattering, which is mostly responsible for the interaction of γ-rays with carbon fibres and causes ionization and atomic displacements, as described in [29].

In the case of the VGCF, the irradiated sample also shows a transient regime before to reach the steady state.

The most remarkable fact is that irradiation decreases the electrical resistivity (see Figures 1-3), according with previous results that shows as the γ-irradiation increases the electrical conductivity [30].

It is also possible to see in the same three figures that beside of the decrease in the electrical resistivity, the irradiated samples show an evolution towards a minor resistivity, as suggested in previous studies that described the healing of the generated defects during γ-irradiation [31].

The process of desorption of hydrogen is studied by comparing the Figures 4-6, corresponding to the outgassing process of the fibres in a 1-bar hydrogen atmosphere. The results for the high strength ex-PAN fibres, shown in Figure 4, point out that pristine and irradiated samples reach the steady state very soon. In contrast, the high modulus ex-PAN carbon fibres show a clear difference between pristine and irradiated samples. The irradiated sample reaches very soon the steady state, but the pristine fibre increases its resistivity very slowly, suggesting a slow outgassing of hydrogen.

In the case of VGCF, in Figure 6, it is observed the opposite. This means that the pristine sample reach steady state very fast, but the irradiated samples need almost 30 hours to reach stabilisation.

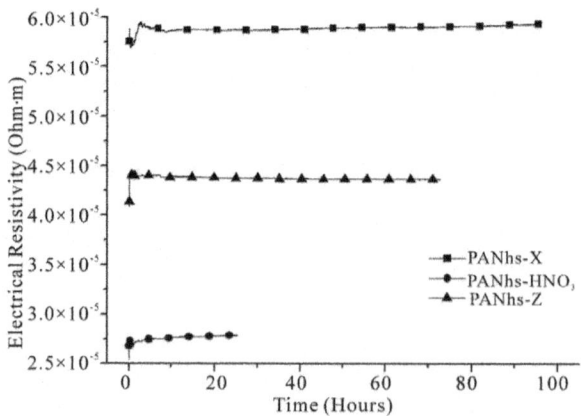

Figure 4: Desorption of hydrogen in high strength ex-PAN fibres.

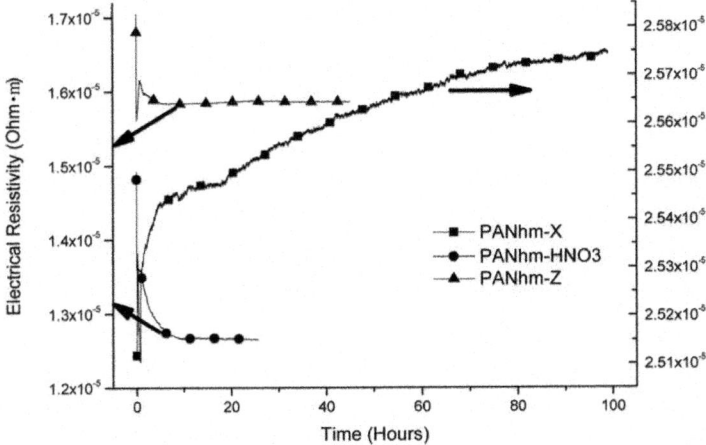

Figure 5: Desorption of hydrogen in high modulus ex-PAN fibres.

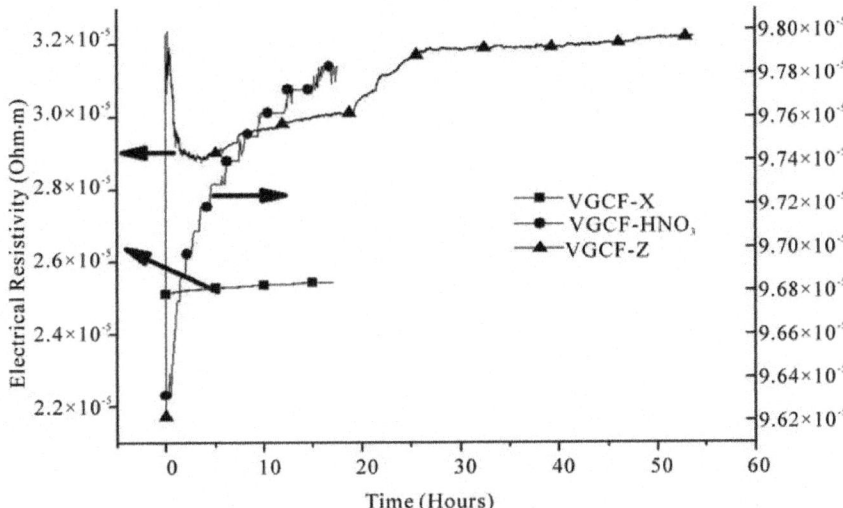

Figure 6: Desorption of hydrogen in vapour grown carbon fibres.

The behaviour of activated fibres is rather different. Comparing Figures 1-3, it can be seen that after 5 hours, the adsorption is stabilised. In all cases, the resistivity is much lower than pristine and irradiated samples.

In the results of the desorption of activated VGCF shown in Figure 6 the evolution of the activated fibre is to reach an electrical resistivity value higher than for ex-PAN fibres. That means that the fibre emitted a large amount of hydrogen, because high electrical resistivity means scarcity of hydrogen [32].

The measurement of the surface energy gives interesting information, shown in Table 1, where it is possible to see similar changes in high strength ex-PAN fibres and VGCF, but different in high modulus ex-PAN fibres. In the first case, the action of irradiation is to decrease the surface energy of the pristine samples, but in the second one the irradiation increases the surface energy. Just the opposite happens if we compare the influence of the acid etching on the energy surface. In the high modulus ex-PAN fibres and VGCF, the action of the activation is to decrease the surface energy, but in the case of high strength ex-PAN fibres, the effect of the activation is an increase of the surface energy.

Apparently, there is no parallelism between the γ-irradiation process and the acid etching effects in the surface of these fibres. This is due to the huge difference in the mechanism of the interaction. The Compton Effect

Table 1: Energy surface of the fibres by means of contact angle measurement

Fibre Sample	Energy surface measurement variables			
	Liquid tested	Contact Angle (°)	Component Surface Energy (mJ/m²)	Total Surface Energy (mJ/m²)
PANhs-X	Coconut Oil[a]	54	16	25
	Glycerol[b]	75	9	
PANhs-HNO₃	Coconut Oil[a]	50	18	34
	Glycerol[b]	63	16	
PANhs-Z	Coconut Oil[a]	51	17	22
	Glycerol[b]	82	4	
PANhm-X	Coconut Oil[a]	61	14	38
	Glycerol[b]	59	24	
PANhm-HNO₃	Coconut Oil[a]	60	15	27
	Glycerol[b]	72	13	
PANhm-Z	Coconut Oil[a]	61	14	44
	Glycerol[b]	52	30	
VGCF-X	Coconut Oil[a]	37	21	52
	Glycerol[b]	38	31	
VGCF-HNO₃	Coconut Oil[a]	43	20	34
	Glycerol[b]	63	14	
VGCF-Z	Coconut Oil[a]	60	15	27
	Glycerol[b]	72	13	

[a]Conout Oil: Non-polar component; [b]Glycerol: Polar component.

of the irradiation mainly changes crystalline structures, whereas acid attack to a carbon is more pronounced if the carbon is amorphous. As an example, it is possible to say that when nanotubes are fabricated, the current practice to eliminate the amorphous soot is a treatment with acid, which totally annihilates the amorphous carbon, but the crystalline nanotubes remain intact [33].

Finally, we used Raman spectroscopy because it is known that γ-irradiation effect affects bands D and G of the carbons [34].

In the particular case of this study, for the samples used, it is observed the influence of the irradiation in the D and G peaks morphology. As an example, in Figure 7 it is shown the Raman spectra of the two samples of VGCF presented as pristine and irradiated.

Besides, we have used Raman spectroscopy to determine the crystallite size, L_a [27] and the hydrogen content [28], as it is shown in Figure 8.

As it can be seen, the values of L_a for the high modulus ex-PAN fibre are much higher than the ones for

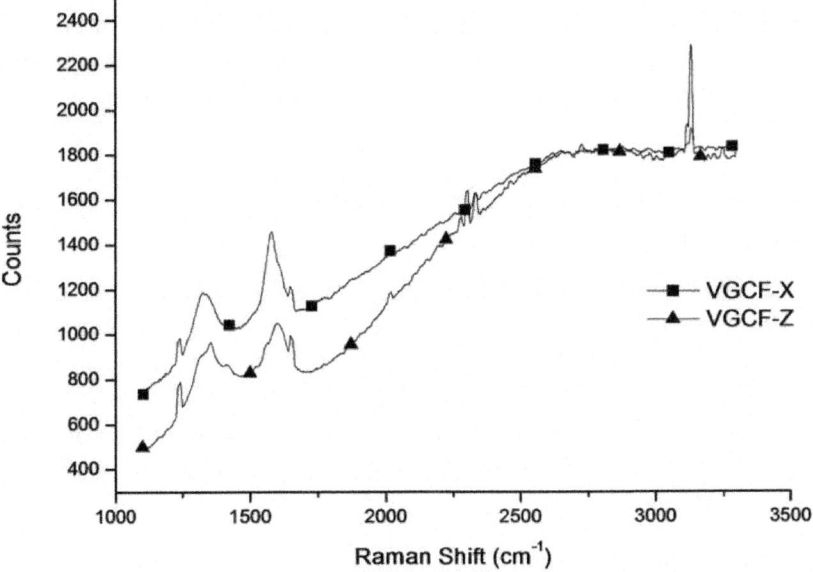

Figure 7: Raman spectra showing changes in crystallinity due to the irradiation.

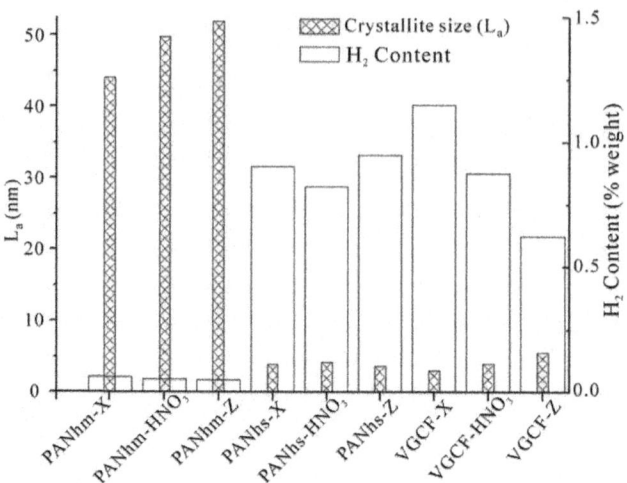

Figure 8: Hydrogen content and crystallite size obtained by means of Raman spectroscopy.

the high strength ex-PAN fibre and the VGCF. Also, it is observed that the more L_a is, the bigger the effect of the acid etching and irradiation treatments. Thus, both treatments would have more influence in a sample with a bigger value of L_a.

In addition to this, for the high strength ex-PAN fibre and VGCF there is an increase of L_a when the sample is irradiated. In the case of the high strength ex-PAN fibre, L_a does not change so remarkably.

Also, in Figure 8, it is interesting to highlight that the values of the hydrogen content for the high strength ex-PAN fibres and VGCFs are much higher than those for the high modulus ex-PAN fibres. This is equivalent to say that crystallite size is inversely proportional to hydrogen content. This is in agreement with published results [35] that explain that the hydrogen is mainly stored in the grain boundaries. The hydrogen retention increases with decreasing grain size.

CONCLUSIONS

As a summary of these results, we can conclude:

1) It is possible to track the adsorption and desorption of hydrogen in the carbon fibres studied in this work by measuring the electrical resistivity.

2) The numbers of surface energy are not directly related to the hydrogen content, what means that hydrogen is not only stored on the surface, but also in the bulk material.

3) The irradiation treatment has influenced over the adsorption and desorption processes of hydrogen. In general, it decreases the electrical resistivity due to the changes produced in the crystalline quality.

4) For hydrogen storage purposes, it is better to use materials with a small crystallite size, because hydrogen tends to anchor into the grain boundaries.

5) The two treatments used in this study, the chemical activation and the γ-irradiation, are supplementary. The chemical treatment works on the outer surface of the material, whereas the irradiation works also within the bulk material.

ACKNOWLEDGEMENTS

We would like to acknowledge support from the Spanish Ministry of Finances and Competitiveness for support via the Programme ConsoliderIngenio 2010 (project CSD 2010-00044).

REFERENCES

1. W. McDowall and M. Eames, "Towards a Sustainable Hydrogen Economy: A Multi-Criteria Sustainability Appraisal of Competing Hydrogen Futures," International Journal of Hydrogen Energy, Vol. 32, No. 18, 2007, pp. 4611-4626.doi:10.1016/j.ijhydene.2007.06.020

2. K. L. Lim, H. Kazemian, Z. Yaakob and W. R. W. Daud, "Solid-State Materials and Methods for Hydrogen Storage: A Critical Review," Chemical Engineering and Technology, Vol. 33, No. 2, 2010, pp. 213-226. doi:10.1002/ceat.200900376

3. R. Ströbel, J. Garche, P. T. Moseley, L. Jörissen and G. Wolf, "Hydrogen Storage by Carbon Materials," Journal of Power Sources, Vol. 159, No. 2, 2006, pp. 781-801.doi:10.1016/j.jpowsour.2006.03.047

4. E. Raymundo-Piñero, D. Cazorla-Amorós, A. LinaresSolano, S. Delpeux, E. Frackowiak, K. Szostak and F. Béguin, "High Surface Area Carbon Nanotubes Prepared by Chemical Activation," Carbon, Vol. 40, No. 9, 2002, pp. 1597-1617. doi:10.1016/S0008-6223(02)00134-3

5. D. Qu, "Investigation of Hydrogen Physisorption Active Sites on the Surface of Porous Carbonaceous Materials," Chemistry—A European Journal, Vol. 14, No. 3, 2008, pp. 1040-1046. doi:10.1002/chem.200701042

6. M. A.Obolensky, A. V. Basteev and L. A. Bazyma, "Hydrogen Storage in Irradiated Low-Dimensional Structures," Fullerenes, Nanotubes and

Carbon Nanostructures, Vol. 19, No. 1-2, 2011, pp. 133-136. doi:10.108 0/1536383X.2010.490134

7.   S. Gupta, B. L. Weiss, B. R. Weiner, L. Pilione, A. Badzian and G. Morell, "Electron Field Emission Properties of Gamma Irradiated Microcrystalline Diamond and Nanocrystalline Carbon Thin Films," Journal of Applied Physics, Vol. 92, No. 6, 2002, pp. 3311-3317.doi:10.1063/1.1499996

8.   K. Hanada, H. Shiono and K. Matsuzaki, "Hydrogen Uptake of Carbon Nanofiber under Moderate Temperature and Low Pressure," Diamond and Related Materials, Vol. 12, No. 3-7, 2003, pp. 874-877. doi:10.1016/ S0925-9635(02)00360-6

9.   F. Suárez-García, J. Nauroy, A. Martínez-Alonso and J. M. D. Tascón, "Thermogravimetric Studies on the Activation of Nanometric Carbon Fibers," Journal of Thermal Analysis and Calorimetry, Vol. 79, No. 3, 2005, pp. 525- 528. doi:10.1007/s10973-005-0573-1

10.   J. Q. Li , Y. D. Huang, S. Y. Fu, L. H. Yang, H. Qua and G. Wu, "Study on the Surface Performance of Carbon Fibres Irradiated by G-ray under Different Irradiation Dose," Applied Surface Science, Vol. 256, No. 7, 2010, pp. 2000- 2004.doi:10.1016/j.apsusc.2009.09.035

11.   P. Voit, E. A. Evard and I. E. Gabis, "Effect of Sorbed Hydrogen on the Conductivity of Nanoporous Carbon," Materials Science, Vol. 38, No. 4, 2002, pp. 570-575.doi:10.1023/A:1022970818229

12.   Madroñero, "Possibilities for the Vapour-Liquid-Solid Model in the Vapour-Grown Carbon Fibre Growth Process," Journal of Materials Science, Vol. 30, No. 8, 1995, pp. 2061-2066.doi:10.1007/BF00353034

13.   Ph. Serp, A. Madroñero and J. L. Figueiredo, "Production of Vapour-Grown Carbon Fibres: Influence of the Catalyst Precursor and Operating Conditions," Fuel, Vol. 78, No. 7, 1999, pp. 837-844. doi:10.1016/S0016-2361(98)00216-6

14.   Madroñero and M. Verdú, "Hydrogen Content Evaluation in Vapour-Grown Carbon Fibres by SIMS," Carbon, Vol. 33, No. 3, 1995, pp. 247-251. doi:10.1016/0008-6223(94)00139-Q

15.   L. H. Peebles Jr., "Carbon Fibres: Structure and Mechanical Properties," International Materials Reviews, Vol. 39, No. 2, 1994, pp. 75-92. doi:10.1179/095066094790326248

16.   P. Vinke, M. van der Eijk, M. Verbree, A. F. Voskamp and H. van Bekkum, "Modification of the Surfaces of a Gas-Activated Carbon and a Chemically Activated Carbon with Nitric Acid, Hypochlorite, and Ammonia," Carbon, Vol. 32, No. 4, 1994, pp. 675-686.doi:10.1016/0008-6223(94)90089-2

17. H. Takagi, Y. Soneda, H. Hatori, Z. H. Zhu and G. Q. Lu, "Effects of Nitric Acid and Heat Treatment onHydrogen Adsorption of Single-Walled Carbon Nanotubes," Australian Journal of Chemistry, Vol. 60, No. 7, 2007, pp. 519-523. doi:10.1071/CH06409

18. http://www.csn.es/images/stories/actualidad_datos/ofin_11/ccnn_11/ain_cie_191_11.pdf

19. G. U. Sumanasekera, C. K. V. Adu, G. Chen, H. E. Romero and P. C. Elkund, "Thermoelectric Study of Hydrogen Storage in Carbon Nanotubes," Physical Review B, Vol. 65, No. 3, 2001, Article ID: 035408. doi:10.1103/PhysRevB.65.035408

20. M. Hájek, J. Veselýa and M. Cieslara, "Precision of Electrical Resistivity Measurements," Materials Science and Engineering A, Vol. 462, No. 1-2, 2007, pp. 339-342.doi:10.1016/j.msea.2006.01.175

21. Madroñero and C. Merino, "Some Geometrical Singularities in the Characterization of Vapor Grown Carbon Fibers Using Laser Diffraction Technique," Materials Research Bulletin, Vol. 33, No. 10, 1998, pp. 1503-1515. doi:10.1016/S0025-5408(98)00144-5

22. K. Holmberg, D. O. Shah and M. J. Schwuger, "Handbook of Applied Surface and Colloid Chemistry," Wiley, New York, 2002.

23. F. M. Fowkes, "Attractive Forces at Interfaces," Industrial and Engineering Chemistry, Vol. 56, No. 12, 1964, pp. 40-52. doi:10.1021/ie50660a008

24. Q. Shu, J. F. Wang, B. X. Peng, D. Z. Wang and G. R. Wang, "Predicting the Surface Tension of Biodiesel Fuels by a Mixture Topological Index Method, at 313 K," Fuel, Vol. 87, No. 17-18, 2008, pp. 3586-3590. doi:10.1016/j.fuel.2008.07.007

25. W. C. Jones and M. C. Porter, "A Method for Measuring Contact Angle on Fibres," Journal of Colloid and Interface Science, Vol. 24, No. 1, 1967, pp. 1-3. doi:10.1016/0021-9797(67)90269-X

26. Madroñero, A. Asenjo, C. Gil, M. Jaafar and A. López, "Reconnaissance of the Specific Surface of Vapour Grown Carbon Micro and Nanofibres as a Main Controller of the Sorption of Hydrogen," Applied Surface Science, Vol. 256, No. 20, 2010, pp. 5797-5802.doi:10.1016/j.apsusc.2010.02.056

27. L. G. Cançado, K. Takai, T. Enoki, M. Endo, Y. A. Kim, H. Mizusaki, A. Jorio, L. N. Coelho and R. MagalhãesPaniago, "General Equation for the Determination of the Crystallite Size L_a of Nanographite by Raman Spectroscopy," Applied Physics Letters, Vol. 88, No. 16, 2006, Article ID: 163106. doi:10.1063/1.2196057

28. M. Culebras, A. Madroñero, A. Cantarero, J. M. Amo, C. Domingo and A. Lopez, "Confident Methods for the Evaluation of the Hydrogen Content in Nanoporous Carbon Microfibers," Nanoscale Research Letters, Vol. 7, 2012, pp. 588-592. doi:10.1186/1556-276X-7-588

29. Z. W. Xu, Y. D. Huang, C. Y. Min, L. Chen and L. Chen, "Effect of Gamma-Ray Radiation on the Polyacrylonitrile Based Carbon Fibers," Radiation Physics and Chemistry, Vol. 79, No. 8, 2010, pp. 839-843. doi:10.1016/j.radphyschem.2010.03.002

30. V. Skakalova, U. Dettlaff-Weglikowska and S. Roth, "Gamma-Irradiated and Functionalized Single Wall Nanotubes," Diamond and Related Materials, Vol. 13, No. 2, 2004, pp. 296-298. doi:10.1016/j.diamond.2003.11.003

31. S. Suzuki and Y. Kobayashi, "Healing of Low-Energy Irradiation-Induced Defects in Single-Walled Carbon Nanotubes at Room Temperature," Journal of Physical Chemistry C, Vol. 111, No. 12, 2007, pp. 4524-4528. doi:10.1021/jp067398r

32. P. Voit, E. A. Evard and I. E. Gabis, "Effect of Sorbed Hydrogen on the Conductivity of Nanoporous Carbon," Materials Science, Vol. 38, No. 4, 2002, pp. 570-575.doi:10.1023/A:1022970818229

33. S. Lyu, D. Jung, K. Ahn, H. Lee, N. Lee, Y. Park and J. Sok, "Purification of Single-Walled Carbon Nanotubes by HCl Treatment and Analysis of the Field Emission Property," Journal of Korean Institute of Metals and Materials, Vol. 48, No. 4, 2010, pp. 335-341.doi:10.3365/KJMM.2010.48.04.335

34. M. Hulmana, V. Skákalová, S. Roth and H. Kuzmany, "Raman Spectroscopy of Single-Wall Carbon Nanotubes and Graphite Irradiated by γ Rays," Journal of Applied Physics, Vol. 98, No. 2, 2005, Article ID: 024311. doi:10.1063/1.1984080

35. Sh. Michaelson, O. Ternyak and A. Hoffman, "Correlation between Diamond Grain Size and Hydrogen Retention in Diamond Films Studied by Scanning Electron Microscopy and Secondary Ion Mass Spectroscopy," Applied Physics Letters, Vol. 90, No. 3, 2007, Article ID: 031914. doi:10.1063/1.2432996

Chapter 6

HYDROGEN ECONOMY: MODERN CONCEPTS, CHALLENGES AND PERSPECTIVES

Vladimir A. Blagojević [1], Dragica M. Minić[1], Dejan G. Minić[2] and Jasmina Grbović Novaković[3]

[1]University of Belgrade, Faculty for Physical Chemistry, Serbia
[2]Kontrola LLC, Austin, TX, USA
[3]Laboratory for Material Sciences, Institute for Nuclear Science Vinča, University of Belgrade, Belgrade, Serbia

INTRODUCTION

Identifying and building a sustainable energy system are two of the most critical issues for any modern society. Ideally, current energy system, based mostly on fossil fuels, which have limited supply and considerable negative environmental impact, would be replaced with a system based on a renewable fuel. Hydrogen, as an energy carrier primarily derived from water, can address the issues of sustainability, environmental emissions and energy security. If one assumes a full hydrogen economy the size of United States, the amount of hydrogen for just purposes of transportation would be about 150 million tons per year, which would amount to consuming, with current production efficiency, between 2 and 5 billion tons of water. As a comparison, current water consumption in United States for purpose of thermoelectric power generation in power plants is around 300 billion tons, with another 1.2 billion tons consumed for gasoline production. Therefore, rather than consume, a hydrogen economy would most likely significantly reduce water consumption for purposes of energy generation (Turner, 2004).

Hydrogen is the most abundant element in the universe, burns clean, producing only water and has the highest energy density per unit mass. This is why hydrogen is considered most suitable to replace fossil fuels as the primary energy material for the mobile industry (Šušić, 1997c). However, hydrogen is not an energy source, only an energy carrier, and it is not freely available in nature and needs to be produced, either from water or other compounds. If it is produced from water, it costs more energy to produce it than one would recover burning it. This is why, ideally, a hydrogen cycle would include hydrogen produced by splitting water using electrolysis with solar energy and stored reversibly in a solid. However, there are considerable difficulties associated with efficient hydrogen production, storage and use in fuel cells. Among them, hydrogen storage for mobile applications is currently the most difficult obstacle.

Gasoline has very high energy density of 31.6 MJ/L, compared to 4.4 MJ/L of compressed hydrogen and 8.8 MJ/L of liquid hydrogen. In addition, gasoline tank has extremely short filling time, is capable of providing energy at low temperatures and provides excellent control of energy discharge, allowing rapid acceleration, high sustained speed and considerable range. These are the challenges that a successful hydrogen tank has to meet, too. US Department of Energy (DOE) target requirements for a hydrogen tank require hydrogen gravimetric density of 7.5 mass% and volumetric density of 70g/L, operating temperature between 233 and 358K, minimum delivery pressure of 12bar (1.2MPa) and fueling time of 3min. In addition, the storage system should be safe, durable (1500 operational cycle life) and cost effective. None of the existing systems meet these requirements.

In order to achieve the hydrogen economy, there are some obstacles that need to be overcome to make hydrogen a viable energy carrier. They are characterized by four main aspects of hydrogen use and some of these will be addressed here:

- Production – since hydrogen needs to be produced, ideally from water, it is necessary to develop production methods that would consume the least amount of energy and provide ability to produce hydrogen renewably on a large scale.

- Storage – fuel needs to be easily stored for use and transport, where one of the main requirements is that it is readily available, which requires not just short charge/discharge times, but also excellent control of charge/discharge process coupled with sufficient energy and gravimetric/volumetric density.

- Power generation – once hydrogen is ready to be consumed, it is necessary to do so in the most effective way: the power generation system that uses hydrogen needs to be both efficient and, for mobile application, lightweight.

- Safety – hydrogen use and storage comes with some risks (flammability) which necessitate certain precautions and safety measures; another aspect related to this is environmental impact of the hydrogen cycle, which depends on the methods used to produce, store and use it.

Since hydrogen is thought to be a renewable fuel for the future, it is only appropriate that, when we consider all the challenges associated with its production, storage and use, we keep in mind that when we consider proposed systems, efficiency is only one of the factors that will determine the success of these systems. Other important aspects are production cost (both financial and in resource), durability, stability of operation and safety, and these can, more than efficiency, determine the success or failure of any of the proposed solutions for a part of the hydrogen cycle.

HYDROGEN PRODUCTION

There are several potential sources of hydrogen on our planet, although these are exclusively hydrogen compounds, necessitating extraction of hydrogen at energy cost. The most abundant is water, while hydrogen can also be obtained from hydrocarbons, either fossil fuels or biomass. While production from water is clean and renewable, with no CO_2 emission, production from fossil fuels generates similar or even higher levels of CO_2 emissions as burning of coal and gasoline. Hydrogen production from biomass is carbon-neutral, since plants and organisms used during the process sequester approximately the same amount of CO_2 during their growth as it is emitted during the process of extraction of hydrogen from them. However, their negative environmental impact is considerable due to the fact that they require large land surfaces for growth.

From water. Although many technologies have been explored for production of hydrogen by splitting water into hydrogen and oxygen, these processes have yet to achieve the necessary efficiency and scalability for industrial application. The main advantage of hydrogen production from water is that it is clean, renewable and has little or no negative environmental impact, although the energy cost of its production is currently too high. There are several processes of interest, like electrolysis, catalytic thermolysis, photocatalytic water splitting and sulfur-iodine cycle.

Electrolysis of water is used today to produce around 5% of all industrial hydrogen. There are several types of cells in use: solid oxide electrolysis cell, alkaline electrolysis cell and polymer electrolyte membrane cell. These cells operate at elevated temperatures (350-570K) and contain high electrolyte concentrations and catalysts (typically yttrium-stabilized zirconium oxide mixed with nickel). Efficiency of typical electrolysis processes is usually between 50-80%, when inefficiencies of production of power used for electrolysis are not taken into account. With these taken into account, the energy efficiency would decrease to 30-45% for a typical nuclear or thermal power plant used as the power source, or even lower for a typical solar cell or array (Hauch et al., 2008).

Water thermolysis is thermal dissociation of water, which occurs spontaneously around 2800K. Although this temperature is too high for practical applications, significant effort has been invested into research of catalysts to reduce water thermolysis temperature and make it an industrially viable process. The goal is to use water thermolysis either in solar concentrators or in nuclear power plants to produce hydrogen directly using thermal energy. Solar concentrators can produce very high temperatures (over 1800K) by concentrating sunlight using a system of mirrors. Next generation nuclear power plants will be operating at lower temperatures (1000-1300K), but it is hoped that new catalysts will make it possible to use them for direct hydrogen generation using water thermolysis.

Photocatalytic water splitting is a process of directly producing hydrogen using solar energy. It relies on use of photocatalyst to capture the solar energy and use for water dissociation (Ni et al., 2007). There are two principal types of catalysts: photoelectrochemical and photobiological (discussed in biohydrogen production section). Photoelectrochemical (PEC) systems can be divided further into four groups:

- Type 1 – a single electrolyte-filled reactor containing a colloidal suspension of PEC nanoparticles producing a mixture of H_2 and O_2 gas;
- Type 2 – dual electrolyte-filled reactor beds containing a colloidal suspension of PEC nanoparticles, each carrying out half of the reaction process (one producing oxygen, the other hydrogen from H^+ produced in the first reactor bed);
- Type 3 – fixed PEC planar array using multi-junction photovoltaic/PEC cells immersed in electrolyte reservoir;
- Type 4 – PEC solar concentrator system, using reflectors to focus the solar flux at 10:1 intensity onto multi-junction photovoltaic/PEC cell receivers immersed in electrolyte reservoir and pressurized to 2MPa.

Estimated hydrogen production costs for these systems are: 1.63-10.36\$/ kgH_2 depending on the type of system. DOE target for 2015 price of hydrogen was originally set at 1.50\$/kgH_2, but it was increased in 2005 to 2-3\$ per kgH_2, which means that some of these systems already meet that modified requirement. The main issues in these systems remain separation of O_2 and H_2 from the product for Type 1, ionic conduction (through ionic bridge) between two reactor beds for Type 2, improvement of PEC cell structure to reduce cost for Type 3 and new composite concentrator structure and high pressure PEC operation for Type 4. Potential benefits are clean and renewable direct hydrogen production with relatively low cost, although efficiency, depending on the system, is in range of 5-25% for the entire system.

Sulfur-iodine cycle is a thermochemical process which produces hydrogen from hydrogen iodide at much higher efficiency than water splitting, and at lower temperature (700K). One of its advantages is that sulfur and iodine are recovered and reused during the process and not consumed. It is usually coupled with concentrating solar power systems to produce hydrogen using solar energy, providing clean and renewable source of energy.

From fossil fuels. Fossil fuels are the dominant source of industrial hydrogen today. Hydrogen can be produced from natural gas with efficiency of around 80% and from other hydrocarbon sources with a varying degree of efficiency.

Most widely used method of hydrogen production today is steam reforming of methane or natural gas. At high temperatures (1000-1300K), water vapor reacts with methane to yield syngas (mixture of hydrogen and carbon monoxide), which can be used to produce more hydrogen through reaction of water and carbon monoxide (also known as water gas shift reaction, performed around 400K). The drawback of this process is that it produces CO_2 waste (Lee & Lee, 2001).

Other methods of hydrogen production from fossil fuels are partial oxidation of hydrocarbons, which includes partial combustion of fuel-air mixture at high temperatures or in a presence of a catalyst, plasma reforming (Kvaerner process), which produces hydrogen and carbon black from hydrocarbons (no CO_2 waste), and coal gasification, where coal is converted to syngas and methane.

Biohydrogen production. Biological H_2 production represents an effort to harness biological processes to generate hydrogen on the industrial scale. Although they have found no industrial application, there are a number of processes for conversion of biomass and waste streams into biohydrogen. Some of them are the same as the ones described above for fossil fuels, except they use biomass in place of fossil fuel (biomass gasification, steam reforming), while others use biological conversion of solar energy (Tao et al., 2007). Biological conversion is process where biological organisms (usually plants) convert sunlight into hydrogen through their metabolic processes (Melis, 2002).

There are variety of pathways for biological conversion that include unicellular green algae, cyanobacteria, photosynthetic bacteria and dark fermentative bacteria. Photobiological production of hydrogen offers a perspective of operating with solar energy conversion efficiency to H_2 as high as 10%, if some of the barriers could be overcome, like slow H_2 production rate, or discontinuity of H_2 production due to co-generation of O_2 (Maness, 2002). Another challenge represents system engineering for cost-effective photobiological H_2 production. Because of an excellent variety of different bacteria, which absorb light in different spectral regions, it is hoped that, ultimately, a mixture of bacteria tailored to maximize sunlight absorption would be used to improve efficiency. This level of adaptability is one of the advantages of photobiological hydrogen production. There are no photobiological hydrogen production systems that are even close to being competitive with other methods of hydrogen generation, while relatively low overall efficiency would require large surface areas for harvesting and conversion of sunlight. On the other hand, biological production would probably have positive impact on environmental pollution and potentially serve as a source of high value bio-products, which could be useful in the food and synthetic chemistry industries. However, the main limitation of biological production is that it ultimately depends on availability of land for biomass production. This means that it cannot provide the amounts of hydrogen needed by an entire civilization, especially taking into account the fact that we live in a food-limited world with increasing population.

HYDROGEN STORAGE

Hydrogen storage is a problem that has been a focus of scientific research for decades (Minić & Šušić, 2002). During this time a variety of methods have been investigated, although, none of these have accomplished the required performance level so far. Current methods for hydrogen storage can be broadly separated into:

- mechanical storage: storage in a tank of compressed gas or liquid hydrogen;
- physisorbtion: storage in a solid material through physisorbtion; includes:
- graphene and other carbon structures
- metals and metallic nanocrystals and composites
- metal-organic frameworks, zeolites;
- hydrogen: storage in solid or liquid material of chemically bound hydrogen that is released on decomposition; includes:

- light metal hydrides (e.g. alkaline hydrides, alanates, alane)
- borohydrides
- amines and imides
- amino borane.

Each of these methods has its advantages and disadvantages, but all on-board storage technologies have to meet the following requirements:

- safety
- performance
- cost
- technical adaptation for the infrastructure
- scalability (application in both small and large vehicles).

Mechanical Storage

Low-pressure gaseous form of hydrogen is preferable in terms of efficiency. However, since vehicles cannot store enough hydrogen in this form, compression or liquefaction of hydrogen represents a relatively straightforward way of increasing vehicle on-board capacity. Mechanical storage methods store hydrogen by confining its gas or liquid form in a mechanical tank, similar to how gasoline is stored. The advantages are relatively good charge/discharge time and durability, but the capacity of the tanks is limited by relatively low energy density of hydrogen gas and liquid. In addition, the weight of tank is considerably larger than the gasoline tank due to demands of safe hydrogen storage.

There are currently three broad methods of mechanical hydrogen storage:

- high-pressure tank systems
- hydrogen-absorbing alloy tank system
- liquid hydrogen tanks.

Most current vehicles using hydrogen are equipped with a composite high-pressure tank, due to its simple structure and ease of charge-discharge cycle. Standard high-pressure tanks store hydrogen at 35MPa (350bar) pressure, which provides vehicle autonomy of 300-350km. Tanks have been developed recently to store hydrogen at 70MPa (700bar) pressure.

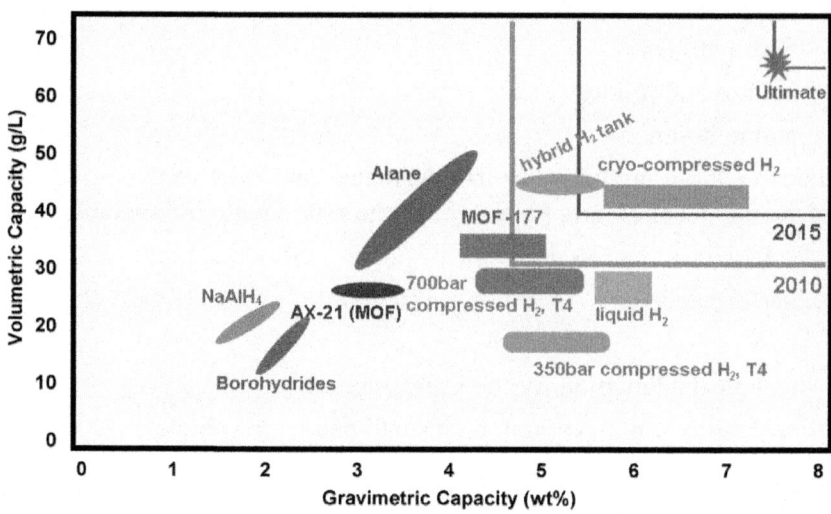

Figure 1: DOE set requirements (solid lines) and overview of existing developed hydrogen storage methods with respect to tose requirements

However, pressure in this range makes the relationship between hydrogen pressure and amount non-linear. Therefore, doubling of pressure only leads to an increase of 40-60% more hydrogen, increasing vehicle autonomy to around 500km (Mori & Hirose, 2009).

Table 1: Different high-pressure (35MPa) tank types for hydrogen storage (GFRP – glass fiber reinforced polymer, CFRP – carbon fiber reinforced polymer)

Tank type	Type 1	Type 2	Type 3	Type 4
composition	all metal (carbon steel)	metal liner + GFRP layer	metal liner + CFRP layer	plastic liner + CFRP layer
thickness ratio (cylinder part)	1.0	0.7	0.32	0.28

In order to improve storage capacity of a high-pressure tank, hydrogen absorbing metal alloy is added to the tank to produce a hybrid high-pressure hydrogen absorbing tank system. The alloy increases capacity of the tank from around 3kg per 180L tank to 7-10kg per 180L tank, but hydrogen absorption on charging releases considerable amount of heat, requiring heat exchanger and radiator to be fitted with the tank, resulting in an increase in tank weight from around 100kg to over 400kg. High-pressure (35MPa) hydrogen environment helps metal alloy absorb hydrogen quickly (around 80% charge in 5min) and improves discharge speed, which is a common difficulty of classical metal hydrides, although high hydrogen pressure means that this system has the same safety issues as a regular high-pressure tank system. This system is also capable of operating at temperatures as low as 243K, which is not possible with low-pressure metal hydride systems. These hybrid systems offer a lot of promise offering good charge/discharge performance, but they still don't achieve the required hydrogen capacity and they retain the same safety issues associated with conventional high-pressure systems.

Another improvement of high-pressure system represents the use of cryo-compressed storage, where hybrid high-pressure tank is equipped with heat-transfer system, allowing it to maintain hydrogen gas at temperature around 50K and 0.4MPa pressure (around 6mass%). As the tank is emptied during use, temperature is gradually increased using the heat released during discharge to maintain a minimum 0.4MPa pressure. Figure 2 shows a comparison of gravimetric and volumetric densities of hydrogen for different high-pressure tank systems. Projected capacity of cryo-compressed hybrid system would meet DOE interim requirements for 2015, the first system so far to achieve this. However, this system is still well short of ultimate goals set by DOE.

Liquid hydrogen exists at atmospheric pressure at temperatures below 20K and its density is much higher than compressed gas. From the point of use in vehicles and infrastructure, liquid hydrogen is very attractive, because it offers an opportunity to easily transport and store large amounts of hydrogen. Some studies show that the extra energy consumption of liquefaction process can be compensated by the ease of delivery and storage. However, due to low temperatures involved, liquid hydrogen tanks require double wall construction for thermal insulation, minimizing heat conductivity with vacuum multi-layer insulation. This consists of a thin metal layer on the spacer material which prevents both radiation and thermal irradiation between individual layers and heat intrusion from irradiation and gaseous convection. The most advanced liquid hydrogen tanks have limited heat flow to a few watts per second, which results in hydrogen evaporation and, therefore, loss of a few percent per day. Further developments in thermal insulation, advanced liquefaction and charge/discharge strategy are necessary in order to make liquid hydrogen a viable commercial alternative to a gasoline tank.

Carbon Materials

There have been a number of studies of use of different carbon materials for hydrogen storage. Although initial reports suggested high hydrogen storage capacity of carbon nanotubes and other related nanostructures, more recent results have shown that the maximum sorption capacity of carbon nanostructures is around 2mass% (Zuettel et al., 2004) and that it is dependent only on the surface area of individual carbon nanostructure. Of those nanostructures, graphene sheets have exhibited the highest surface area and represent the most promising carbon material for hydrogen storage. However, since its capacity in pure form is about 2mass%, recent work has been focused on improving both performance and capacity of graphene. One of the issues that need to be overcome is that binding energy of atomistic hydrogen to carbon is 0.8eV, much lower than the energy of H-H bond in H_2 molecule (2.3eV per hydrogen atom). That is why recent research has been focused on catalyst assisted hydrogen sorption on carbon through spillover effect. Hydrogen spillover refers to transport of an active species generated on one substance (activator) to another (receptor) that would not normally adsorb it. In this case, activator is commonly a metal or a metal oxide, while carbon acts as a receptor, overcoming the energy barrier associated with dissociation of hydrogen molecule. Enhancement of hydrogen adsorption via spillover effect is much more pronounced at lower hydrogen pressure (below 0.5MPa), while it saturates at high pressures. This suggests existence of two distinct mechanisms corresponding to different spillover behavior, which can be controlled by activation of catalyst (Tsao et al., 2010).

Recent first-principle calculations have suggested that use of graphene double-layer and multi-layer structures could increase capacity for hydrogen storage (3-4mass%) (Patchkovskii et al., 2005), while recent experiments of Pd-loaded single-wall carbon nanotubes show improved performance after Pd-loading and thermal treatment at temperatures around 700K (Kocabas et al., 2008). Additional theoretical calculations of spillover on graphene suggests that hydrogen atoms on graphene surface should create compact clusters so that the lowest-energy luster is composed of closed six-hydrogen rings, which would correspond, if entire surface area of graphene is used, to hydrogen storage capacity of about 7.7mass% (Lin et al., 2008). However, recent studies of kinetics of hydrogen adsorption/desorption kinetics in Pd/Pt/Ni/Ru doped ordered mesoporous carbon indicated that there is no difference in the kinetics of doped and pure carbon (Saha & Deng, 2009), suggesting that doping of carbon materials using transition metals might not be able to achieve significant increase in capacity and charge/discharge rates.

Although work on carbon materials is ongoing, it is unlikely that carbon materials will, in the foreseeable future, achieve the necessary performance levels required to replace fossil fuels. These types of materials have exhibited encouraging hydrogen capacity at low temperatures (20-80K), but their performance has regularly diminished at room temperature and this remains the biggest challenge in research of carbon materials for hydrogen storage.

Zeolites and Metal-Organic Frameworks

Zeolites are crystalline microporous materials, usually alumosilicates or aluminophosphates. They consist of microporous framework, which, in principle, appears highly suitable for hydrogen adsorption, as the adsorption energies in the narrow pores are very low, allowing thermal cycling to be used for adsorption and desorption of hydrogen. Initial reports of their storage capacities (Weitkamp et al., 1992,1995, 1997) indicated very low capacities at room temperature, but these substantially increased at 77K, to 1.5mass% (Jhung et al., 2007). However, these fall well short of technical requirements. Projected maximum capacities for zeolite systems (assuming hydrogen packing density equal to liquid hydrogen) are a maximal 2.5 mass%, indicating that the currently known systems would not be able to meet the requirements to serve as hydrogen storage materials in mobile applications, although new materials might offer better performance.

Metal-organic framework (MOF) materials are composed of metal ions as vertices connected by organic molecules, often polyvalent carboxylic acids, to create a porous material of exceedingly high surface area (Rowsell & Yaghi, 2005). Reported hydrogen storage capacities of these materials have been encouraging: material MOF-177 was reported to have saturation uptake of 7.5 mass% at 77K and atmospheric pressure (Wong-Foy et al., 2006), although this declines to 1.62 mass% at 0.1MPa pressure (Rowsell et al., 2004). These materials exhibit an interesting feature, which could be of great importance for hydrogen adsorption – the so-called gated adsorption (Kitaura et al., 2003; Zhao et al., 2004). This process relies on flexibility of the framework of some MOFs allowing the structure to expand upon adsorption of guest species and shrink back upon desorption. This leads to a pronounced hysteresis, which can be exploited to load the materials at high pressure and still be able to capture hydrogen at lower pressures or higher temperature. However, these materials have some disadvantages as the loading of the material has to be performed at low temperatures, which consumes additional energy, and the binding of hydrogen to MOF materials is stronger than in carbon materials like graphite and carbon nanotubes (Rosi et al., 2003). However, this is still a new class of materials and most of them have yet to be investigated as hydrogen storage materials. In addition, in the foreseeable future, there should be many more new materials of this type, therefore, this class of materials shows great promise when it comes to hydrogen storage capabilities and offers genuine prospect of a hydrogen storage system that could meet all the requirements necessary for mobile applications.

Metal Hydrides

Many metals and alloys are capable of absorbing considerable amounts of hydrogen according to the reaction (1):

$$Me + \frac{x}{2}H_2 \Leftrightarrow MeH_x$$

(1)

Here MeMe is a metal, a solid solution, or an intermetallic compound and MeHxMeHx is a hydride (x is the ratio of hydrogen to metal H/Me). In most cases this reaction is exothermic and reversible. Heating of the hydride causes hydrogen desorption. Charging can be done using molecular hydrogen gas or hydrogen ions from an electrolyte. If hydrogen is loaded from gas phase, several reaction stages of hydrogen with metal would occur, as shown in Fig.2. The metal is usually in form of powder, and can be amorphous or crystalline (Minić et al., 1996). Repeated thermal treatment during hydrogen absorption and desorption causes structural changes in amorphous metals and intermetallic compounds, leading to crystallization (Minić et al., 1995; Šušić et al., 1996).

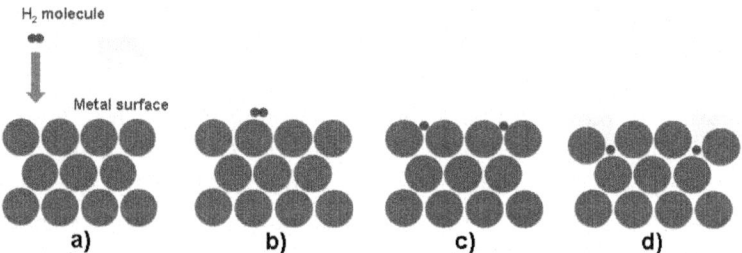

Figure 2: Reaction of H_2 molecule with metal surface: a) H_2 molecule move toward the metal surface. b) physisorbtion of H_2 molecule through Van der Waals interaction with metal surface c) chemisorption of hydrogen after dissociation d) occupation of subsurface sites and diffusion into bulk.

Most of the known metal hydrides exhibit unsatisfactory charge/discharge kinetics, which led to devotion of significant research effort on improving it using surface catalysts and taking advantage of spillover mechanism of hydrogen absorption (Fig. 3). This mechanism includes adsorption and subsequent dissociation of hydrogen molecule on surface catalyst, and migration and subsequent diffusion of adsorbed hydrogen atoms from surface catalyst into the metal (Minić et al., 1997). Since catalyst is a metal or intermetallic compound with superior hydrogen adsorption, it serves as a gateway for hydrogen absorption. This improves charging kinetics of the metal, while using relatively small amounts of catalyst (most commonly Pd) (Šušić, 1997a, 1997b).

The reaction of hydrogen gas with a metal can be described by one-dimensional Lennard-Jones potential curve, Figure 4 (Lennard-Jones, 1932). Far from the metal surface, the potential energy difference of a hydrogen molecule and that of 2 individual hydrogen atoms is equal to dissociation energy ($H_2 \rightarrow 2H$, E_D = 435.99 kJ/molH$_2$). The molecular hydrogen initially exhibits Van der Waals attractive interaction during approach to metal surface, leading to the physisorbed state ($E_{phys} \approx$ -5 kJ/molH) at a distance from the metal surface approximately equal to hydrogen molecule radius ($\approx$0.2 nm).

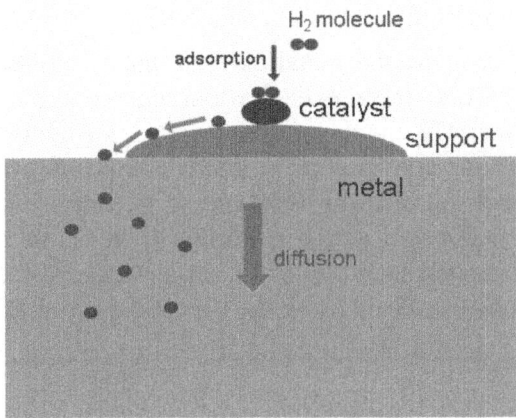

Figure 3: Illustration of spillover mechanism of absorption of hydrogen into a metal using a catalyst

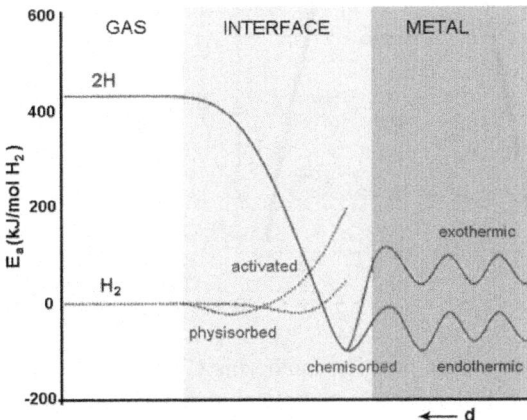

Figure 4: One-dimensional potential energy curve (one-dimensional Lennard-Jones potential for a hydrogen metal system.

Closer to the metal surface, hydrogen has to overcome an activation barrier for dissociation and formation of the hydrogen-metal bond (crossing point of dashed blue and solid red line). The height of the activation barrier depends on the chemical composition of the surface. When hydrogen atom becomes chemisorbed ($E_{chem} \approx -50$ kJ/molH$_2$) it shares its electron with the metal atoms at the surface. These hydrogen atoms have high surface mobility, interacting with each other and forming surface phases. In the next step chemisorbed hydrogen atom can migrate into the subsurface layer and, finally, diffuse into the interstitial sites through the host metal lattice, contributing their electrons to the band structure of the metal (Zuettel, 2003).

After dissociation on the metal surface, the H atoms generally diffuse rapidly through the bulk metal even at room temperature to form Me-H solid solution or α-phase. The thermodynamic aspects of hydride formation from gaseous hydrogen are described by pressure–composition isotherms (Fig. 5). After the maximum solubility of hydrogen in the α-phase is reached, hydride phase (β-phase) will begin to form. Further increase in hydrogen pressure will result in substantial increase in the amount of absorbed hydrogen. This phenomenon can be explained using the Gibbs phase rule (2)

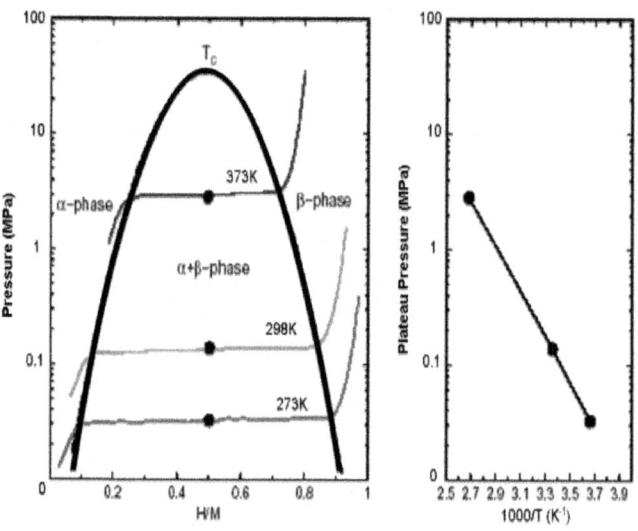

Figure 5: Left: Pressure-Composition-Isotherms (PCI) for a hypothetical metal hydride. Right: Van't Hoff plot for a hypothetical metal hydride derived from the measured pressures at plateau midpoints from the PCI's.

$$F = 2 - \pi + N \tag{2}$$

where FF is the degree of freedom, π is the number of phases and N is the number of components.Hydrogen pressure at which this transformation takes place is called the plateau pressure, where α- and β-phase co-exist. When the stoichiometric hydride is formed, completely depleting the α-phase, an additional degree of freedom is regained and the additional absorption of hydrogen would now require a substantial pressure increase, corresponding to the solid solution of hydrogen in β-phase.

The plateau pressure, described by Van't Hoff equation (3), gives us valuable information about reversible storage capacity. Width of the plateau and the position of the plateau at a given temperature give us a sense of the stability of the hydrides. Stable hydrides (enthalpy of formation, $H_f \ll 0$) would require higher temperatures than less stable hydrides ($H_f < 0$) to achieve a certain plateau pressure. Recording a series of PCI›s at different temperatures makes it possible to construct a phase diagram from the end points of the plateaus in the individual PCI›s.

$$\ln p = \frac{\Delta H}{RT} - \frac{\Delta S}{R} \tag{3}$$

where $\Delta H \Delta H$ is enthalpy, $\Delta S \Delta S$ is entropy, R is the gas constant and T is temperature.Figure 6 represents a Van't Hoff diagram showing the dissociation pressures and temperatures of a number of hydrides. Light elements, such as Mg, have shown promising levels of stored hydrogen (about 7 wt% hydrogen), but require higher temperature for dehydrogenation. Conventional metal hydrides, which have been well characterized and their capacity for interstitial hydrogen storage is well-established, include type AB, AB_2, AB_5, A_2B (TiFe, $ZrMn_2$, $LaNi_5$, and Mg_2Ni) intermetallic compounds, and body-centered cubic metals. These materials typically store between 1.4 and 3.6 wt% hydrogen. (Table 2). However, the requirements for gravimetric capacity, fast kinetic and high storage capacity is barely satisfied, so development of new lightweight materials presents many scientific and technical challenges. Among metal hydrides, magnesium hydride appears to be the most promising, because of its high storage capacity and relatively low cost.

Table 2: Hydrogen storage properties of some intermetallic compounds

Type	Intemetallic compound	H/M	H_2capacity (mass%)	Temperature (K) for 0.1MPa $_{Pdesorption}$
A_2B	Mg_2Ni	1.33	3.6	528
AB	TiFe	0.975	1.86	265
AB	ZrNi	1.4	1.85	565
A_B2	ZrM_n2	1.2	1.77	440
A_B5	LaN_i5	1.08	1.49	285
A_B2	$TiV_{0.62}M_{n1.}5$	1.14	2.15	267

Magnesium Hydride

High gravimetric (7.6 wt.%) (Hanada et al., 2004) and volumetric density (130 kg H_2/m^3) and relatively low price make magnesium hydride (MgH_2) an attractive hydrogen storage material. However the wide industrial application is still not feasible, due to its high enthalpy of formation ($\Delta H° = -75$ kJ/molH_2) and dehydrogenation temperature (720 K), as well as slow kinetics of hydrogenation reactions/dehydrogenation reaction. For example, there is no detectable hydrogen desorption at temperature of 573 K, while at 623 K it takes more than 3000s for complete decomposition of MgH_2(Varin et al., 2009). Furthermore, in order to allow for the formation of MgH_2, it is necessary to perform an activation process of Mg metal by consecutive heating and cooling of metal in vacuum and in hydrogen atmosphere, which makes the material permeable to hydrogen. The efforts to overcome these deficiencies have made MgH_2 one of the most investigated materials in last two decades.

At moderate hydrogen pressure the only hydride phase existing in equilibrium with Mg is magnesium hydride, β-MgH_2. Pure Mg has a hexagonal crystal structure, while its hydride has a tetragonal lattice unit cell (rutile type) (Zeng et al., 1999; Yu & Lam, 1988). Figure 7 shows the crystal structure of β-MgH_2 ($P4_2/mnm$ space group), were each Mg atom is coordinated with six H atoms forming an irregular slightly distorted octahedron (Noritake et al., 2003). Each H atom is coordinated with three planar Mg atoms. Synchrotron X-ray diffraction gives the parameters of crystal $a = 0.45180(6)$ nm and$c = 0.30211(4)$ nm (Lide, 2006; Lide, 2007) while the powder diffraction file (JCPDS 12–0697) provides similar values for $a = 0.4517$ nm and $c = 0.30205$ nm. Density of MgH_2 is 1.45 g/cm^3(Bastide, 1980).

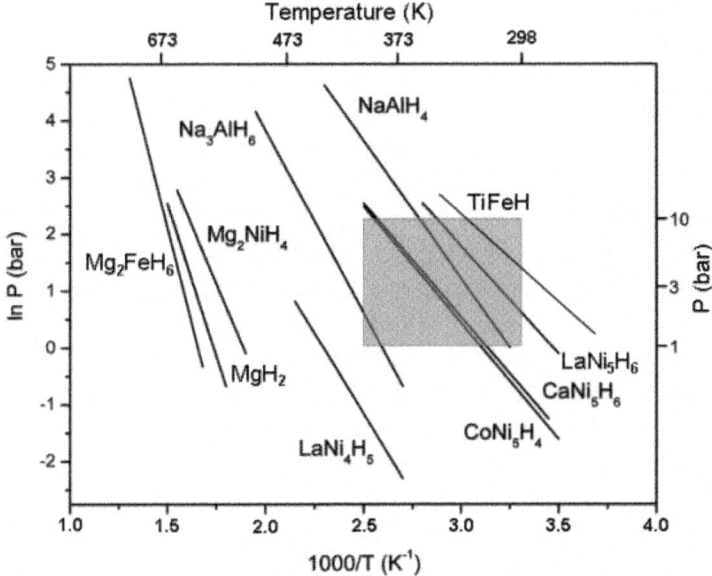

Figure 6: Van't Hoff plots of various metal hydrides, showing hydrogen dissociation pressures and temperatures (red shaded area represents desirable operating conditions).

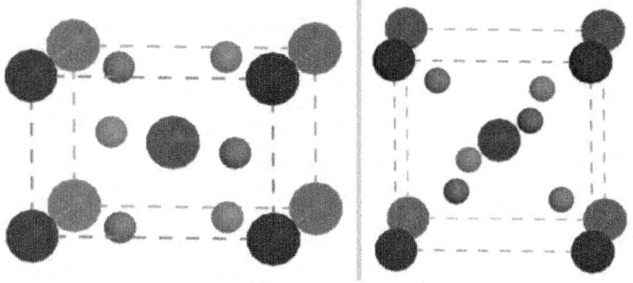

Figure 7: Two perspectives of crystal structure of β-MgH$_2$ showing the positions of Mg (blue) and H atoms (red)

Although there is considerable literature dealing with structure and properties (Novaković et al., 2009;Schimmel et al., 2004, 2005; Kelkar & Pal, 2009; Basseti et al., 2005) of MgH$_2$, there are still some uncertainties about H-desorption kinetics, since the Mg–H interaction strongly depends on method of synthesis and presence of additives. For instance, ball milling causes mechanical deformation, surface modification, and metastable phase formation which generally promote the solid–gas reaction. It also introduces defect zones which may accelerate the diffusion of hydrogen (Shan et al., 2004; Jensen et al., 2006; Bobet et al., 2001; Montone et al., 2006, 2007; Aguey-Zinsou et al., 2007). Addition of transition metals, metal oxides or ternary hydrides to mechanically milled MgH_2 decreases its thermal stability and decomposition temperature. It has also been established that nanosized powders provide a possible solution for the problem of hydrogen desorption kinetics (Varin et al., 2006). *Ab initio* DFT calculations (Kurko et al., 2011) have been used to determine possible ways of improving the performance of MgH_2. They have shown that destabilization temperature of MgH_2 reduces with decreasing cluster size, and that a crystallite size of 0.9 nm could result in a desorption temperature below 500 K (Wagemans et al., 2005). However ball milling can also cause agglomeration and cold welding of particles (Montone et al., 2005), therefore experimentally obtained decrease in hydrogen adsorption temperature in ball-milled samples is relatively low, keeping MgH_2 far from practical application (Milovanović et al., 2008). Based on *ab initio* calculations, Novakovic *et al.* (Novaković et al., 2009) examined the possible pathways for adsorption/desorption of hydrogen from MgH_2, showing that different H-filling patterns influence initial hcp-Mg structure in different ways, qualifying wurtzite MgH as a probable intermediate phase between the hcp-Mg and MgH_2 at 1:1 stoichiometry. Du et al. suggested that surface desorption is rate-determining since the activation barrier computed for H-vacancy diffusion from the surface into sublayer is less than 0.70 eV, much smaller than the activation energy for desorption of hydrogen on the $MgH_2(110)$ surface (1.78-2.80 eV/H_2) (Du et al., 2007). Furthermore, the first principle calculations of MgH_2-*TM* (*TM*=Al, Ti, Fe, Ni, Cu, Nb, Co) were carried out to investigate their influence on the stability of the magnesium hydride (Novaković et al., 2010; Song et al., 2004). It was found that TM-H bonding is stronger than the Mg–H bond, but at the same time it weakens other bonds in the second and third coordination around TM atom, which leads to overall destabilization of the MgH_2 compound. Due to a higher number of d-electrons, this effect is more pronounced in late transition metals. In case of Co doping, spin polarization has an additional stabilizing influence on the compound structure.

Alanates

Alanates, or aluminohydrides, are a family of compounds containing aluminum and hydrogen, with $NaAlH_4$ being the most popular material in this family (Bogdanović & Schwickardi, 1997; Downs & Pulham, 1994; Fichter et al., 2003; Jain et al., 2010). One of the attractive features of alanates is their easy accessibility, since sodium and lithium alanate are commercially available and magnesium alanate can be easily synthesized. They have relatively high gravimetric hydrogen capacity (5-9 mass%), however, these materials usually undergo multi-step thermal decomposition (Eq. 4) to release hydrogen and these reactions require relatively high temperatures and considerable reaction time (NaAlH$_4$ releases 3.7 mass% H$_2$ in 3h at 483K).

$$NaAlH_4 \rightarrow NaH + Al + \frac{1}{2}H_2$$

(4)

Additionally, some of the alanates are meta-stable, making their first decomposition (dehydrogenation) step exothermic and irreversible. This makes their direct rehydrogenation impossible, which would rule them out as candidates for on-board hydrogen storage applications.

In order to improve their charge/discharge performance and reduce decomposition temperatures, alanates have been doped with transition metals, with Ti- and Ni-doping exhibiting promising results thus far in improving kinetics and dehydrogenation rates, while preserving very high hydrogen capacity.

Table 3: Overview of characteristics of some of the alanates for hydrogen storage [18]

Material	H$_2$ capacity (mass%)	Dehydogenation temperature (K)	Dissociation enthalpy (kJ/mol H$_2$)
NaAl$_H$4	5.6	480-490 (I step) "/>525 (II step)	37 (I step, 3.7 mass% H$_2$) 42 (II step, 1.9 mass% H$_2$)
LiAl$_H$4	7.9	430-450 (I step) 450-490 (II step)	-10 (I step, 5.3 mass% H$_2$) 25 (II step, 2.6 mass% H$_2$)
Mg(AlH$_4$)2	9.3	380-470 (I step) 510-650 (II step)	41 (I step, 7 mass% H$_2$) 76 (II step, 2.3 mass% H$_2$)
KAl$_H$4	5.7	570 (I step) 610 (II step) 650 (III step)	55 (I step, 2.9 mass% H$_2$) 70 (II step, 1.4 mass% H$_2$)
Ca(AlH$_4$)2	5.9	400 (I step) 520 (II step)	-7 (I step, 2.9 mass% H$_2$) 28 (II step, 2.9 mass% H$_2$)

Alanates exhibit very high hydrogen storage capacity, although they cannot be charged and discharged easily and their working temperatures are too high. There is some hope that doping could alleviate some of these difficulties, but it is clear that if these materials become utilized for hydrogen storage, it will probably be in a hybrid high-pressure system, which could additionally improve their charge/discharge performance and be adapted to account for relatively slow discharge rates, while taking advantage of high hydrogen capacities.

Borohydrides

Borohydride, or tetrahydroborate, refers to a group of complex hydride compounds in which hydrogen in covalently bonded to the central atom in $[BH_4]^-$ complex anion (Eq. 5).

$$NaBH_4 + (2+x)H_2O \rightarrow 4H_2 + NaBO_2 \cdot xH_2O \tag{5}$$

They exhibit high gravimetric and volumetric hydrogen capacity, making the materials of interest for hydrogen storage research (Schlesinger et al., 1953). One of the issues that would have to be resolved is the fact that they exhibit exothermic desorption reactions (30-300 kJ/mol), making their rehydrogenation more difficult than that of other materials and making direct rehydrogenation thermodynamically impossible. In addition, they exhibit high dehydrogenation temperatures (520-670°C), which practically means that only part of their hydrogen capacity is accessible at normal working temperatures.

Table 4: Overview of characteristics of some of the borohydrides for hydrogen storage [18]

Material	H_2 capacity (mass%)	Dehydogenation temperature (K)	Dissociation enthalpy (kJ/mol H_2)
NaB_H4	10.8	670	-217 to -270
LiB_H4	13.4	650	-177
$Mg(BH_4)2$	13.7	530-670	-39.3 to -50
$Ca(BH_4)2$	9.6	620	32

Therefore, most of the work associated with borohydrates is focused on use of doping and catalysts to improve charge/discharge kinetics, lower dehydrogenation temperature and increase hydrogen discharge capacity and cycling ability of these materials (Soler et al., 2007, Pena-Alonso et al., 2007; Lee et al., 2008; Demirci et al., 2009).

Amides and Imides

Amides and imides have attracted a lot of interest, due to their high hydrogen capacity and relatively low operating temperature (Chen et al., 2003; Hu et al., 2003). However, their poor absorption kinetics limits their current practical application (Eq. 6). Therefore there has been a lot of focus on overcoming this by doping with a catalyst, usually through mechanical ball milling (Liu et al., 2008).

$$LiNH_2 + LiH \rightarrow Li_2NH + H_2 \tag{6}$$

Another direction of research has been combining alkaline metal amides with other hydride materials using ball milling. Although $LiNH_2$ and $LiAlH_4$ mixture released approximately 4 atom equivalents of hydrogen, its rehydrogenation was unsuccessful at H_2 pressures up to 80bar (8MPa) (Xiong et al., 2006).

Table 5: Overview of characteristics of some of the amida/imide systems for hydrogen storage [18]

Metal ion of amide/imide system	H_2 capacity (mass%)	Dehydogenation temperature (K)	Dissociation enthalpy (kJ/mol H_2)
Li	5.4 6.5	550	148 45
Mg	7.4	470	3.5
Ca	3.5 2.1	620 770	
Li-Mg	4.5	470	39
Li-Al	4 2 2.1	360 440 470	

Amino Borane

Ammonium borane, NH_3BH_3, is considered a promising candidate for chemical storage of hydrogen, due to its low molecular weight and high hydrogen content of 19.6 mass%. Its thermal decomposition occurs in three stages (Eqs. 7 and 8), around 363, 420 and 970K, respectively, with one molarequivalent (6.5 mass%) of hydrogen released during each step (Gutowska et al. 2005; Baitalow et al., 2002).

$$NH_3BH_3 \rightarrow (BN)_n + 3nH_2 \tag{7}$$

$$NH_3BH_3 + H_2O \rightarrow NH_4^+ + BO_2^- + 3H_2 \tag{8}$$

Dehydrogenation rates are very slow due to long induction period (around 200min at 353K) (Chandra & Xu, 2006a, 2006b). The enthalpy is -21 kJ/mol, which is in the range that is suitable for on-board application. Therefore, most of the scientific effort has been focused on overcoming the slow reaction kinetics using additives like silica scaffolds, metal catalysts and ionic liquids (Umegaki et al., 2009).

Alane

Aluminum hydride (alane, AlH_3) has an average enthalpy of formation of -11.4 kJ/mol and freshly synthesized non-solvated alane is reported to desorb around 10 mass% of H_2 at temperatures below 373K (Graetz & Reilly, 2006). However, while the dehydrogenation reaction occurs readily (Eq. 9), the reverse process does not, requiring 2.5GPa pressure of H_2 to rehydride (Baranowski & Tkacz, 1983).

$$AlH_3 \rightarrow Al + \frac{3}{2}H_2$$

(9)

In addition, dehydrogenation kinetics is too slow for practical applications, as it is limited by nucleation and growth of aluminum particles. Therefore, research has been focused on decreasing the dehydrogenation temperature and improving its kinetics using additives like alkali metal hydride (Sandrock et al., 2006) and particle size reduction using ball milling (Orimo et al., 2006). Best results have been achieved by doping AlH_3 with small amounts of LiH, NaH or KH.

HYDROGEN POWER GENERATION – FUEL CELLS

As early as 1839, William Grove discovered the basic operating principle of a fuel cell by reversing the electrolysis of water to generate electricity from hydrogen and oxygen. This principle remains unchanged today, that a fuel cell is an electrochemical device which continuously converts chemical energy into electric energy (and heat) for as long as fuel and oxidant are supplied. A fuel cell does not need recharging, operates quietly and efficiently and, with hydrogen as fuel, generates only power and water. This is why it is called a zero-emission engine. Unlike thermal engines, it is not limited by the Carnot efficiency; therefore, while thermal engines typically achieve efficiency of 24-32%, fuel cells achieve efficiencies of 35-60% (Minh & Takahashi, 1995).

One of the major obstacles in fuel cell application so far has been their insufficient lifetime: most fuel cells exhibit major performance decay after 1000 hours of operation, while DOE target for 2015 is 5000 hours of operation at 60% efficiency for mobile applications. Additionally, their price per kW is almost twice that of an internal combustion engine (Hoogers, 2003). There is a wide variety of fuel cell types, which can be distinguished by the electrolyte used, but they all function in the same basic way. At the anode a fuel (usually hydrogen) is oxidized into electrons and protons, and at the cathode, oxygen is reduced to oxide species. Depending on the electrolyte, either protons or oxide ions are transported through the electrolyte (ion-conducting, but electronically insulating) to combine with oxide species or protons, respectively, to generate water and electric power. Alkaline fuel cells do not tolerate presence of CO_2 in the system, which forms alkaline carbonates, meaning that they can only use carbon-free fuel.

Table 6: Overview of different types of fuel cells and their characteristics

Fuel Cell Type	Electrolyte	Charge Carrier	Operating temperature (K)	Fuel	Electric Efficiency (system)	Power Range (kW)
Alkaline	KOH	$OH-$	330-390	H_2	35-55%	<5kW
Proton Exchange Membrane	Solid polymer	$H+$	320-373	H_2 (tolerates CO_2)	35-45%	5-250
Phosphoric acid	Phosphoric acid	$H+$	~590	H_2 (tolerates CO_2)	40%	200
Molten carbonate	Li and K carbonate	CO_3^{2-}	~1020	H_2, CO, CH_4 (tolerates CO_2)	"/>50%	200-1000
Solid oxide	Solid oxide electrolyte (Y, Zr)	O_2-	~1270	H_2, CO, CH_4 (tolerates CO_2)	"/>50%	2-1000

Of the currently developed types of fuel cells, proton exchange membrane fuel cells (PEMFC) and alkaline fuel cells (AFC) satisfy the required range of operating temperature and provide sufficient current density for mobile applications, although the power range for AFC is sufficient only for specialty applications (like power generation of space crafts). PEMFC, also known as solid polymer fuel cell, takes its name from the special plastic membrane it uses for electrolyte, combining all the key parts (anode, cathode, electrolyte) in a very compact unit, not thicker than few hundreds microns. The membrane requires presence of liquid water, limiting its operating temperature to below 373K, which means that, to achieve a good performance, this fuel cell requires a good catalyst (Wang et al., 2011).

Two high-temperature types of fuel cells (molten carbonate and solid oxide) are mainly considered for large scale stationary power generation. They achieve higher efficiencies than low-temperature systems and high operating temperatures allow for direct internal processing of fuels like natural gas, reducing the system's complexity (Tucker, 2010). However, this also means that they cannot easily be turned off.

In addition to using pure hydrogen of hydrocarbons and carbon monoxide as fuel, in recent times, fuel cells have been coupled with biomass-derived fuel processors. This makes it possible to use biomass-derived fuels, such as ethanol, methanol, biodiesel or biogas, and feed them to a fuel processor as a raw fuel, which would, after reforming, be used by a fuel cell (Xuan et al., 2009).

Although all of types of fuel cells described above operate on gas fuel, recently, PEM fuel cells have been constructed to operate using liquid fuel – methanol (Ismail et al., 2011). This has advantages for mobile applications and fuel transportation, but these direct methanol fuel cells are not yet advanced enough to generate high enough current and power density to compete with gas fuel cells.

CONCLUSION

The limitations of our world's natural resources mean that our civilization has to find an alternative energy source and energy carrier to replace fossil fuels. Taking into account available fossil fuel reserves and that such a conversion of energy source/carrier historically takes 75-100 years, it is clear that this is crucial time for alternative energy research. Hydrogen is the obvious candidate for the renewable energy carrier for the future, due to its availability in form of water, high energy density and lack of negative environmental impact. However, current state of technology is such that significant advances are needed in the next decade in order to make hydrogen economy viable. There is still lack of an effective large-scale hydrogen production process from water, while hydrogen storage still falls well behind the requirements set by gasoline. When it comes to power generation, hydrogen fuel cells are still lagging behind internal combustion engines and conventional batteries in performance and power/weight ratio. On the other hand, significant strides have been made in recent times, which give hope that hydrogen will, in the foreseeable future, be able to challenge fossil fuels as the primary energy carrier, but only through a sustained and focused effort.

ACKNOWLEDGEMENT

The research has been supported by the by the Serbian Ministry of Education and Science under grants 172015 and III45012.

REFERENCES

1.   K. Aguey-Zinsou-F, Fernandez. J. R. Ares, T. Klassen, R. Bormann, 2007Effect of Nb2O5 on MgH2 properties during mechanical milling, International Journal of Hydrogen Energy, 321324002407

2.   F. Baitalow, J. Baumann, G. Wolf, K. Jaenicke-Rößler, G. Leitner, decomposition. Thermal, B. of-N-H, investigated. compounds, using. by, thermoanalytical. combined, Thermochimica. methods, vol. Acta,

159168

3. B. Baranowski, M. Tkacz, 1983The Equilibrium Between Solid Aluminium Hydride and Gaseous Hydrogen, Zeitschrift für Physikalische Chemie N.F., 1352738

4. A. Bassetti, E. Bonetti, L. Pasquini, A. Montone, J. Grbovic, Antisari. M. Vittori, 2005Hydrogen desorption from ball milled MgH2 catalyzed with Fe, European Physics Journal B, 4311928

5. J. P. Bastide, B. Bonnetot, J. M. Letoffe, P. Claudy, 1980Materials Research Bulletin, 151215

6. J. Bobet-L, E. Akiba, B. Darriet, 2001Study of Mg-M (M=Co, Ni and Fe) mixture elaborated by reactive mechanical alloying: hydrogen sorption properties, International Journal of Hydrogen Energy, 265493501

7. B. Bogdanovic, M. Schwickardi, 1997Ti-doped alkali metal aluminium hydrides as potential novel reversible hydrogen storage materials, Journal of Alloys and Compounds, 253-25419

8. U. B. Demirci, O. Akdim, P. Miele, 2009Aluminum chloride for accelerating hydrogen generation from sodium borohydride, Journal of Power Sources, 19223103 15

9. M. Chandra, Q. Xu, 2006A high-performance hydrogen generation system: Transition metal-catalyzed dissociation and hydrolysis of ammonia-borane, Journal of Power Sources, 1562190194

10. M. Chandra, Q. Xu, 2006Dissociation and hydrolysis of ammonia-borane with solid acids and carbon dioxide: An efficient hydrogen generation system, Journal of Power Sources, 1592855860

11. P. Chen, Z. Xiong, J. Z. Luo, J. Y. Lin, K. L. Tan, 2003Interaction between Lithium Amide and Lithium Hydride, Journal of Physical Chemistry B, 1071096710970

12. Downs A.J. & Pulham C.R.1994The hydrides of aluminium, gallium, indium, and thallium: a re-evaluation, Chemical Society Review, 233175184

13. A. J. Du, S. C. Smith, G. Q. Lu, 2007First-Principle Studies of the Formation and Diffusion of Hydrogen Vacancies in Magnesium Hydride, Journal of Physical Chemistry C, 1112383608365

14. M. Fichtner, O. Fuhr, O. Kircher, 2003Magnesium alanate-a material for reversible hydrogen storage?, Journal of Alloys and Compounds, 356-357418422

15. A. Gutowska, L. Li, Y. Shin, C. M. Wang, X. S. Li, J. C. Linehan, R. S. Smith, B. D. Kay, B. Schmid, W. Shaw, M. Gutowski, T. Autrey,

2005Nanoscaffold Mediates Hydrogen Release and the Reactivity of Ammonia Borane, Angewandte Chemie. International Edition, 442335783582

16. J. Graetz, J. J. Reilly, 2006Thermodynamics of the α, β and polymorphs of AlH3, Journal of Alloys and Compounds, 424262265

17. N. Hanada, T. Ichikawa, S. Orimo-I, H. Fujii, 2004Correlation between hydrogen storage properties and structural characteristics in mechanically milled magnesium hydride MgH2, Journal of Alloys and Compounds, 366269273

18. A. Hauch, S. D. Ebbesen, S. H. Jensen, M. Mogensen, 2008Highly Efficient High Temperature Electrolysis, Journal of Materials Chemistry, 1823312340

19. Hoogers G. (editor), Fuel Cell Technology Handbook, CRC Press, New York, U.S.A., 2003

20. I. P. Jain, P. Jain, A. Jain, 2010Novel hydrogen storage materials: A review of lightweight complex hydrides, Journal of Alloys and Compounds, 503303339

21. Y. H. Hu, E. Ruckenstein, 2003Ultrafast Reaction between LiH and NH3 during H2 Storage in Li3N, Journal of Physical Chemistry A, 1074697379739

22. A. Ismail, S. K. Kamarudin, W. R. W. Daud, S. Masdar, M. R. Yosfiah, 2011Mass and heat transport in direct methanol fuel cells, Journal of Power Sources, 19698479855

23. T. R. Jensen, A. Andreasen, T. Vegge, J. W. Andreasen, K. Ståhl, A. S. Pedersen, M. M. Nielsen, A. M. Molenbroek, F. Besenbacher, 2006Dehydrogenation kinetics of pure and nickel-doped magnesium hydride investigated by in situ time-resolved powder X-ray diffraction, International Journal of Hydrogen Energy, 311420522062

24. S. H. Jhung, J. W. Yoon, J. S. Lee, J. S. Chang, 2007Low-Temperature Adsorption/Storage of Hydrogen on FAU, MFI, and MOR Zeolites with Various Si/Al Ratios: Effect of Electrostatic Fields and Pore Structures, Chemistry A European Journal, 132265026507

25. T. Kelkar, S. Pal, 2009A computational study of electronic structure, thermodynamics and kinetics of hydrogen desorption from Al- and Si-doped α-,-, and β-MgH2, Journal of Materials Chemistry, 192543484355

26. R. Kitaura, K. Seki, G. Akiyama, S. Kitagawa, 2003Porous Coordination-Polymer Crystals with Gated Channels Specific for Supercritical Gases, Angewandte Chemie International Edition, 424428431

27. S. Kocabas, T. Kopac, G. Dogu, T. Dogu, 2008Effect of thermal treatments and palladium loading on hydrogen sorption characteristics of single-walled carbon nanotubes, International Journal of Hydrogen Energy 3316931699

28. S. Kurko, Mamula. B. Paskaš, Lj. Matovic, Novakovic. J. Grbovic, N. Novakovic, 2011The Influence of Boron Doping Concentration on MgH2 Electronic Structure, Acta Physica Polonica A, 1202238241

29. Lee W-J. & Lee Y-K.2001Internal Gas Pressure Characteristics Generated during Coal Carbonization in a Coke Oven, Energy & Fuels, 153618623

30. Y. Lee, Y. Kim, H. Jeong, M. Kang, 2008Hydrogen production from the photocatalytic hydrolysis of sodium borohydride in the presence of In-, Sn-, and Sb-TiO2s, Journal of Industrial Engineering and Chemistry, 145655660

31. Lennard-Jones J.E.1932Processes of adsorption and diffusion on solid surfaces, Transactions of Faraday Society, 28333

32. Lide D.R.editor),Physical constants of inorganic compounds, CRC Handbook of Chemistry and Physics, Editor: 87th on-line edition (2006-2007CRC Press, New York, USA

33. Y. Lin, F. Ding, B. I. Yakobson, 2008Hydrogen storage by spillover on graphene as a phase nucleation process, Physical Review B 78041402R)

34. Y. Liu, K. Zhong, M. Gao, J. Wang, H. Pan, Q. Wang, 2008Chemistry of Materials, 2035213527

35. P. Maness-C, S. Smolinski, A. C. Dillon, M. J. Heben, P. F. Weaver, 2002Characterization of the oxygen tolerance of a hydrogenase linked to a carbon monoxide oxidation pathway in Rubrivivax gelatinosus, Applied Environmental Microbiology, 6826362636

36. A. Melis, 2002Green alga hydrogen production: progress, challenges and prospects. International Journal of Hydrogen Energy, 2712171228

37. S. Milovanović, Lj. Matović, M. Drvendžija, Novaković. J. Grbović, 2008Hydrogen storage properties of MgH2- diatomite composites obtained by high energy ball milling, Journal of Microscopy, 2323522525

38. N. Minh, T. Takahashi, 1995Science and Technology of Ceramic Fuel Cells, Elsevier, The Netherlands

39. D. Minić, M. Šušić, 1995Thermal Behaviour of 82Ni-18P Amorphous Powder Alloy in Hydrogen atmosphere, Materials Chemistry and Physics, 40281284

40. D. Minić, M. Šušić, A. Maričić, 1996Absorption of Hydrogen by Amorphous and Crystalline 89Fe-11P Powder. Deformation of the

Powder under Pressure and Relaxation on Heating, Materials Chemistry and Physics, 45280283

41. D. Minić, M. Šušić, Ž. Tešić, R. Dimitrijević, 1997Investigation of the Thermal Behaviour of Ag-Pd Intermetallic Compounds in Hydrogen Atmosphere, Studies in surface and catalysis, 112447456

42. D. Minić, M. V. Šušić, 2002Modern Concepts of Conversion and Storage of Energy bz Disperse Material Absorption, Science of Sintering, 34247259

43. A. Montone, J. Grbović, A. Bassetti, L. Mirenghi, P. Rotolo, E. Bonetti, L. Pasquini, Antisari. M. Vittori, 2005Role of Organic Additives in Hydriding Properties of Mg-C Nanocomposites, Materials Scence Forum, 494137142

44. A. Montone, J. Grbović, Lj. Stamenković, A. L. Fiorini, L. Pasquini, E. Bonetti, Antisari. M. Vittori, 2006Desorption Behaviour in Nanostructured MgH2-Co, Materials Science Forum, 5187984

45. A. Montone, Novaković. J. Grbović, Antisari. M. Vittori, A. Bassetti, E. Bonetti, A. L. Fiorini, L. Pasquini, L. Mirenghi, P. Rotolo, 2007Nano-micro MgH2-Mg2NiH4 composites: Tayloring a multichannel system with selected hydrogen sorption properties, International Journal of Hydrogen Energy, 321429262934

46. D. Mori, K. Hirose, 2009Recent challenges of hydrogen storage technologies for fuel cell vehicles, International Journal of Hydrogen Energy 3445694574

47. M. Ni, M. K. H. Leung, D. Y. C. Leung, K. Sumathy, 2007A review and recent developments in photocatalytic water-splitting using TiO2 for hydrogen production, Renewable and Sustainable Energy Reviews, 11401425

48. T. Noritake, S. Towata, M. Aoki, Y. Seno, Y. Hirose, E. Nishibori, M. Takata, M. Sakata, 2003Charge density measurement in MgH2 by synchrotron X-ray diffraction, Journal of Alloys and Compounds 356-3578486

49. N. Novaković, Lj. Matović, Novaković. J. Grbović, M. Manasijević, N. Ivanović, 2009Ab initio study of MgH2 formation, Materials Science and Engineering B, 1653235238

50. N. Novaković, Novaković. J. Grbović, Lj. Matović, M. Manasijević, I. Radisavljević, Mamula. B. Paskas, N. Ivanović, 2010Ab initio calculations of MgH2, MgH2:Ti and MgH2:Co compounds, International Journal of Hydrogen Energy, 352598608

51. S. Orimo, Y. Nakamori, T. Kato, C. Brown, C. M. Jensen, 2006Intrinsic and mechanically modified thermal stabilities of α-, β- and-aluminum trihydrides AlH3, Applied Physics A, 83158

52. S. Patchkovskii, J. S. Tse, S. N. Yurchenko, L. Zhechkov, T. Heine, G. Seifert, 2005Graphene nanostructures as tunable storage media for molecular hydrogen, Proceedings of National Academy of Sciences, 102301043910444

53. R. Pena-Alonso, A. Sicurelli, E. Callone, G. Carturan, R. Raj, 2007A picoscale catalyst for hydrogen generation from NaBH4 for fuel cells, Journal of Power Sources, 1651315323

54. N. L. Rosi, J. Eckert, M. Eddaoudi, D. T. Vodak, J. Kim, M. O'Keeffe, O. M. Yaghi, 2003Hydrogen Storage in Microporous Metal-Organic Frameworks, Science, 300562211271129

55. J. L. C. Rowsell, A. R. Millward, K. S. Park, O. M. Yaghi, 2004Hydrogen Sorption in Functionalized Metal−Organic Frameworks, Journal of the American Chemical Society 1261856665667

56. Rowsell J.L.C. & Yaghi O.M.2005Strategies for Hydrogen Storage in Metal-Organic Frameworks, Angewandte Chemie International Edition, 443046704679

57. D. Saha, S. Deng, 2009Hydrogen Adsorption on Ordered Mesoporous Carbons Doped with Pd, Pt, Ni, and Ru, Langmuir 25211255012560

58. G. Sandrock, J. J. Reilly, J. Graetz, W. Zhou-M, J. Johnson, J. Wegrzyn, 2006Alkali metal hydride doping of α-AlH3 for enhanced H2 desorption kinetics, Journal of Alloys and Compounds, 421185189

59. H. G. Schimmel, M. R. Johnson, G. J. Kearley, A. J. Ramirez-Cuesta, J. Huot, F. M. Mulder, 2004The vibrational spectrum of magnesium hydride from inelastic neutron scattering and density functional theory, Materials Science and Engineering B, 1081-23841

60. H. G. Schimmel, M. R. Johnson, G. J. Kearley, A. J. Ramirez-Cuesta, J. Huot, F. M. Mulder, 2005Structural information on ball milled magnesium hydride from vibrational spectroscopy and ab-initio calculations, Journal of Alloys and Compounds, 39314

61. H. I. Schlesinger, C. H. Brown, A. E. Finholt, J. R. Gilbreath, H. R. Hoekstra, E. K. Hyde, 1953Sodium Borohydride, Its Hydrolysis and its Use as a Reducing Agent and in the Generation of Hydrogen, Journal of the American Chemical Society, 751215219

62. C. X. Shan, M. Bououdina, Y. Song, Z. X. Guo, 2004Mechanical alloying and electronic simulations of (MgH2+M) systems (M=Al, Ti, Fe, Ni, Cu

and Nb) for hydrogen storage, International Journal of Hydrogen Energy, 2917380

63. L. Soler, J. Macanás, M. Munoz, J. Casado, 2007Synergistic hydrogen generation from aluminum, aluminum alloys and sodium borohydride in aqueous solutions, International Journal of Hydrogen Energy, 321847024710

64. Y. Song, Z. X. Guo, R. Yang, 2004Influence of selected alloying elements on the stability of magnesium dihydride for hydrogen storage applications: A first-principles investigation, Physical Review B, 699094205

65. M. Šušić, D. Minić, A. Maričić, B. Jordović, D. Krsmanović, 1996Structural Changes During Heating a Cold Sintered Amorphous Powder of 82Ni and 18P Alloy, Science of Sintering, 28105110

66. M. V. Šušić, 1997Hydriding and dehydriding of a graphite powder doped with palladium, Journal of Serbian Chemical Society, 621211831186

67. M. V. Šušić, 1997Hydriding and dehydriding of palladium-doped charcoal, Journal of Serbian Chemical Society, 628631634

68. M. V. Šušić, 1997Kinetics of the process of isothermal hydriding and dehydriding of hydrogen absorbers, International Journal of Hydrogen Energy, 226585589

69. Y. Tao, Y. Chen, Y. Wu, Y. He, Z. Zhou, 2007High hydrogen yield from a two-step process of dark- and photo-fermentation of sucrose, International Journal of Hydrogen Energy, 322200206

70. C. S. Tsao, Y. R. Tzeng, M. S. Yu, C. Y. Wang, H. H. Tseng, T. Y. Chung, H. C. Wu, T. Yamamoto, K. Kaneko, S. H. Chen, 2010Effect of Catalyst Size on Hydrogen Storage Capacity of Pt-Impregnated Active Carbon via Spillover, Journal of Physical Chemistry Letters, 110601063

71. J. Turner, 2004Sustainable hydrogen production, Science 305972974

72. Tucker M.C.2010Progress in metal-supported solid oxide fuel cells: A review, Journal of Power Sources, 19545704582

73. T. Umegaki, J. M. Yan, X. B. Zhang, H. Shioyama, N. Kuriyama, Q. Xu, 2009Boron- and nitrogen-based chemical hydrogen storage materials, International Journal of Hydrogen Energy, 34523032311

74. R. A. Varin, T. Czujko, Z. Wronski, 2006Particle size, grain size and-MgH2 effects on the desorption properties of nanocrystalline commercial magnesium hydride processed by controlled mechanical milling, Nanotechnology, 171538563865

75. R. A. Varin, T. Czujko, Z. S. Wronski, 2009Nanomaterials for Solid State Hydrogen Storage, Springer Science+Business Media, LLC, 978-

0-38777-711-5

76. R. W. P. Wagemans, J. H. van Lenthe, P. E. de Jongh, A. J. van Dillen, K. P. de Jong, 2005Hydrogen Storage in Magnesium Clusters: Quantum Chemical Study, Journal of the American Chemical Society, 127471667516680

77. Y. Wang, K. S. Chen, J. Mishler, S. C. Cho, X. C. Adroher, 2011A review of polymer electrolyte membrane fuel cells: Technology, applications, and needs on fundamental research, Applied Energy, 889811007

78. J. Weitkamp, M. Fritz, S. Ernst, 1992Zeolithe als Speichermaterialien für Wasserstoff, Chemie Ingenieur Technik, 641211061109

79. J. Weitkamp, M. Fritz, S. Ernst, 1995Zeolites as media for hydrogen storage, International Journal of Hydrogen Energy, 2012967970

80. J. Weitkamp, S. Ernst, F. Cubero, F. Wester, W. Schnick, 1999Nitridosodalite Zn6[P12N24] as a material for reversible hydrogen encapsulation, Advanced Materials 93247248

81. A. G. Wong-Foy, A. J. Matzger, O. M. Yaghi, 2006Exceptional H2 Saturation Uptake in Microporous Metal−Organic Frameworks, Journal of the American Chemical Society 1281134943495

82. Z. T. Xiong, G. T. Wu, J. J. Hu, P. Chen, 2006Investigation on chemical reaction between LiAlH4 and LiNH2, Journal of Power Sources, 1591167170

83. J. Xuan, M. K. H. Leung, D. Y. C. Leung, M. Ni, 2009A review of biomass-derived fuel processors for fuel cell systems, Renewable and Sustainable Energy Reviews, 1313011313

84. R. Yu, P. Lam, 1988Electronic and structural properties of MgH2, Physical Revew B 371587308737

85. X. Zhao, B. Xiao, A. J. Fletcher, K. M. Thomas, D. Bradshaw, M. J. Rosseinsky, 2004Hysteretic Adsorption and Desorption of Hydrogen by Nanoporous Metal-Organic Frameworks, Science 306569810121015

86. K. Zeng, T. Klassen, W. Oelerich, R. Bormann, 1999Critical assessment and thermodynamic modeling of the Mg-H system, International Journal of Hydrogen Energy, 24109891004

87. A. Zuettel, 2003Materials for hydrogen storage, Materials Today, 692433

88. A. Zuettel, P. Wenger, P. Sudan, P. Mauron, S. Orimo, 2004Hydrogen density in nanostructured carbon, metals and complex materials, Materials Science and Engineering B 108918

Chapter 7

SMALL SCALE HYDROGEN PRODUCTION FROM METAL-METAL OXIDE REDOX CYCLES

Doki Yamaguchi[1], Liangguang Tang[1], Nick Burke[1], Ken Chiang[1], Lucas Rye[2], Trevor Hadley[3] and Seng Lim[3]

[1]CSIRO Earth Science and Resource Engineering,, Australia

[2]CSIRO Marine and Atmospheric Research,, Australia

[3]CSIRO Process Science and Engineering,, Australia

INTRODUCTION

The industrial production of hydrogen by reforming natural gas is well established. However, this process is energy intensive and process economics are adversely affected as scale is decreased. There are many situations where a smaller supply of hydrogen, sometimes in remote locations, is required. To this end, the steam-iron process, an originally coal-based process, has been re-considered as an alternative. Many recent investigations have shown that hydrogen (H_2) can be produced when methane (CH_4) is used as the feedstock under carefully controlled process conditions. The chemistry driving this chemical looping (CL) process involves the reduction of metal oxides by methane and the oxidation of lower oxidation state metal oxides with steam. This process utilises oxygen from oxide materials that are able to transfer oxygen and eliminates the need of purified oxygen for combustion. Such a system has the potential advantage of being less energy intensive than reforming processes and of being flexible enough for decentralised hydrogen production from stranded reserves of natural gas. This chapter first reviews the existing hydrogen production technologies then highlights the recent progress made on hydrogen production from small scale CL processes. The development of oxygen carrier materials will also be discussed. Finally, a preliminary economic appraisal of the CL process will be presented.

A BRIEF OVERVIEW ON HYDROGEN PRODUCTION

Hydrogen can be produced from the reaction of feedstock including fossil fuels and biomass with water. Today, 96 % of hydrogen is derived from fossil fuels of which 48 %, 30 % and 18 % originates from natural gas, higher hydrocarbons and coal, respectively and the remaining 4 % comes from electrolysis.Fossil fuel based hydrogen production processes are mature technologies and are currently the most economic routes for large scale hydrogen production. Because coal, natural gas and biomass all contain carbon, carbon dioxide is inevitably produced as a by-product of the energy released.A pictorial overview of the available hydrogen production processes is given in Figure 1. The basics of two commercialised processes, namely steam methane reforming and partial oxidation, are considered in this section. A brief discussion on emerging hydrogen production technology will also be presented.

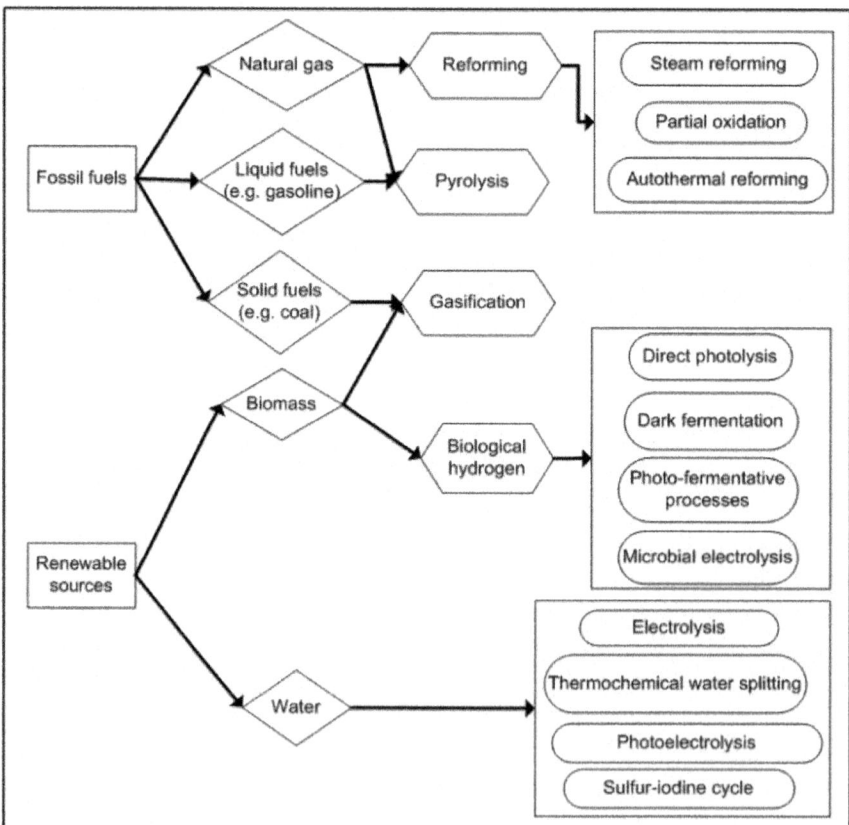

Figure 1: An overview of existing hydrogen production process from different sources.

Steam Methane Reforming

Steam reforming of methane (SMR) is one of the most developed and commercially used technologies.Compared to other fossil fuels, natural gas, which contains mostly methane, is a cost effective feedstock for making hydrogen.This is because methane has a high hydrogen-to-carbon ratio, meaning the yield of hydrogen is higher. Today, almost 48 % of the world's hydrogen is produced from this technology [1].In this process, hydrogen is produced according to the following two reactions:

In the SMR, the natural gas feedstock is first reformed in the presence of steam over a catalyst at elevated temperatures (700 – 925 C) to produce a mixture of carbon monoxide and hydrogen (syngas) as shown in Equation 1.Then, the yield of hydrogen is further increased by reacting the carbon monoxide with make up steam via the water-gas shift reaction (WGS) as shown in Equation 2.Finally, hydrogen is separated and purified by processes such as pressure swing absorption, wet scrubbing or membrane separation. SMR is currently the most cost effective hydrogen production process which offers a minimum energy efficiency of 80 – 85 % in a large scale facility if residual steam is re-used [1].Furthermore, the process is economically viable for large scale operation [2].According to Pardor et al.[3], the price of hydrogen produced from SMR ranges from $5.97/GJ for a 25.4 million Nm³/day plant to $7.46/GJ for a 1.34 million Nm³/day plant.A figure of $11.22/GJ was estimated for hydrogen produced from a small facility (0.27 million Nm³/day).However, the price of hydrogen varies with the price of natural gas feedstock.In general, the price of the natural gas feedstock accounts for 52 – 68 % and 40 % of the total cost for large and small SMR plants, respectively.It can be seen that decreasing the scale of operation would lead to an increase in cost of the hydrogen produced.

Partial Oxidation

Hydrogen can also be produced from the partial oxidation (POX) of hydrocarbons over a catalyst at high temperatures (Equation 3).

The reaction requires the use of high purity oxygen and is mildly exothermic. Similar to the SMR process, the yield and purity of hydrogen may be further increased by the WGS reaction and a subsequent purification process. The reported efficiency of POX is in the range of 66 – 76 % [1]. Mirabal [4] estimated the cost of hydrogen to be $12.43/GJ for a 2.83 million Nm³/day plant, which is higher than that produced from SMR. However, based on the use of coke off-gas and residual oil (both having a price of lower than natural gas), Pardro et al.[3] estimated the price to be in the range of $6.94

– 9.83/GJ for large facilities (1.34 – 2.80 million Nm³/day) and \$10.73/GJ for a small facility (capacity is 0.27 million Nm³/day). Similar to SMR, the economics appears more favourable for large scale operations.

Coal Gasification

Gasification can be used to convert a varied range of solid fuels such as coal and biomass into syngas (Equation 4).

Coal gasification is a mature process and is commercially available. Although the cost of the coal feedstock is generally much cheaper than natural gas, the price of hydrogen produced from coal gasification process is estimated to be \$17.45/GJ.This is higher compared to SMR (\$10.26/GJ) and POX (\$12.43/GJ), and this is due to the higher capital investment required for coal gasification.Coal is an economically viable option for making hydrogen in very large centralised plants where the demand for hydrogen becomes large enough to support an associated large distribution network and establishment costs.It is therefore seen that coal gasification would become more competitive than SMR and POX as the price of natural gas increases [4].Much of the engineering experience accumulated from coal fired power plant is directly useful for coal gasification.

Other Novelroutes for Hydrogen Production

Water splitting is one of the options for producing hydrogen and has received wide attention.The current reported energy efficiency is between 10 – 27 % and the cost of hydrogen is estimated to be 3-10 times of the hydrogen produced from the SMR process [5]. Biological routes for producing hydrogen are also being considered because of the renewable nature and the mild operating conditions of these processes.These alternative routes have yet to become economically competitive with technologies in practice such as SMR and POX that use fossil fuel feedstock.

HYDROGEN PRODUCTION FROM CYCLIC REDOX PROCESSES

There is an ongoing demand for viable processes for producing hydrogen on a small scale for decentralised distribution. For this reason, there is currently much attention being paid to the development of cyclic redox processes or commonly referred as chemical looping (CL) processes for small scale hydrogen production. In addition to the compactness of the process, another advantage is the ability to produce a near sequestration-ready stream of carbon dioxide from the process. The operating concept behind these processes resembles the well-known steam-iron process and is illustrated in Figure 2a.

Some widely reported variations and applications include chemical looping combustion (CLC) for power generation and, chemical looping hydrogen production (CLH2). The schematic diagrams representing these processes are shown in Figure 2b and Figure 2c. A typical chemical looping operation consists of a reduction and an oxidation steps. During the reduction, a metal oxide is used as the oxygen carrier to oxidise carbonaceous fuels (e.g. natural gas, coal or biomass) into carbon dioxide and steam. The reduction can be optimised such that syngas (a mixture of carbon monoxide and hydrogen) can be obtained. Subsequently, the partially or fully reduced metal oxide is oxidised with air or steam to re-generate the original metal oxide and other oxidation products. When steam is used, water is split to produce hydrogen as the main product.

One of the fundamental parameters that determine the overall efficiency of many chemical looping processes is the effectiveness of the oxygen carriers. Therefore many research groups have focused on improving the activity and the stability of oxygen carrying materials. This section reports the latest developments of oxygen carrier materials for CL applications.

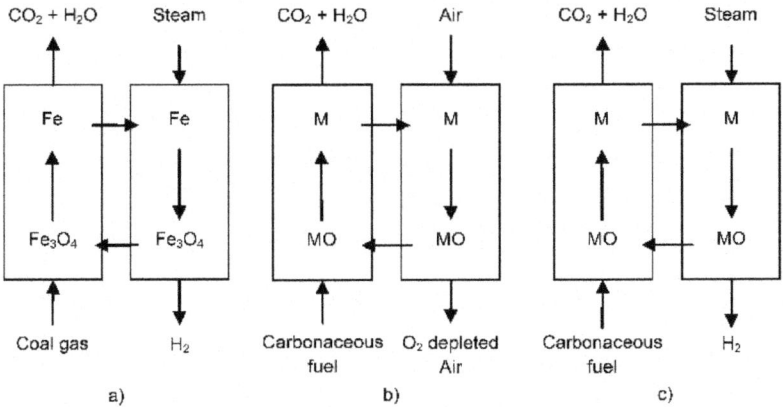

Figure 2: a) The traditional steam-iron process and chemical looping (CL) processes, b) CLC for power generation, and c) CLH2.

Thermodynamic Constraints

The selection of an oxygen carrier requires comprehensive appraisal of the physiochemical properties of the material.Some properties include reaction kinetics, oxygen content, long-term recyclability and durability, attrition resistance, heat capacity, melting points, tendency to form coke, resistance to carbon deposition, cost and toxicity [6, 7].Nevertheless, the most important requirement is the thermodynamic feasibility of oxygen transfer to and from these oxygen carriers.Figure 3 shows the changes in Gibbs free energy (ΔG) of some oxygen carriers commonlystudied for CL applications. Some selected properties are provided in Table 1.

For the current topic, the oxygen carrier can be divided into two groups based on their ability to oxidise methane. The first group contains oxides that are capable of only partially oxidising methane into carbon monoxide and hydrogen. Some representative redox couples are ZnO/Zn, V_2O_5/V and CeO_2/ Ce_2O_3 couples. The second group contains oxides that are able to support the complete oxidation of methane. NiO/Ni, CuO/Cu and Co_3O_4/Co are redox couples that fall into this category. In addition, the oxidations of these reduced oxides are favourable over a wide temperature range as indicated by the negative ΔG values in Figure 3. Therefore these three redox couples are often regarded as good candidates for CL applications.

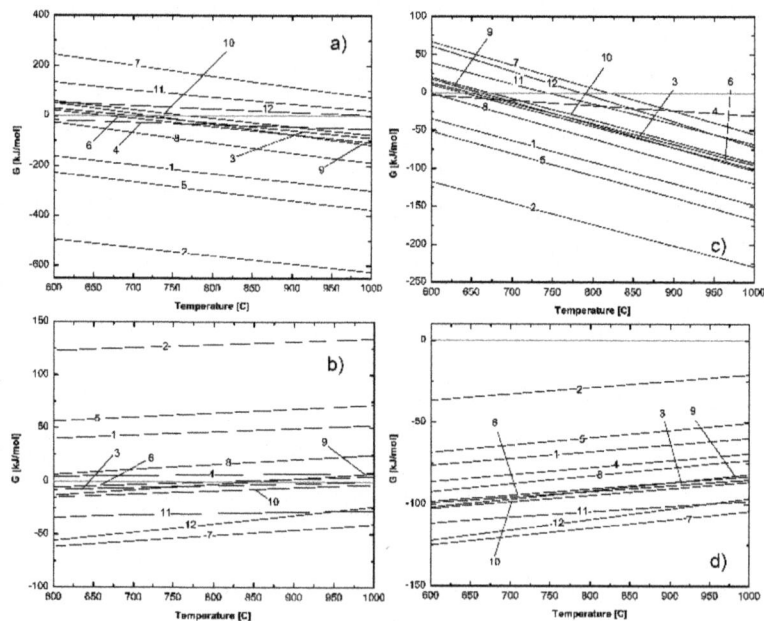

Figure 3: Variation of Gibbs free energy of reactions, a) CH_4 combustion (CH_4 + 4/ $yM_xO_y \rightarrow CO_2$ + $2H_2O$ + 4x/yM) and b) CH_4 partial oxidation (CH_4 + 1/$yM_xO_y \rightarrow$ CO + $2H_2$ + x/yM), c) steam oxidation (xM + $yH_2O \rightarrow M_xO_y$ + yH_2), and d) air oxidation (xM + y/2$O_2 \rightarrow M_xO_y$).See Table 1 for the legends used.

Table 1: Selected properties of oxygen carriers

Number	Redox couple	Melting point [°C]	Oxygen transport capacity [kg/kg-metal]	Price [USD/t]
1	NiO/Ni	1955/1455	0.27	21,800
2	CuO/Cu	1326/1084	0.25	7,680
3	Fe_3O_4/Fe	1597/1538	0.38	100
4	MnO_2/Mn	535/1267	0.58	1,500
5	Co_3O_4/Co	895/1495	0.36	39,700
6	WO_3/W	1472/3407	0.26	27,000
7	ZnO/Zn	1975/420	0.24	2,250
8	SnO/Sn	1080/232	0.13	21,000
9	In_2O_3/In	1913/157	0.21	565,000
10	MoO_2/Mo	1100/2623	0.33	34,900
11	V_2O_5/V	670/1910	0.78	25,600
12	CeO_2/Ce_2O_3	2400/2230	0.06	24,611

Compared to oxidation using molecular oxygen, the ΔG shifts to higher values when steam is used as the oxidising agent.As a result, it is not thermodynamically feasible to produce hydrogenby reacting steam with metallic Ni, Cu or Co.MnO_2/Mn and SnO/Sn couples are also not reactive when they are brought into contact with steam. ZnO/Zn and V_2O_5/V couples react with steam to produce hydrogen, however, their melting points in either the oxide or the metallic form are too low for CL applications in general.Despite the moderate ΔG values associated with Fe_3O_4/Fe, WO_3/W and CeO_2/Ce_2O_3 redox couples, the reported redox kinetics and thermo-mechanicalstrength have made them appealing candidates for CL processes.The Fe_3O_4/Fecouple also possesses a relatively high oxygen content, and is widely available, non-toxic and less costly.When iron oxide is used, it is only possible to oxidise the reduced state to magnetite (Fe_3O_4) due to thermodynamic limitations.

Common Feedstocks for Producing Hydrogen

A number of studies have employed non-gaseous fuels including coal [8-13], biomass [14-17] and pyrolysis oil [18, 19].In a syngas chemical looping (SCL) process, the fuel is first converted into syngas in a separate gasification unit. The syngas generated is then used in the reduction cycle and steam is used to regenerate the oxide and to produce hydrogen.An additional air oxidation cycle may be required to regenerate the oxygen carrier. The SCL process generally has lower efficiency for conversion, owing to the low conversions in the syngas generation step and the steam oxidation step [8].Li et al. [9] examined the cyclic performance of a Fe-based oxygen carrier at 830 C when a simulated syngas was used.They showed that the syngas was completely converted in the reduction half cycle giving an oxygen carrier conversion of 94.6 %.For the steam half cycle, the reduced oxygen carrier was oxidised into Fe_3O_4producing a stream of 99.8 % pure hydrogen.In a separate study, the same group also demonstrated the feasibility of using a moving bed reactor at 900 C for the same reaction[10]. A syngas conversion in excess of 99.5% and an oxygen carrier conversion of 50 % were recorded. A process simulation conducted by Gupta et al. [8] confirmed that the maximum efficiency for the SCL process could reach 74.2 % for hydrogen production which is comparable to or more effective than steam reforming (65-75 %), partial oxidation (50 %) and gasification (43-47 %). Considering the complexity of the SCL, it is clear that footprint of the process would be large because of the large number of unit operations involved in its design.

When coal is used as the feedstock, the solid fuelcan be used to reduce oxygen carriers directly.This process is often referred as the coal direct chemical looping (CDCL) process because a gasification unit, as well as air separation and gas cleaning units, is not required[12]. The CDCL process is reported to be significantly more efficient than the SCL process for hydrogen production [6, 13].Yang et al. [11] investigated the CDCL process using a lignite-derived char in a fluidised bed reactor. The complete gasification of the char achieved a maximum carbon dioxide concentration of 90% in the presence of a K_2CO_3 catalyst.A high oxygen carrier-to-char ratioimproved the complete gasification to carbon dioxide but this also led to lower hydrogen yields as a result of low conversions of the oxygen carrier.Under the optimum condition, thehydrogen production efficiency was reported to be 50.2 % at an oxygen carrier conversion of 70.2 %. The use of counter-current moving bed reactor was found to improve oxygen carrier conversion, and achieved (due to the significantly low mass required) a char conversion of > 90 % and an overall carbon dioxide capturing efficiency of > 95 %[6].

Biomass has found limited applications for SCL processes. This is because of the high water content generally associated with biomass feedstocks. Sime et al. [14] investigated the use of gases derived from woody biomass gases for SCL and reported that such process was less efficient and more costly than conventional gasification processes for producing hydrogen. Li et al. [16] pointed out that it is critical to reduce the moisture content in the biomass feedstock to less than 5 %in order to achieve a conversion of 56.6 % in gasification.Similar to other solid feedstocks, unreacted biomass must be separated before the oxygen carrier is circulated to the steam reactor.

Otherwise, the unreacted biomass could be gasified and lower the purity of the hydrogen produced.

Natural gas is an efficient feedstock for CL processes since it is fed to the process in gaseous form. This minimises the need of solid handling and improvesmass transfer processes [20]. Cormos (2011) recently assessed and compared hydrogen production from a natural gas CL process and a coal/lignite based SCL [21]. It was concluded that when natural gas was used to produce hydrogen, the recorded efficiency was 78.1 %. This value was higher compared to the values of 65.7 % and 63.3 % recorded for the coal- or lignite-based SCL processes, respectively. In addition, the separation and capturing of CO_2were said to be more effective when natural gas was used. Another clear advantage of using natural gas as the feedstock is that no additional up-stream unit operations are required for producing syngas.

Oxygen Carriers for Cyclic Redox Processes

As mentioned previously, redox kinetics and thermal stability are the two main issues associated with the use of oxide-based oxygen carriers for CL processes. In order to improve their performance, support and/or promoting materials to assist in material stabilisation are often added to improve the performance of the metal oxide. A comprehensive list of oxygen carriers developed for various CL applications in the last decade can be found in an excellent review published by Adanez et al. [7]. This section highlights some recent studies on developing novel oxygen carriers.

Effects of Metal Addition to Oxide Carriers

Otsuka et al. [22, 23] investigated the effects of 26 different metal dopants on iron oxide.It was found that some metal dopants were more effective in preventing the iron oxide from sintering and some were more effective in facilitating the splitting of water.Among these 26 metals, Mo and Cr were found to improve the thermal stability of iron oxide in the cyclic process.The improved redox stability after the introduction of Mo metal (5 mol%) was also reported by Wang et al. [24], and Liu and Wang [25].Despite the fact that Cr addition could improve the sintering resistance of iron oxide, temperature programmed analysis revealed that a temperature of ca. 500 C is required to split water when compared to a temperature of 420 Cas required by iron oxide modified with Mo [22].In addition, no oxidation of methane was observed when the temperature was lower than 700 C [26].It was proposed that the main role of Cr and Mo dopants was to partially transform the iron oxide into the ferrite structure ($M_xFe_{3-x}O_4$, M = Mo and Cr) [22, 26] and therefore inhabited the agglomeration of neighbouring particles.

Some metals including Ru, Rh, Pd, Ag, Ir and Pt have been shown to improve reaction kinetics by facilitating the dissociation of hydrogen, methane and water.Otsuka et al. [22] reported that the improvement on splitting of water into hydrogen by metal in a CLprocess increased inthe order of Rh > Ir > Ag > Pd > Ru.Ryu et al. [27] also found that Rh was more effective than Pb, Pt and Ru in enhancing the hydrogen production step in a chemical looping process. Therole of Rh was to decrease the onset temperature for the water splitting reaction.A XANES/EXAFS study on Rh-Cr-added iron oxide revealed that Rh was also able to form Rh-Fe alloy upon reductions[26]. However, Rh segregated in the alloy structure when it contactedsteam and thus accelerated the sintering of iron oxide.This led to the observed deterioration in redox activity after repeated redox operation.Although Ni- and Cu-ferrites also exhibited an enhancing effect on redox kinetics, Ni and Cu were shown not to be effective in improving sintering resistance [28, 29].

The addition of a second and a third metal have been shown to further improve the redox activity [22,24-27, 30, 31]. Common choices of metal combinations often consisted of a first metal such as Rh, Pt, Ni and Cu which is thought to catalytically activates the reducing gas (e.g. hydrogen, carbon monoxideormethane), and a second metal such as Mo and Cr which exhibitsa structural stabilising effect.Otsuka et al [22] examined the addition of Rh and Mo to iron oxide for the chemical storage of hydrogen and observed an enhancement in reaction kinetics and a reduction in reaction temperature for hydrogen formation.Most importantly, the Mo provided good stabilising effect and largely mitigated the sintering of the oxygen carrier. The effect of bimetal addition on iron oxide was also investigated under methane oxidation at a temperature range of 200 – 800 C by Takenaka et al. [30]. The methane conversion was found to increase by adding a second metal and the performance increased in the order of Rh-Cr > Ir-Cr > Pt-Cr > Ni-Cr > Pd-Cr > Cu-Cr = Co-Cr.Other research groups also reported similar findings [24, 25, 27, 31]. Despite the improvement in reactivity and thermal stability, most of the bimetallic modified oxygen carriers produce carbon upon methane oxidation. The production of carbon usually leads to a rapid deterioration of the oxygen carrier and is the source of carbon oxides (CO_x) contamination.

Supported Oxygen Carriers

Another approach to improve the thermal stability of oxygen carriers is to introduce inert support materials such as Al_2O_3, SiO_2, TiO_2 and ZrO_2.Adanez et al. [32] assessed the reactivity of 240 different types of oxygen carriers composed of Cu, Fe, Mn or Ni supported on SiO_2, TiO_2, ZrO_2, Al_2O_3 or sepiolite ($Mg_4Si_6O_{15}(OH)_2 \cdot 6H_2O$) over a temperature range of 950 – 1300 C.The best

Fe-based oxygen carriers were those supported on Al_2O_3 or ZrO_2.It was also found that the formation of aluminate ($NiAl_2O_4$ and $CoAl_2O_4$) lowered the oxygen transport capacity and hence reduced the redox activity [33].SiO_2 was found to be the most suitable support for Cu-based oxygen carrier because it remained inert at high temperatures and did not form $Cu-SiO_2$ composites. However, Fe-based oxygen carriers showed a strong tendency to form unreactive iron silicates with SiO_2[34].ZrO_2 and TiO_2 were suggested as the best supports for Mn- and Ni-based oxygen carriers, respectively.In terms of the cyclic redox activity, however, TiO_2 supported Ni-based oxygen carriers showed lower reactivities, compared to Ni supported on Al_2O_3. This is because NiO is more prone to react with TiO_2 and form $NiTiO_3$which is known to be less reducible than NiO. It also exhibits a high carbon formation tendency. Therefore, Al_2O_3 supported Ni-based oxides were considered to be the most promising oxygen carrier for a large scale CLC applications.

Some metal doped iron oxide oxygen carriers were also supported on ZrO_2 for CL processes [29, 35-37]. Kodama et al. [35, 36] showed improved thermal resistance for the Ni- and Co-ferrites when ZrO_2support was introduced. The reported methane conversion and carbon monoxide selectivity by using $Ni_{0.39}Fe_{2.61}O_2$ (33 wt%)/ZrO_2 were 46-58% and 44-48%, respectively. However, since Fe and Ni are excellent catalysts for methane decomposition, the material was severely deactivated by coke and the subsequent carbide species formed. Because Cu has lower activity for methane decomposition, $CuFe_2O_4$ was used to produce syngas from methane [29]. The results showed that no CO_x was formed during the operation. The same group also found beneficial effects of ZrO_2 and CeO_2 supports for $CuFe_2O_4$ (20 wt%) [38]. Compared to the methane conversion obtained for $CuFe_2O_4$ (34–56 %), the methane conversions achieved by $CuFe_2O_4/CeO_2$ and $CuFe_2O_4/ZrO_2$ were 89-92 % and 74-83 %, respectively. From these results, CeO_2 was found to be more active in promoting methane oxidation while ZrO_2 was considered to be a more effective stabiliser against thermal sintering. Since CeO_2 is known to be able to oxidise soot through lattice oxygen transfer [39, 40], it is thought that this property could help to minimise carbon formation when $CuFe_2O_4/CeO_2$ is used. Cha et al. [37] also confirmed that CeO_2 modified $CuFe_2O_4/ZrO_2$ was a more effective oxygen carrier than Ni- modified $CuFe_2O_4/ZrO_2$ for chemical looping syngas and hydrogen productions.

A recent study conducted by Yamaguchi et al. [41] also demonstrated the improved performance of CeO_2/ZrO_2 modified $Fe_2O_3$for producing hydrogen from methane-steam cycles.Some results obtained from temperature programmed analysis and isothermal reduction are shown in Figure 4 and are summarised in Table 2.Figure 4a shows that CeO_2 and ZrO_2 altered the redox properties of Fe_2O_3 with the most significant enhancement observed for the reducibility at low temperatures (< 600 C) (see Table 2).The isothermal reduction analysis (Figure4b) further confirmed the accelerated reduction kinetics after the introduction of CeO_2 and ZrO_2.The observed overall enhancement was derived from the combined effects of CeO_2 and ZrO_2. CeO_2 improved the reducibility of Fe_2O_3 while ZrO_2 provided thermal stability and helped to suppress the reduction of FeO to metallic Fe.The latter was supported by the incomplete reduction of $Fe_{15}Ce_{10}Zr_{75}$ and $Fe_{40}Zr_{60}$ (Table2). Similar observations were also reported when WO_3 was modified with CeO_2 and ZrO_2[42].The synergic effect provided by CeO_2 and ZrO_2 effectively defined the redox window of the oxygen carriers.An immediate consequence is the minimisation of carbon and carbide formation during repeated redox cycles. This can be demonstrated by the fact that CO_x free hydrogen was produced by using CeO_2-ZrO_2 modified WO_3 in a methane-steam CL process [42]. The addition of a small amount of Mo or Cr could further improve the thermal stability of this type of oxygen carrier.Galvita et al. [43]showed the addition of 2 wt% of Mo to $Fe_2O_3/Ce_{0.5}Zr_{0.5}O_2$ could maintain a stable level of hydrogen production over 100 cycles in a cyclic water-gas shift process.In this reaction, the main role of Mo is to improve the dispersion of Fe-Mo oxide material and minimise the migration of material across the boundary of adjacent particles [44].

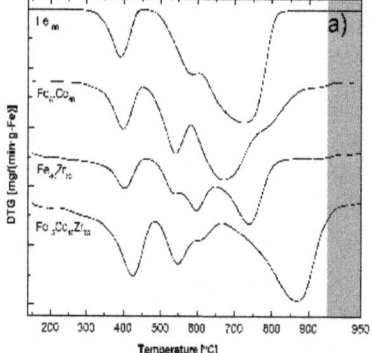 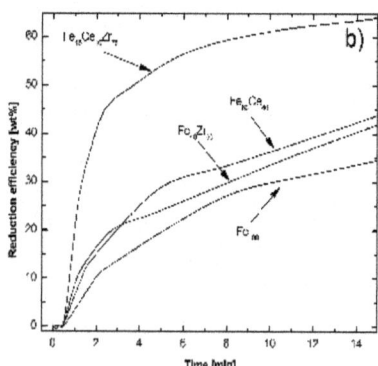

Figure 4: Effect of CeO_2 and/or ZrO_2 addition on Fe_2O_3 reducibility during a) temperature programmed and b) isothermal reduction with H_2[41].

Table 2: A summary of oxides used in methane-steam redox cycle [41]. [1]The oxygen removal represents a cumulative weight reduction at temperatures < 600 C during the TPR analysis (Figure4a). [2]The overall reduction efficiency represents a final reduction efficiency obtained during isothermal reduction analysis at 750 C for 240 min

Oxygen carrier	Oxygen removal[1] [mg-O/g-Fe]	Overall reduction efficiency[2] [wt%]	H_2 yield [μmol/g-Fe]	H_2 purity [%]
$Fe_{10}O$	125	98.1	15	11.5
$Fe_{60}Ce_{4}O$	169	97.1	368	49.1
$Fe_{40}Zr_{6}O$	153	66.7	88	17.4
$Fe_{15}Ce_{10}Zr_{75}5$	255	77.2	6283	97.5

Naturally Occurring Oxide Materials

Recently, many naturally occurring minerals and ashy waste produced from industry have been considered for use as oxygen carriers. These materials include natural ilmenite (Fe and Ti mixed oxide often denoted as $FeTiO_3$), iron ore, manganese ore and oxide scales. An advantage of using these materials is the low cost compared to many synthetic oxygen carriers. In addition, naturally occurring oxides usually contain Si, Al, Mg, and many other metals which have been shown to modify the physiochemical properties of the materials to various degrees. Leion et al. [45] investigated the feasibility of using ilmenite, iron ores, oxide scales from steel industry and manganese ores as oxygen carriers in a fluidised bed reactor. They concluded that many Fe based oxides, particularly ilmenite, were suitable for CLC application. However, the Mn-based oxides showed poor mechanical stability and fluidising properties, and were determined to be non-ideal candidates for this application. In a separate study, Leion et al. [46] also proved the feasibility of using ilmenite to completely capture carbon dioxide upon its reaction with syngas and reported a moderate conversion when methane is used. Adanez et al. [47] observed increases in ilmenite, and syngas and methane conversions with increasing the time on stream and the number of redox cycles. Another important finding was the enhanced activation of ilmenite when the raw ilmenite material was subjected to an oxidation pre-treatment. The authors also found the redox properties of ilmenite changed with the temperature of oxidative pre-treatment. However, the positive effect only became apparent when the ilmenite was first oxidised to pseudobrookie (Fe_2TiO_5) which is usually formed above 1000 C.

Table 3: Oxygen transfer capacity and major phase of various ilmenite samples before and after pre-oxidation. [1] Phases were identified by XRD analysis

Pre-oxidation temperature [°C]	**Major crystalline phase[s]1**	Oxygen transfer capacity [wt%]
Raw	$FeTiO_3$, Ti_O2	1.1
800	Fe_2O_3, Ti_O2	1.0
1000	Fe_2TiO_5, Ti_O2	1.8

Leion et al. [46] also reported that an ilmenite sample remained active with minimum carbon formation after a continuous operation for three days at 975 C.Furthermore, natural ilmenite is known to react just as well with petroleum coke, syngas and methane as synthetically prepared Fe_2O_3/ $MgAl_2O_4$[48].Lorente et al. [49] reported a better hydrogen storage capacity and redox stability when iron ore samples was used instead of pure Fe_2O_3. The improvement in the overall redox performance was due to the presence of impurities including SiO_2, Al_2O_3, MgO and CaO.Among these impurities, Al_2O_3 and SiO_2 are considered to be good stabilisers against sintering, while CaO and MgO are able to facilitate kinetics of water splitting.

Deactivation of Oxygen Carriers

The life time of the oxygen carrier is a critical factor in determining the efficiency and viability of CL processes.In general, the efficacies of oxygen carriersdecrease over time because of material alternation by sintering and/or coking.

Generally, for the CLH2 application, a relatively high temperature is required for driving the reduction reaction in order to achieve satisfactory conversion and kinetics. As a result, the high temperature environment irreversibly alters the structure and the morphology of oxygen carriers, and lowers the activity during the cyclic operation. The sintering process starts as two spherical particles adhere to one another. The process involves the diffusion of metal cations between neighbouring spheres. Figure 5shows the SEM images of a pure Fe_2O_3 sample and the same material recovered after six methane-steam redox cycles performed at 750 C. Severe sintering is clearly evident. The heat generated from the redox reactions could accelerate the rate of sintering. When oxygen carriers sinter and agglomerate inside a fluidised bed reactor, bed defluidisation may occur. The change in solid circulation and the subsequent occurrence of gas by-pass would significantly lower the gas-solid contact and hence the overall conversion efficiency.

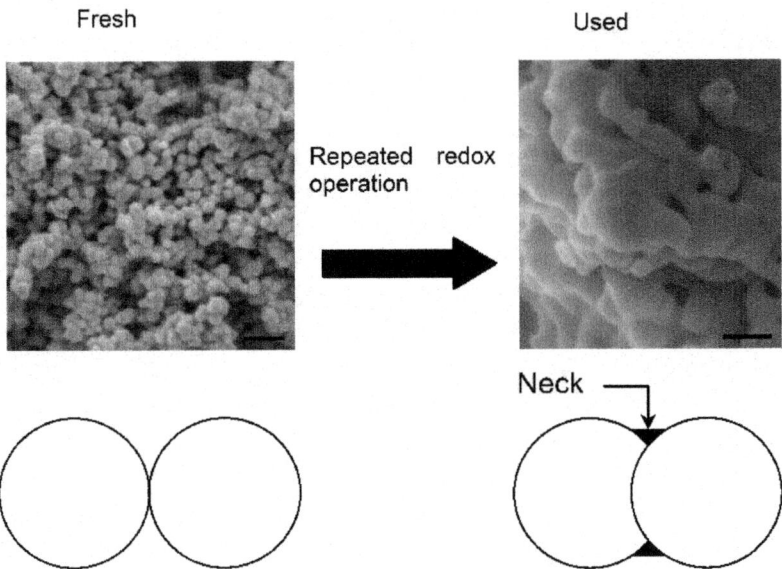

Figure 5: SEM images of Fe_2O_3 sample before and after six methane-steam redox cycles at 750 Cand representative schematics of neck growth between two particles[41].5

One of the approaches to minimise material sintering is to inhibit the diffusion in the solid particle.The complete reduction of the oxygen carrier to the corresponding zero valent metal is also a main cause of sintering since most metals agglomerates easily under elevated temperature conditions.Fukase and Suzuka [50] reported that the formation and accumulation of FeO during CL operation was mainly responsible for deactivation when iron oxide was used as the oxygen carrier.They also pointed out the importance of balancing the stoichiometry of reduction and oxidation of iron oxide and to avoid the formation of FeO by controlling reduction and oxidation temperatures. It is also important that the reduced iron species were completely oxidised to Fe_3O_4 phase.This mitigates the crystallite growth of the iron oxide and effectively prevents it from any structural changes.

Carbon is a common by-product of the CL process when a carbonaceous fuel is used as the feedsstock.Two possible routes for carbon formation are the decomposition of methane (Eq. 5) and the Boudouard reaction (Eq. 6). Methane decomposition is an endothermic reaction, and it is thermodynamically favourable at a high temperature, while the Boudourard reaction is favourable at a low temperature.These reactions could become significant in the presence of catalysts.Upon reduction, many metal oxides such as NiO, CuO and Fe_2O_3 could give rise to active metal centres which are able to rapidly produce carbon on the oxygen carrier surfaces.Once the solid carbon is formed, it will be carried over to the subsequent oxidation cycle where it is gasified to produce CO_x.When this happens, the purity of the hydrogen produced will be inevitably lowered.

In general, as the oxygen ratio in the system decreases, there is a higher tendency towards carbon formation. The oxygen ratio is defined as the actual amount of oxygen contained in the metal oxide to the stoichiometric amount of oxygen required for complete oxidation of the fuel. It is also clear that carbon formation becomes more favourable as the oxygen in the oxygen carrier is depleted through the reaction with fuel. Cho et al. [51] reported that when more than 80 % of the available oxygen in the Ni-based oxygen carrier was consumed, the rate of carbon formation increased rapidly. This was accompanied by a drastic decrease in the fuel conversion because of the decreasing oxygen content available for oxidation. Galvita and Sundmacher [43] reported that a maximum Fe reduction of 60 % largely minimised carbon formation and a high purity hydrogen stream (< 20 ppm CO) could be obtained.

PROCESS ECONOMICS

In view of the lack of information on the cost of hydrogen produced from the CLprocess, the preliminary economic analysis and greenhouse gas footprint (GHG equivalent emissions in terms of carbon dioxide) of a methane-steam redox process will be provided in this section. A simple design for hydrogen production via a two-reactor layout wasfirst obtained by considering the mass and energy balances as well as the overall pressure balance in order to establish a circulation of solids between the two reactors. The means of exchanging heat (direct, indirect, counter-current, available surface area, approach temperatures etc) has been considered, but has not been addressed further in this study. The pressure balance was affected by variables including the physical properties of the solid and gas, fluid velocity, solids recirculation rate as well as the geometry of the system.The pressure balance was solved using a one-dimensional model[52].The basis of the design was a hydrogen production rate of 49 kg/h (or 547 Nm³/h). This process considered the use of iron oxide as the oxygen carrier. Because the reduction of the iron oxide was much slower than its oxidation, a bubbling fluidised bed was chosen for the fuel reactor and a riser for the steam reactor.A particle size and density of the iron oxide particles were assumed to be 160µm and 5850kg/m³, respectively. Other assumptions made for the operation are listed in Table 4.A high solids (i.e. the iron oxide) flow rate was required through the riser in order to meet the mass balance.This resulted in a high pressure drop across the riser, which was reduced by increasing the excess steam used for oxidation of the reduced iron oxide in the riser (at constant superficial gas velocity). The resultant mass balance is given inFigure6 and the CLH2 design is presented in Figure 5.

Table 4: Assumptions used in the design of a CLH2process

1.	Steam Reactor (riser)	Down-comer	Fuel Reactor (bubbling fluidised bed)	Loop Seal to Steam Reactor
Superficial gas velocity [m/s]	6.0	0.1	0.15	0.1
Temperature [C]	750	700	750	700
Feed gas	Steam	Steam	Natural Gas	Steam
Feed gas temperature [C]	240	240	500	240
Feed gas pressure [bar]	5	5	5	5
Conversion	100% FeO to $Fe_{30}4$	None	20% F_3O_4 to FeO 100% conversion of NG	None
Residence time required	2.	3.	1 minute	4.

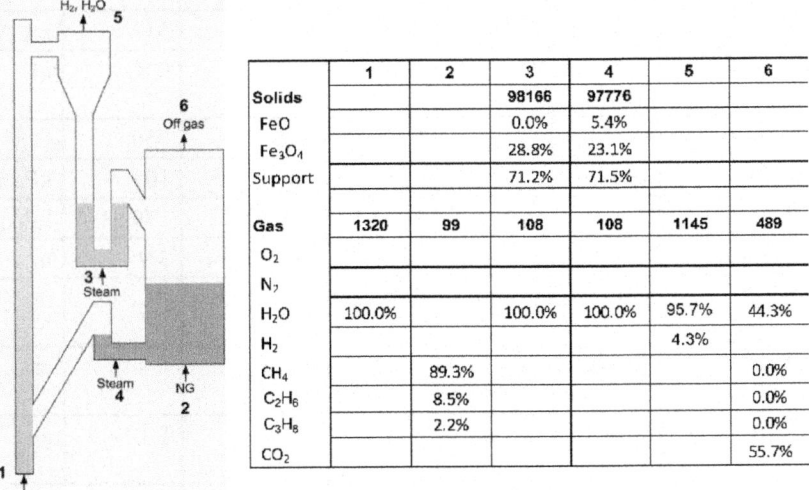

	1	2	3	4	5	6
Solids			98166	97776		
FeO			0.0%	5.4%		
Fe_3O_4			28.8%	23.1%		
Support			71.2%	71.5%		
Gas	1320	99	108	108	1145	489
O_2						
N_2						
H_2O	100.0%		100.0%	100.0%	95.7%	44.3%
H_2					4.3%	
CH_4		89.3%				0.0%
C_2H_6		8.5%				0.0%
C_3H_8		2.2%				0.0%
CO_2						55.7%

Figure 6: A schematic of CLH2processand the mass balance used for hydrogenproduction.Flow rates are represented in kg/hr and compositions in mass percentage.

The process flow diagram including the major peripheral equipment is shown inFigure7. The heat from the exothermic reaction in the riser is used to raise superheated steam at 20 bar and 400 C. This is used to generate electricity, with the steam let down to 5 bar and 240 C. 25% of the steam is used as feed to the steam reactor and to fluidise the two loop seals. The water vapour content in the hydrogen product stream is due to the excess steam fed to the riser as well as from steam used to fluidise the loop seals. This is condensed out and returned with the water from the steam turbine to the boiler, in order to reduce the fresh water requirement. The heating required for the endothermic reaction in the fuel reactor is reduced by pre-heating the natural gas using the waste heat from the off gas from the fuel reactor. For the current heat balance purpose it is assumed that there are different ways of supplying this remaining heat. One of the possible ways of supplying direct heat is by including a third combustion loop operated at higher temperature, which is outside the scope of this study.

Table 5: Reactor configuration for CLH2process

5.		Steam reactor	Steam down-comer	Fuel re-actor	Units
Gas flow	Entering	957	16	159	Nm³/h
6.	Exiting	457	14	137	Nm³/h
Superficial gas velocity	7.	6.0	0.1	0.15	m/s
$_G^S$	Entering	346	346	51	kg/m²s
Internal di-ameter	8.	0.32	0.32	0.83	m
Temperature	9.	750	700	750	C
Pressure	Bottom	113	98	107	kPa,g
10.	Top	102.45	179.31	179.31	kPa,g
Height	Total internal	15	-	3.9	m
11.	Gas exit (from top of riser)	0.8	-	-	m
12.	Downcomer (not including cyclone)	-	7	-	m
13.	Bubbling bed / loop seal	-	0.7	1.1	m
Height rela-tive to datum	14.	15.	16.	17.	18.
19.	Bottom	0.0	5.1	1.9	m

20.	Loop seal entrance to riser	1.2	-	-	m
Solids void-age (ε)	21.	0.88	0.47	0.53	22.

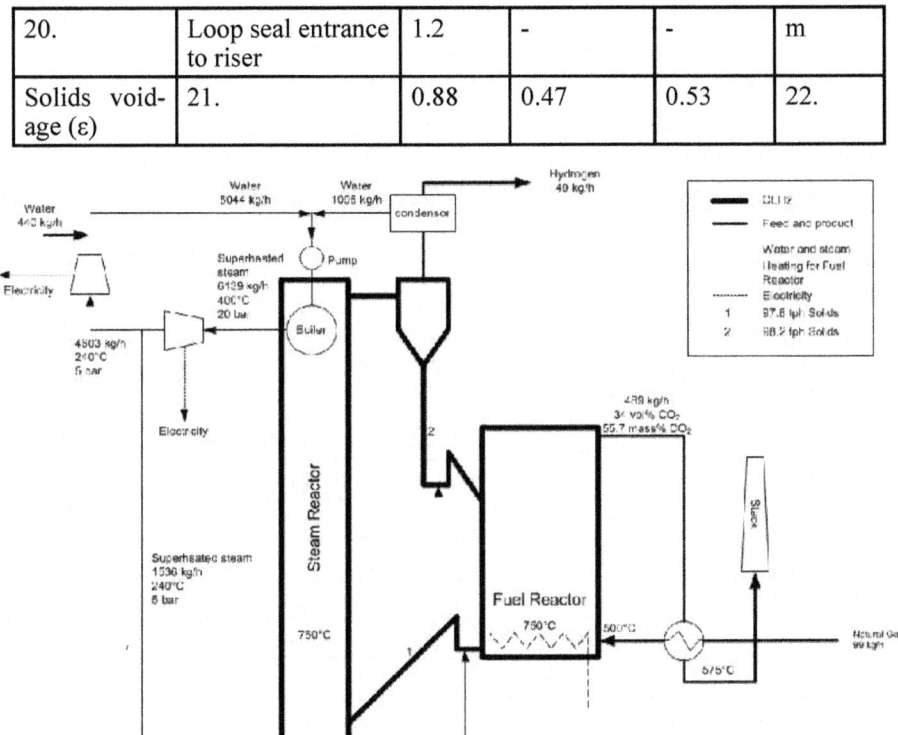

Figure 7: Proposed flow diagram of CLH2 process, showing peripheral equipment.

The greenhouse gas emissions associated with the production of a unit of hydrogen were calculated using lifecycle assessment (LCA) techniques. Principally, LCA is a technique used to assess the environmental impacts of all stages associated with the production, use and disposal of a product or delivery of a service (product life from cradle to grave).In the case of a fossil fuel for example, this includes not only the combustion emissions associated with the fuel's use, but also includes pre-combustion or upstream emissions resulting from the extraction, production, transportation, processing, conversion and distribution of the fuel.The international standards contained in the ISO 14040 series [53] provide a basic framework in which to undertake LCA.A more general introduction to LCA may be found in Horne et al. [54]and Weidema et al. [55].In this study, all fuel production and feedstock supply processes, as specified in Figure 7, were included in the LCA. The analysis is therefore limited to processes upstream of the refinery gate and thus does not include the delivery and combustion of hydrogen.Emission results are reported using the concept of a global warming potential (GWP), which enables different greenhouse gases to be compared and expressed using an equivalent carbon dioxide (gCO_2e) value.Data used for the analysis are summarised in Table 6 based on an hourly hydrogen production rate of 49 kg.The GHG impact of the CLH2 processunder consideration is 18,690 gCO_2e/kg H_2 produced or 154 gCO_2e/MJ H_2.The impact is dominated by the need to supply process heat to the fuel reactor(redox heater emissions:9,628 gCO_2/kg H_2) as shown in Figure 8.

Table 6: LCA inputs/outputs (per hour) for the CLH2 process

Inputs	Value	Units	Comments
Resources	23.	24.	25.
Natural gas	99	kg	Natural gas for reaction
Oxide material	1.26	kg	Yearly make-up (per hour)
Water	440	kg	Make-up water (reaction and cooling)
Energy	26.	27.	28.
Natural gas	151	kg	Fuel reactor heat requirement
Electricity	189	kW	Net electricity requirement
Outputs	29.	30.	31.
Hydrogen	49	kg	Compressed hydrogen output
Emissions	32.	33.	34.
H_2O	217	kg	Fuel reactor (stack emissions)
C_02	272	kg	Fuel reactor (stack emissions)
C_02	419	kg	Fuel reactor (heater emissions)

Preliminary results demonstrate the need to optimise the delivery of heat to the fuel reactor. The introduction of a third combustion loop operated at higher temperature is one such means to reduce upstream emissions. However, this may negatively influence total capital expenditure.The literature reports hydrogen production through current steam reforming technology produces between 9,830 gCO_2e/kg H_2 (24,000 kg H_2/day; midsized facility) and 12,130 gCO_2e/kg H_2 (480 kg H_2/day; distributed facility), and thus are higher than the direct redox process emissions [56], although significantly lower than the total CLH2 emissions.The literature only considered electricity and natural gas related emissions and thus total upstream emissions of existing technologies maybe higher than the reported values.

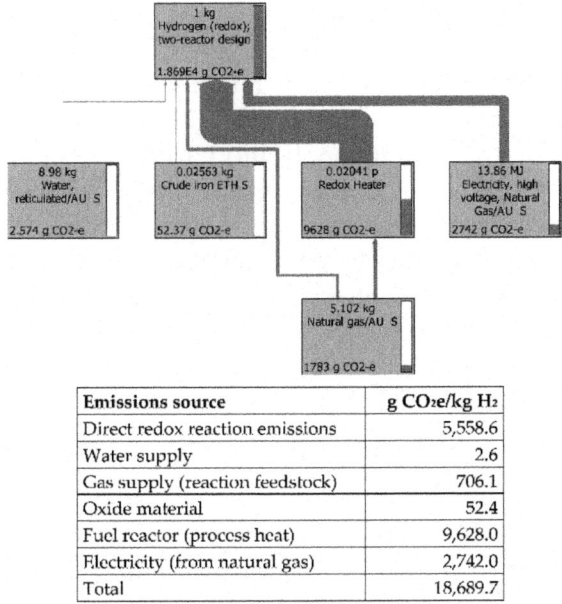

Emissions source	g CO₂e/kg H₂
Direct redox reaction emissions	5,558.6
Water supply	2.6
Gas supply (reaction feedstock)	706.1
Oxide material	52.4
Fuel reactor (process heat)	9,628.0
Electricity (from natural gas)	2,742.0
Total	18,689.7

Figure 8: Redox emissions breakdown (per kg H$_2$).

The commercial viability of the redox process was estimated using cost estimate practices outlined in the literature [56, 57]. Results are reported in $/kg H$_2$. Material and fuel operating expenditure was calculated using the inputs identified in Figure 7, as summarised in the lifecycle analysis section (Table 6). Fixed operating and maintenance costs were calculated based on the total capital expenditure. Battery limit capital expenditure (e.g. redox process) is based on the engineering judgment of the authors, with capital build-up (facilities, engineering, permitting, start-up, contingencies, working capital and land) estimated using a percentage of the battery limit cost. Capital charges are calculated using a percentage of total capital expenses. Importantly, although the estimates may look precise, they are simply estimates based on the judgment of the authors. There remains significant uncertainty about the actual cost of the redox process as it has not been commercially demonstrated. A breakdown of cost data is provided in Table 7.

Initial costing estimates show that the redox process may produce hydrogen at $8.93/kg ($9.36/kg, including carbon tax). The cost breakdown demonstrates that onsite storage of compressed hydrogen represents a significant expense. However, this arises from the conversion of stranded methane. If demand for hydrogen is identified close to a stranded gas reserve, storage costs will decrease significantly. Delivery of compressed hydrogen represents an additional cost that has not been considered in this analysis. Literature cost estimates for at gate hydrogen production via steam reforming, using current technology, range between \$1.51/kg (midsize facility: 24,000 kg H_2/day) to \$3.68/kg (distributed facility: 480 kg H_2/day facility). Hence the hydrogen at gate cost for the CLH2 process is higher than steamreforming technology. Electrolysis production of hydrogen ranges between \$4.94 and \$6.82 per kg for a midsize and distributed facility respectively and thus is closer to CLH2production costs [56]. Experience gained through the commercialisation and deployment of the redox technology is expected to reduce costs, particularly capital build-up costs. However the stranded nature of the product may significantly increase total delivered hydrogen cost.

Table 7: Redox process cost estimates

Expense	\$M/yr	Comment
OpEX	35.	Variable (fuel and materials)
Oxide material	0.55	\$50/kg
Natural gas	0.20	Reaction feed and reducer heating
Electricity	0.12	Net electricity demand
Water	0.01	Make-up supply
Total OpEX	**0.88**	36.
37.	38.	39.
CapEX	\$M	40.
CLH2 reactor	10.0	41.
H_2Compression	0.44	\$3,000/kW capacity
H_2 Storage	5.97	\$26,417/m³ capacity; 5 days storage
Total process units	16.42	42.
General facilities	3.28	20 % of process unit CapEX
Engineering	2.46	15 % of process unit CapEX
Contingencies	1.64	10 % of process unit CapEX
Working capital	0.82	5 % of process unit CapEX
Total CapEX	24.62	43.
44.	45.	46.
Balance	\$M/yr	47.
OpEx (variable)	0.88	48.
OpEx (fixed)	0.49	2 % of total CapEX
Capital charge	2.46	10 % of total CapEX

Carbon Tax	0.18	$23/T C$_O$2
Total ($M/yr)	**4.02**	49.
Total ($/kg H$_2$)	**8.93**	**(ex. carbon tax)**
	9.36	**(inc. carbon tax)**

CONCLUSION

The feasibility of producing hydrogen from the metal/metal oxide redox process has been demonstrated in the literature. This process offers several advantages including the ability to produce hydrogen of high purity and a concentrated stream of carbon dioxide. Most importantly this process eliminates the need for a supply of high purity oxygen and a water gas shift process that are generally required by commercial processes. However, this redox process is not regarded as a fully developed technology and further R&D development is required for commercialisation.

In view of the literature, much research effort has been devoted to formulating novel oxygen carrier materials. Although several types of improved oxygen carrier materials have been identified, full appraisals of their performance and further optimisation studies are required.Iron oxidesand nickel oxides appear to be attractive candidates for this application in terms of their activity. However, their thermal stabilities need further improvement. Current practices include doping, introducing a diffusional barrier provided by a second oxide, and/or adding a second oxide with higher oxygen storage capacity. There are also a limited number of studies that investigate the life time of oxygen carriers. Apart from chemical stability, the changes in the physical properties such as size and attrition of the carrier particles during fluidisation have received little attention and should be addressed in future research. It is viewed strongly that improvement in these areas would significantly increase process efficiency and economic viability of the cyclic redox process.

The lack of pilot scale studies also impedes the commercialisation of cyclic redox and chemical looping processes. Limited data are available for process design, scale-up and optimisation. For example, the transfer of the oxygen carrier particles between oxidation and reduction is a critical issue when it comes to process design.Fixed bed, moving bed and circulating fluidised bed have been proposed, and the choice of reactor will depend on the reaction kinetics and the required flow dynamics of the process. Because the cyclic redox process is considered as an unsteady process, the definition of the operation window of the process will be determined by limiting the upper and the lower oxidation states of the metal/metal oxide couple. This parameter has a direct impact on the overall conversion efficiencies, process designs and economics. Since the redox reactions usually take place at temperatures above 600 C, most of the sensible heat stored in the gas existing from the oxidation and reduction reactors can be used to generate power with a steam generator.The co-production of excess electricity would reduce the cost of the hydrogen produced and increase overall process viability. Hence, the issue of heat management requires much closer examination when it comes to process optimisation.

Finally, the current preliminary LCA-Economic study has made the first attempt to provide an indicative price of hydrogen produced from the redox process. Although the cost of hydrogen produced from the redox process is higher than hydrogen produced from other commercial processes, several design parameters have been identified as the areas for future improvement. It is seen that the LCA techniques are valuable tools for process optimisation.

ACKNOWLEDGEMENT

The authors acknowledge the support from CSIRO Petroleum and Geothermal Research Portfolio in conducting this study

REFERENCES

50. IEA 2005 Small Scale Hydrogen Production from Metal-Metal Oxide Redox Cycles. OECD Publishing

51. Mueller-Langer, F. , E. Tzimas, M. Kaltschmitt, S. Peteves, 2007 Small Scale Hydrogen Production from Metal-Metal Oxide Redox Cycles. Int. J. Hydrogen Energy 32 37973810 .

52. Pardro, C.E.G. and Putsche, V., Survey of the Economics of Hydrogen Technologies. 1999 National Renewable Energy Laboratory: Golden, Colorado.

53. Mirabal, S. T. , 2003 Small Scale Hydrogen Production from Metal-Metal Oxide Redox Cycles, University of Florida.

54. Raissi, A. T. , D. L. Block, 2004 Hydrogen- automotive fuel of the future. IEEE Power Energ. Mag. November/December 4045 .

55. Fan, L. S. , F. X. Li, 2010 Small Scale Hydrogen Production from Metal-Metal Oxide Redox Cycles. Ind. Eng. Chem. Res. 49 1020010211 .

56. Adanez, J. , A. Abad, F. Garcia-Labiano, P. Gayan, L. F. de Diego, 2012 Small Scale Hydrogen Production from Metal-Metal Oxide Redox Cycles. Prog. Energy Combust. Sci. 38 215282 .

57. Gupta, P. , L. G. Velazquez-Vargas, L. S. Fan, 2007 Small Scale Hydrogen Production from Metal-Metal Oxide Redox Cycles. Energy Fuels 21

29002908 .

58. Li, F. , H. R. Kim, D. Sridhar, F. Wang, L. Zeng, J. Chen, L. S. Fan, 2009 Syngas Chemical Looping Gasification Process: Oxygen Carrier Particle Selection and Performance. Energy Fuels 23 41824189 .

59. Li, F. X. , L. Zeng, L. G. Velazquez-Vargas, Z. Yoscovits, L. S. Fan, 2010 Small Scale Hydrogen Production from Metal-Metal Oxide Redox Cycles. AlChE J. 56 21862199 .

60. Yang, J. B. , N. S. Cai, Z. S. Li, 2008 Small Scale Hydrogen Production from Metal-Metal Oxide Redox Cycles. Energy Fuels 22 25702579 .

61. Fan, L. , F. Li, S. Ramkumar, 2008 Small Scale Hydrogen Production from Metal-Metal Oxide Redox Cycles. Particuology 6 131142 .

62. Gnanapragasam, N. V. , B. V. Reddy, M. A. Rosen, 2009 Small Scale Hydrogen Production from Metal-Metal Oxide Redox Cycles. Int. J. Hydrogen Energy 34 26062615 .

63. Sime, R. , J. Kuehni, L. D'Souza, E. Elizondo, S. Biollaz, 2003 Small Scale Hydrogen Production from Metal-Metal Oxide Redox Cycles. Int. J. Hydrogen Energy 28 491498 .

64. Kobayashi, N. , L. S. Fan, 2010 Small Scale Hydrogen Production from Metal-Metal Oxide Redox Cycles. Biomass Bioenergy 35 12521262 .

65. Li, F. X. , L. Zeng, L. S. Fan, 2010 Small Scale Hydrogen Production from Metal-Metal Oxide Redox Cycles. Fuel 89 37733784 .

66. Hacker, V. , G. Faleschini, H. Fuchs, R. Fankhauser, G. Simader, M. Ghaemi, B. Spreitz, K. Friedrich, 1998 Usage of biomass gas for fuel cells by the SIR process. J. Power Sources 71 226230 .

67. Bleeker, M. F. , S. R. A. Kersten, H. J. Veringa, 2007 Small Scale Hydrogen Production from Metal-Metal Oxide Redox Cycles. Catal. Today 127 278290 .

68. Bleeker, M. F. , H. J. Veringa, S. R. A. Kersten, 2010 Small Scale Hydrogen Production from Metal-Metal Oxide Redox Cycles. Ind. Eng. Chem. Res. 49 5364 .

69. Galvita, V. V. , H. Poelman, G. B. Marin, 2011 Hydrogen Production from Methane and Carbon Dioxide by Catalyst-Assisted Chemical Looping,. Top. Catal. 54907 .

70. Cormos, C. , C. , 2011 Small Scale Hydrogen Production from Metal-Metal Oxide Redox Cycles. Int. J. Hydrogen Energy 36 59605971 .

71. Otsuka, K. , T. Kaburagi, C. Yamada, S. Takenaka, 2003 Small Scale Hydrogen Production from Metal-Metal Oxide Redox Cycles. J. Power Sources 122 111121 .

72. Otsuka, K. , C. Yamada, T. Kaburagi, S. Takenaka, 2003 Small Scale Hydrogen Production from Metal-Metal Oxide Redox Cycles. Int. J. Hydrogen Energy 28 335342 .

73. Wang, H. , G. Wang, X. Wang, J. Bai, 2008 Small Scale Hydrogen Production from Metal-Metal Oxide Redox Cycles. J. Phys. Chem. C 112 56795688 .

74. Liu, X. , H. Wang, 2010 Small Scale Hydrogen Production from Metal-Metal Oxide Redox Cycles 183 10751082 .

75. Otsuka, K. , S. Takenaka, 2004 Small Scale Hydrogen Production from Metal-Metal Oxide Redox Cycles. J. Jpn. Pet. Inst. 47 377386 .

76. Ryu, J. C. , D. H. Lee, K. S. Kang, C. S. Park, J. W. Kim, Y. H. Kim, 2008 Small Scale Hydrogen Production from Metal-Metal Oxide Redox Cycles. J. Ind. Eng. Chem. 14 252260 .

77. Kodama, T. , Y. Watanabe, S. Miura, M. Sato, Y. Kitayama, 1996 Reactive and selective redox system of Ni(II)-ferrite for a two-step CO and H2 production cycle from carbon and water. Energy 21 11471156 .

78. Kang, K. S. , C. H. Kim, W. C. Cho, K. K. Bae, S. W. Woo, C. S. Park, 2008 Small Scale Hydrogen Production from Metal-Metal Oxide Redox Cycles. Int. J. Hydrogen Energy 33 45604568 .

79. Takenaka, S. , N. Hanaizumi, V. T. D. Son, K. Otsuka, 2004 Small Scale Hydrogen Production from Metal-Metal Oxide Redox Cycles. J. Catal. 228 405416 .

80. Urasaki, K. , N. Tanimoto, T. Hayashi, Y. Sekine, E. Kikuchi, M. Matsukata, 2005 Small Scale Hydrogen Production from Metal-Metal Oxide Redox Cycles. Appl. Catal., A 288 143148 .

81. Adanez, J. , L. F. de Diego, F. Garcia-Labiano, P. Gayan, A. Abad, J. M. Palacios, 2004 Selection of Oxygen Carriers for Chemical-looping Combustion. Energy Fuels 18 371377 .

82. Cho, P. , T. Mattisson, A. Lyngfelt, 2004 Small Scale Hydrogen Production from Metal-Metal Oxide Redox Cycles. Fuel 83 12151225 .

83. Zafar, Q. , T. Mattisson, B. Gevert, 2005 Integrated hydrogen and power production with CO2 capture using chemical-looping reforming-redox reactivity of particles of CuO, Mn2O3, NiO, and Fe2O3 using SiO2 as a support. Ind. Eng. Chem. Res. 44 34853496 .

84. Kodama, T. , T. Shimizu, T. Satoh, M. Nakata, K. I. Shimizu, 2002 Stepwise production of CO-RICH syngas and hydrogen via solar methane reforming by using a Ni(II)-ferrite redox system. Sol. Energy 73 363374 .

85. Kodama, T. , Y. Kondoh, R. Yamamoto, H. Andou, N. Satou, 2005

Thermochemical hydrogen production by a redox system of ZrO2-supported Co(II)-ferrite. Sol. Energy 78 623631 .

86. Cha, K. S. , B. K. Yoo, H. S. Kim, T. G. Ryu, K. S. Kang, C. S. Park, Y. H. Kim, 2010 Small Scale Hydrogen Production from Metal-Metal Oxide Redox Cyclesnergy Res. 34 422430 .

87. Kang, K. S. , C. H. Kim, K. K. Bae, W. C. Cho, S. H. Kim, W. J. Kim, Y. H. Kim, C. S. Park, 2010 Small Scale Hydrogen Production from Metal-Metal Oxide Redox Cycles. Int. J. Hydrogen Energy 35 568576 .

88. Li, K. Z. , H. Wang, Y. G. Wei, D. X. Yan, 2009 Small Scale Hydrogen Production from Metal-Metal Oxide Redox Cycles. J. Phys. Chem. C 113 1528815297 .

89. Tang, L. , D. Yamaguchi, N. Burke, D. Trimm, K. Chiang, 2010 Methane decomposition over ceria modified iron catalysts. Catal. Commun. 11 12151219 .

90. Yamaguchi, D. , L. Tang, L. Wong, N. Burke, D. Trimm, K. Nguyen, K. Chiang, 2011 Small Scale Hydrogen Production from Metal-Metal Oxide Redox Cycles. Int. J. Hydrogen Energy 36 66466656 .

91. Sim, A. , N. W. Cant, D. L. Trimm, 2010 Small Scale Hydrogen Production from Metal-Metal Oxide Redox Cycles. Int. J. Hydrogen Energy 35 89538961 .

92. Galvita, V. , K. Sundmacher, 2005 Small Scale Hydrogen Production from Metal-Metal Oxide Redox Cycles. Appl. Catal., A 289 121127 .

93. Datta, P. , L. K. Rihko-Struckmann, K. Sundmacher, 2011 Small Scale Hydrogen Production from Metal-Metal Oxide Redox Cycles. Mater. Chem. Phys. 129 10891095 .

94. Leion, H. , T. Mattisson, A. Lyngfelt, 2009 Small Scale Hydrogen Production from Metal-Metal Oxide Redox Cycles. Energy Fuels 23 23072315 .

95. Leion, H. , A. Lyngfelt, M. Johansson, E. Jerndal, T. Mattisson, 2008 Small Scale Hydrogen Production from Metal-Metal Oxide Redox Cycles. Chem. Eng. Res. Des 86 10171026 .

96. Adanez, J. , A. Cuadrat, A. Abad, P. Gayan, L. F. de Diego, F. Garcia-Labiano, 2010 Ilmenite Activation during Consecutive Redox Cycles in Chemical-Looping Combustion. Energy Fuels 24 14021413 .

97. Leion, H. , T. Mattisson, A. Lyngfelt, 2008 Small Scale Hydrogen Production from Metal-Metal Oxide Redox Cycles. Int. J. Greenhouse Gas Control 2 180193 .

98. Lorente, E. , J. A. Pena, J. Herguido, 2011 Small Scale Hydrogen

Production from Metal-Metal Oxide Redox Cycles. Int. J. Hydrogen Energy 36 70437050 .

99. Fukase, S. , T. Suzuka, 1993 Residual oil cracking with generation of hydrogen- Deactivation of iron-oxide catalyst in the steam iron reaction. Appl. Catal., A 100 117 .

100. Cho, P. , T. Mattisson, A. Lyngfelt, 2005 Small Scale Hydrogen Production from Metal-Metal Oxide Redox Cycles. Ind. Eng. Chem. Res. 44 668676 .

101. Hadley, T. D. , K. Chiang, N. R. Burke, K. S. Lim, 2010 Multiple-loop chemical reactor design with pressure balance consideration. Fluidization XIII 463470 .

102. ISO, ISO/DIS14040, Environmental Management Standard- Life Cycle Assessment, Principlesand Framework. International Standard. Switzerland. 2006

103. Horne, R. , T. Grant, K. Verghese, 2009 Small Scale Hydrogen Production from Metal-Metal Oxide Redox Cycles CSIRO

104. Weidema, B. P. , G. Rebitzer, T. Ekvall, 2004 Scenarios in Life-cycle Assessment Society of Environmental Toxicology and Chemistry

105. NRC 2004 The Hydrogen Economy: Opportunities, Costs, Barriers, and R&D Needs. The National Academies Press.

106. Simbeck, D. , E. Chang, 2002 Hydrogen Supply: Cost Estimate for Hydrogen Pathways Scoping Analysis. NREL, Report: SR-5403252 . National Renewable Energy Laboratory

Chapter 8

NANOMATERIALS FOR HYDROGEN STORAGE APPLICATIONS: A REVIEW

Michael U. Niemann,[1] Sesha S. Srinivasan,[1] Ayala R. Phani,[2] Ashok Kumar,[1] D. Yogi Goswami,[1] and Elias K. Stefanakos[1]

[1]Clean Energy Research Center, College of Engineering, University of South Florida, 4202 East Fowler Avenue, Tampa, FL 33620, USA

[2]Nano-RAM Technologies, 98/2A Anjanadri, 3rd Main, Vijayanagar, Bangalore 5600040, Karnataka, India

ABSTRACT

Nanomaterials have attracted great interest in recent years because of the unusual mechanical, electrical, electronic, optical, magnetic and surface properties. The high surface/volume ratio of these materials has significant implications with respect to energy storage. Both the high surface area and the opportunity for nanomaterial consolidation are key attributes of this new class of materials for hydrogen storage devices. Nanostructured systems including carbon nanotubes, nano-magnesium based hydrides, complex hydride/carbon nanocomposites, boron nitride nanotubes, TiS_2/MoS_2 anotubes, alanates, polymer nanocomposites, and metal organic frameworks are considered to be potential candidates for storing large quantities of hydrogen. Recent investigations have shown that nanoscale materials may offer advantages if certain physical and chemical effects related to the nanoscale can be used efficiently. The present review focuses the application of nanostructured materials for storing atomic or molecular hydrogen. The synergistic effects of nanocrystalinity and nanocatalyst doping on the metal or complex hydrides for improving the thermodynamics and hydrogen reaction kinetics are discussed. In addition, various carbonaceous nanomaterials and novel sorbent systems (e.g. carbon nanotubes, fullerenes, nanofibers, polyaniline nanospheres and metal organic frameworks etc.) and their hydrogen storage characteristics are outlined.

INTRODUCTION

The increase in threats from global warming due to the consumption of fossil fuels requires our planet to adopt new strategies to harness the inexhaustible sources of energy [1, 2]. Hydrogen is an energy carrier which holds tremendous promise as a new renewable and clean energy option [3]. Hydrogen is a convenient, safe, versatile fuel source that can be easily converted to a desired form of energy without releasing harmful emissions. Hydrogen is the ideal fuel for the future since it significantly reduces the greenhouse gas emissions, reduces the global dependence on fossil fuels, and increases the efficiency of the energy conversion process for both internal combustion engines and proton exchange membrane fuel cells [4, 5]. Hydrogen used in the fuel cell directly converts the chemical energy of hydrogen into water, electricity, and heat [6] as represented by

$$H_2 + \frac{1}{2}O_2 \rightarrow H_2O + \text{Electricity} + \text{Heat}.$$

$$(1)$$

Hydrogen storage cuts across both hydrogen production and hydrogen applications and thus assumes a critical role in initiating a hydrogen economy [7–10]. For catering today's fuel cell cars, the onboard hydrogen storage is inevitable and an integral part of the system to be reengineered [11, 12]. The critical properties of the hydrogen storage materials to be evaluated for automotive applications are (i) light weight, (ii) cost and availability, (iii) high volumetric and gravimetric density of hydrogen, (iv) fast kinetics, (v) ease of activation, (vi) low temperature of dissociation or decomposition, (vii) appropriate thermodynamic properties, (viii) long-term cycling stability, and (ix) high degree of reversibility. All the said properties greatly demand from us to understand the fundamental mechanistic behavior of materials involving catalysts and their physicochemical reaction toward hydrogen at an atomic or molecular scale.

Various hydrogen storage systems (see Figure 1), such as metal hydrides, complex hydrides, chemical hydrides, adsorbents and nanomaterials (nanotubes, nanofibers, nanohorns, nanospheres, and nanoparticles), clathrate hydrates, polymer nanocomposites, metal organic frameworks, and so on [13–19], have been explored for onboard hydrogen storage applications. However, none of these materials qualifies and fulfill all hydrogen storage criteria such as (1) high hydrogen content (>6.0 wt.%), (2) favorable or tuning thermodynamics (30–55 kJ/mol H_2) (3) operate below 100°C for H_2 delivery, (4) onboard refueling option for a hydrogen-based infrastructure, (5) cyclic reversibility (~1000 cycles) at moderate temperatures, and so on. Among the various hydrogen storage systems, the concept of nanomaterials and their wide applications for energy storage [20] are discussed in the present paper.

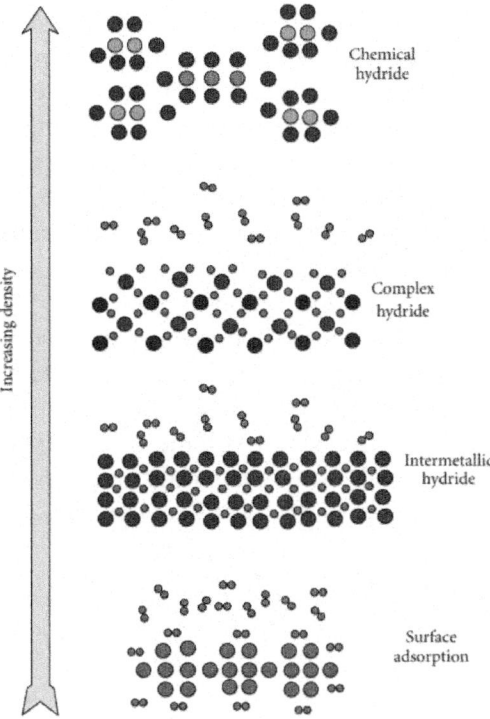

Figure 1: Hydrogen storage density in physisorbed materials, metal/complex, and chemical hydrides.

NANOSTRUCTURED MATERIALS

Nanostructured materials have potential promise in hydrogen storage because of their unique features such as adsorption on the surface, inter- and intragrain boundaries, and bulk absorption [21, 22]. Nanostructured and nanoscale materials strongly influence the thermodynamics and kinetics of hydrogen absorption and dissociation by increasing the diffusion rate as well as by decreasing the required diffusion length. Additionally, the materials at the nanoscale offer the possibility of controlling material tailoring parameters independently of their bulk counterparts. They also lead to the design of light weight hydrogen storage systems with better hydrogen storage characteristics.

Nanocatalyst Doping in Complex Borohydrides

Advanced complex hydrides that are light weight, low cost, and have high-hydrogen density are essential for onboard vehicular storage [8, 23, 24]. Some of the complex hydrides with reversible capacities achieved are Alanates [25, 26], Amides [27, 28], Borohydrides [29–31], and combinations thereof [32, 33]. The challenging tasks to design and develop the complex hydrides mandate an optimization and overcoming of kinetic and thermodynamic limitations [18, 34]. The enhancement of reaction kinetics at low temperatures and the requirement for high hydrogen storage capacity (>6.0 wt%) of hydrogen storage materials could be made possible by catalytic doping. If nanostructured materials with high surface area are used as the catalytic dopants, they may offer several advantages for the physicochemical reactions, such as surface interactions, adsorption in addition to bulk absorption, rapid kinetics, low-temperature sorption, hydrogen atom dissociation, and molecular diffusion via the surface catalyst. The intrinsically large surface areas and unique adsorbing properties of nanophase catalysts can assist the dissociation of gaseous hydrogen and the small volume of individual nanoparticles can produce short diffusion paths to the materials' interiors. The use of nanosized dopants enables a higher dispersion of the catalytically active species [35] and thus facilitates higher mass transfer reactions.

Figure 2 depicts the thermogravimetric weight loss profiles of the new complex borohydride ($LiBH_4 + 1/2\ ZnCl_2$) system undoped and doped with different nanocatalyst (e.g., nano-Ni) concentrations.

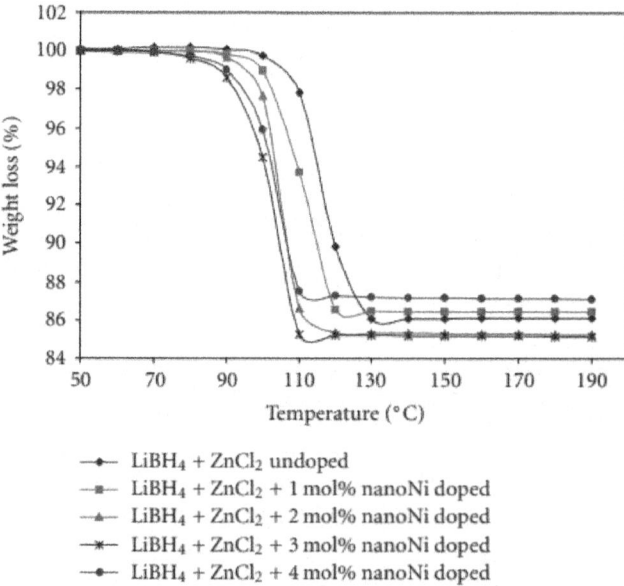

- $LiBH_4 + ZnCl_2$ undoped
- $LiBH_4 + ZnCl_2$ + 1 mol% nanoNi doped
- $LiBH_4 + ZnCl_2$ + 2 mol% nanoNi doped
- $LiBH_4 + ZnCl_2$ + 3 mol% nanoNi doped
- $LiBH_4 + ZnCl_2$ + 4 mol% nanoNi doped

Figure 2: TGA curves for the $LiBH_4 + 1/2\ ZnCl_2$ undoped andXmol% nano-Ni doped system (X=1,2,3,4).

The $Zn(BH_4)_2$, as obtained from the mechanochemical reaction of ($LiBH_4 + 1/2\ ZnCl_2$), exhibits an endothermic melting transition at around 80–90°C (DSC signals are not shown in the figure) and a clear-cut weight loss occurs due to the thermal hydrogen decomposition at around 120°C. Trial experiments were conducted by introducing different nanocatalyst (nano-Ni (particle size of 3–10 nm) obtained from QuantumSphere Inc., Calif, USA) concentrations (X=1,2,3,4 mol%) in this complex system. It is clearly discernible from this figure that nanocatalyst doping helps to lower the temperature of decomposition from 120°C down to 100°C. A concentration of 3 mol% nano-Ni was found to be optimum for the gravimetric weight loss due to thermal decomposition at low temperature. It is also confirmed from the gravimetric analysis that nanocatalyst doping enhances the hydrogen storage characteristics such as reaction kinetics at low-decomposition temperatures (T_{dce})The microstructures of the nanocatalyst doped complex hydride both in imaging mode and EDS mapping (distribution of different elements) mode as obtained from SEM are shown in Figures 3(a) and 3(b).

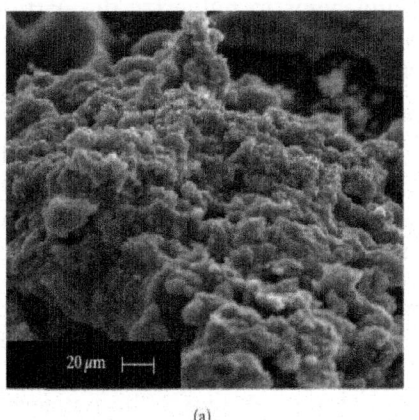 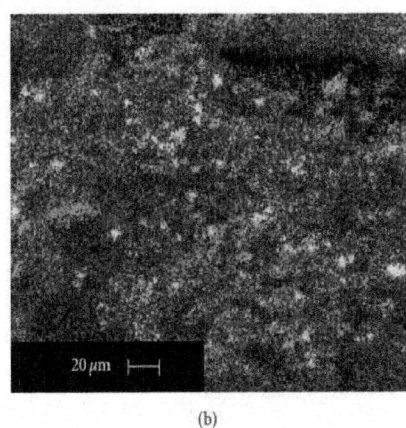

(a)   (b)

Figure 3: (a) SEM imaging and (b) EDS mapping of the sample $LiBH_4 + 1/2$ $ZnCl_2 + 10$ mol% nano-Ni (mapping elements—green: oxygen, yellow: nano-Ni, blue: chlorine, purple: zinc).

Synergistic Behavior of Nanocatalyst Doping and Nanocrystalline Form of MgH_2

It is generally known that pristine MgH_2 theoretically can store ~7.6 wt.% hydrogen [36]. However, so far, magnesium hydride-based materials have limited practical applications because both hydrogenation and dehydrogenation reactions are very slow and, hence, relatively high temperatures are required

[23]. Magnesium hydride forms ternary and quaternary hydride structures by reacting with various transition metals (Fe, Co, Ni, etc.) and thus improved kinetics. Moreover, the nanoscale version of these transition metal particles offers an additional hydrogen sorption mechanism via its active surface sites [36, 37]. In a similar way, the synergistic approach of doping nanoparticles of Fe and Ti with a few mol% of carbon nanotubes (CNTs) on the sorption behavior of MgH_2 has recently been investigated [38]. The addition of CNT significantly promotes hydrogen diffusion in the host metal lattice of MgH_2 due to the short pathway length and creation of fast diffusion channels [39]. The dramatic enhancement of kinetics of MgH_2 has also been explored through reaction with small amounts of $LiBH_4$ [40]. Though the MgH_2 admixing increases the equilibrium plateau pressure of $LiNH_2$ [41] or $LiBH_4$ [29], catalytic doping of these complex hydrides has not yet been investigated.It is generally believed that the role of the CNT/nanocatalyst on either $NaAlH_4$ or MgH_2 is to stabilize the structure and facilitate a reversible hydrogen storage behavior.

The phenomenon of mechanical milling helps to pulverize the particles of MgH_2 into micro- or nanocrystalline phases and thus leads to lowering the activation energy of desorption [42]. The height of the activation energy barrier depends on the surface elements. Without using the catalysts, the activation energy of absorption corresponds to the activation barrier for the dissociation of the H_2 molecule and the formation of hydrogen atoms. The activation energies of the H_2 sorption for the bulk MgH_2, mechanically milled MgH_2 and nanocatalyst-doped MgH_2, are shown in Figures 4(a) and 4(c).

It is undoubtedly seen that the activation barrier has been drastically lowered by nanocatalyst doping which suggests that the collision frequency between the H_2 molecules and transition metal nanoparticles increases with decreasing size of the catalyst. In addition, Figure 4 shows the conceptual model of an MgH_2 nanocluster and the distribution of nanocatalyst over the active surface sites for efficient hydrogen storage.

Recently, we have attempted to establish the above "proof of concept." Commercial MgH_2 exhibits weight loss due to hydrogen decomposition at higher temperature (e.g., 415°C) (see Figure 5).

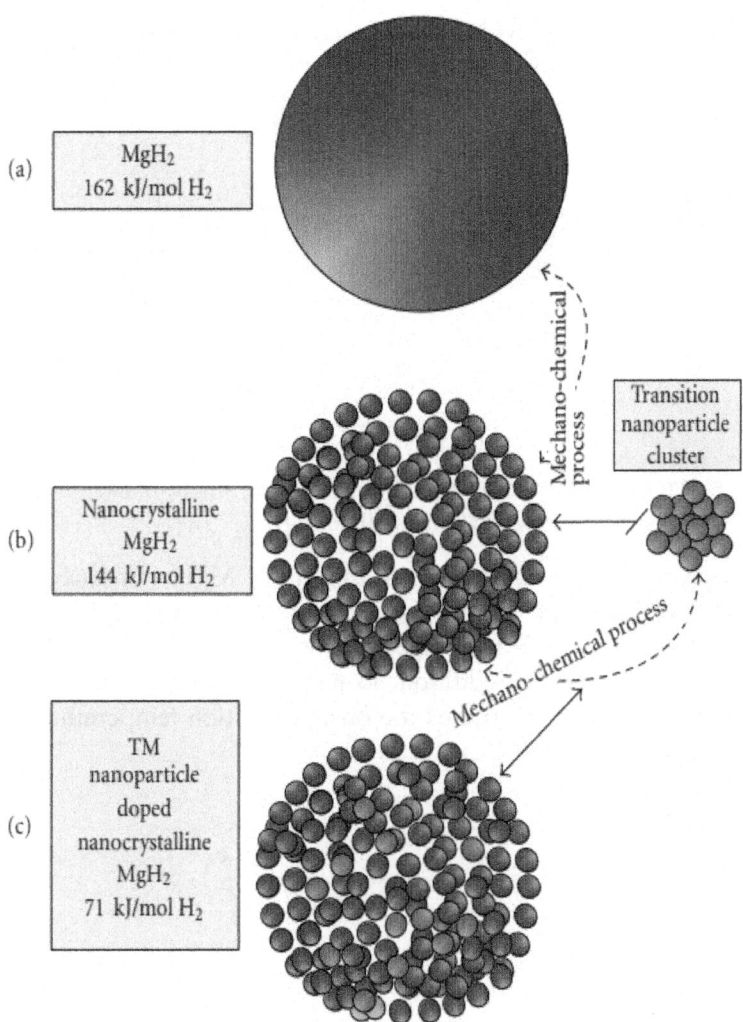

Figure 4: Conceptual model of MgH_2 cluster (a) plain, (b) nanocrystalline, and (c) nanocatalyst-doped materials.

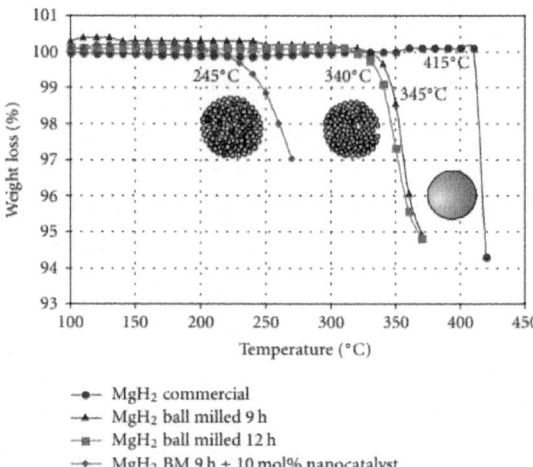

Figure 5: TGA curves for the (a) commercial; (b) micro-/nanocrystalline, and (c) 10 mol% nanocatalyst-doped nanocrystalline MgH_2.

However, the mechanochemical milling of MgH_2 introduces defects and particle size reduction. Thus, obtained micro-/nanocrystalline MgH_2 grains show endothermic hydrogen decomposition (see Figure 6) at an earlier temperature of 340°C. In addition, to nanoscale formation, the doping by a nanocatalyst certainly decreases the onset transition temperature by as much as 100°C (Figures 5 and 6).

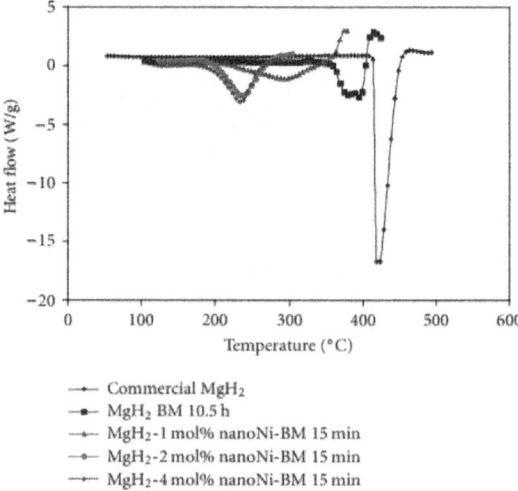

Figure 6: DSC profiles of (a) commercial, (b) ball milled for 10.5 hours, and (c) 10 mol% nano-Ni-doped nanocrystalline MgH_2.

Carbonaceous Nanomaterials (carbon Nanotubes, Fullerenes, and Nanofibers)

Carbonaceous materials are attractive candidates for hydrogen storage because of a combination of adsorption ability, high specific surface area, pore microstructure, and low-mass density. In spite of extensive results available on hydrogen uptake by carbonaceous materials, the actual mechanism of storage still remains a mystery. The interaction may either be based on van der Walls attractive forces (physisorption) or on the overlap of the highest occupied molecular orbital of carbon with occupied electronic wave function of the hydrogen electron, overcoming the activation energy barrier for hydrogen dissociation (chemisorption). The physisorption of hydrogen limits the hydrogen-to-carbon ratio to less than one hydrogen atom per two carbon atoms (i.e., 4.2 mass %). While in chemisorption, the ratio of two hydrogen atoms per one carbon atom is realized as in the case of polyethylene [43–45]. Physisorbed hydrogen has a binding energy normally of the order of 0.1 eV, while chemisorbed hydrogen has C–H covalent bonding, with a binding energy of more than 2-3 eV.

Dillon et al. presented the first report on hydrogen storage in carbon nanotubes [46] and triggered a worldwide tide of research on carbonaceous materials. Hydrogen can be physically adsorbed on activated carbon and be "packed" on the surface and inside the carbon structure more densely than if it has just been compressed. The best results achieved with carbon nanotubes to date confirmed by the National Renewable Energy Laboratory are hydrogen storage density corresponding to about 10% of the nanotube weight [47].

In the present study, carbon nanotubes have been successfully grown by microwave plasma-enhanced chemical vapor deposition (MPECVD), a well-established method [46, 47]. Figure 7(a) represents the as-grown carbon nanotubes on a substrate using optimized processing conditions such as temperature, gas flow, gas concentrations, and pressure. Aligned nanotubes, as seen in Figure 7(b), have been grown to ensure uniformity in the nanotubes' dimensions. Various seed materials have been investigated to grow carbon nanotubes and attempted to determine any effect on the hydrogen sorption capacities.

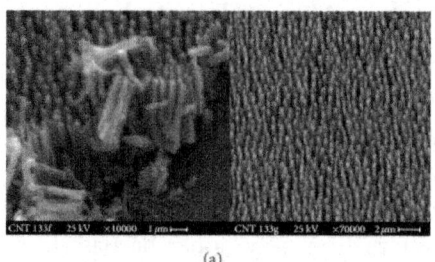 

(a)   (b)

Figure 7: SEM micrographs of (a) carbon nanotubes grown by MPECVD and (b) high density of aligned carbon nanotubes.

Fullerenes, on the other hand, a new form of carbon with close-caged molecular structure were first reported by Kroto et al. in 1985 [48]. It is a potential hydrogen storage material based on the ability to react with hydrogen via hydrogenation of carbon-carbon double bonds. The theory predicts that a maximum of 60 hydrogen atoms can be attached to both the inside (endohedrally) and outside (exohedrally) of the fullerene spherical surface. Thus, a stable $C_{60}H_{60}$ isomer can be formed with the theoretical hydrogen content of ~7.7 wt%. It seems that the fullerene hydride reaction is reversible at high temperatures. The 100% conversion of $C_{60}H_{60}$ indicates that 30 moles of H_2 gas will be released from each mole of fullerene hydride compound. However, this reaction is not possible because it requires high temperature, about 823–873 K [49]. Solid C_{60} has face-centered cubic lattice at room temperature and its density is ~1.69 g/sm^3. Molecules are freely rotating due to weak intermolecular interaction. Fullerene is an allotropic modification of carbon. Fullerene molecules are composed of pentagons and hexagons whose vertexes contain carbon atoms. Fullerene, C_{60}, is the smallest and the most stable structure (owing to high degree of its symmetry).

Hydrogen can be stored in glass microspheres of approximately 50 μm diameter. The microspheres can be filled with hydrogen by heating them to increase the glass permeability to hydrogen. At room temperature, a pressure of approximately 25 MPa is achieved resulting in storage density of 14% mass fraction and 10 kg H_2/m^3 [49]. At 62 MPa, a bed of glass microspheres can store 20 kg H_2/m^3. The release of hydrogen occurs by reheating the spheres to again increase the permeability.

Nanocomposite Conducting Polymers

Nanocomposite material consisting of a polyaniline matrix, which can be functionalized by either catalytic doping or incorporation of nanovariants, is considered to be a potential promise for hydrogen storage. It was reported that polyaniline could store as much as 6–8 wt% of hydrogen [50], which, however, another team of scientists could not reproduce [51]. Yet another recent study reveals that a hydrogen uptake of 1.4–1.7 wt% H_2 has been reported for the polymers of intrinsic microscopy [52]. Polyaniline is a conductive polymer, with conductivity on the order of 10^0 S/cm. This is higher than that of typical nonconducting polymers, but much lower than that of metals [53]. In addition to its conductivity, polyaniline emeraldine base (EB) is very simple and inexpensive to polymerize. It is because of this simplicity that it was chosen as a matrix material for the nanocomposite structure.

Figure 8 represents the hydrogen sorption kinetics of polyaniline nanospheres at room temperature. From this figure, it is discernible that the hydrogen uptake and release of ~4.0 wt.% occurs in the initial run. However, during the consecutive cycles, the hydrogen storage capacity and kinetics were decreased. The SEM microstructure of polyaniline nanospheres are shown in Figure 9(a). Uniform cluster sizes of 50–100 nm are widely distributed over the surface. The microstructures after hydrogen sorption cycles exhibit microchannels or microcrack formation (see Figure9(b)). This correlates very well with effective hydrogenation as observed from sorption kinetic profiles (see Figure 8). Further cyclic reversibility and associated mechanistic behavior for hydrogen uptake and release kinetics are still underway.

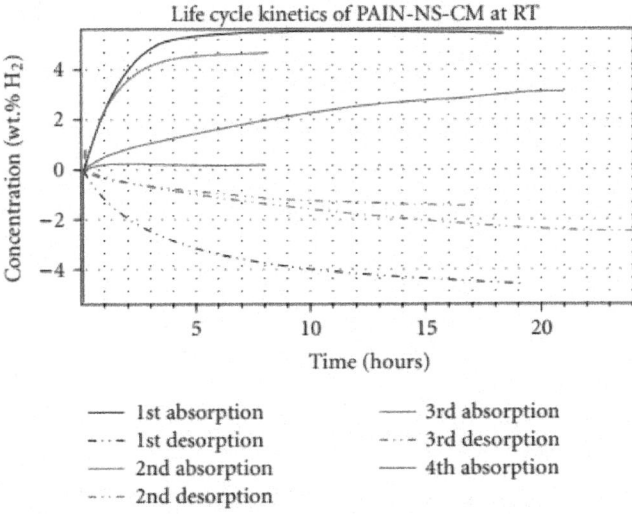

Figure 8: Hydrogen sorption kinetics of polyaniline nanospheres at room temperature showing good reversibility in the initial runs.

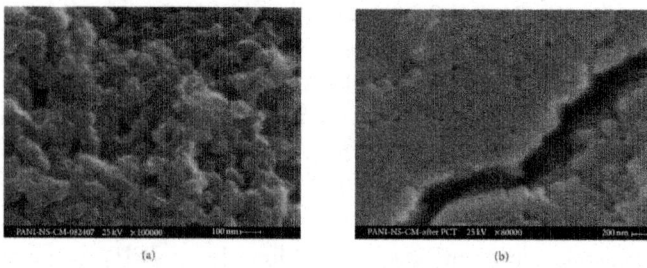

Figure 9: SEM images of polyaniline nanospheres (a) as-prepared and (b) after hydrogen sorption cycles.

Functionalization (see schematic Figure 10) has been carried out by the introduction of chemical groups into a polymer molecule or conversion of one chemical group to another group, which leads to a polymer with chemical, physical, or other functions. Functional polymers act as a catalyst to bind selectively to particular species, to capture and transport electric charge or energy, and to convert light into charge carriers and vice versa.

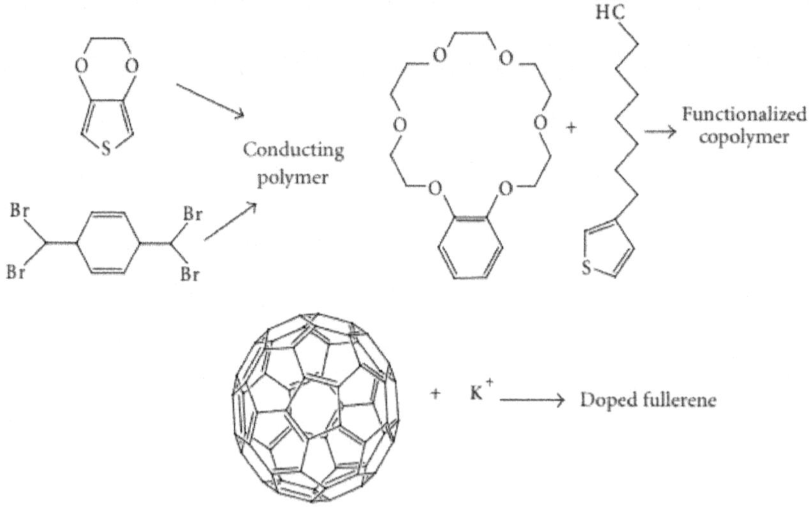

Figure 10: Schematics for the development of functionalized conducting polymer.

High-Surface Area Sorbents and New Materials Concepts

There is a pressing need for the discovery and development of new reversible materials. One new area that may be promising is that of high-surface area hydrogen sorbents based on microporous metal-organic frameworks (MOFs).

Such materials are synthetic, crystalline, and microporous and are composed of metal/oxide groups linked together by organic struts. Hydrogen storage capacity at 78 K ($-195°C$) has been reported as high as 4 wt% via an adsorptive mechanism, with a room temperature capacity of approximately 1 wt% [54]. However, due to the highly porous nature of these materials, volumetric capacity may still be a significant issue.

Another class of materials for hydrogen storage may be clathrates [15], which are primarily hydrogen-bonded H_2O frameworks. Initial studies have indicated that significant amounts of hydrogen molecules can be incorporated into the sII clathrate. Such materials may be particularly viable for off-board storage of hydrogen without the need for high pressure or liquid hydrogen tanks.

SUMMARY

Nanostructured materials such as nanotubes, nanofibers, and nanospheres show potential promise for hydrogen storage due to high-surface area, and they may offer several advantages for the physicochemical reactions, such as surface interactions, adsorption in addition to bulk absorption, rapid kinetics, low-temperature sorption, hydrogen atom dissociation, and molecular diffusion via the surface catalyst. The intrinsically large surface areas and unique adsorbing properties of nanophase materials can assist the dissociation of gaseous hydrogen, and the small volume of individual nanoparticles can produce short diffusion paths to the materials' interiors. The use of nanosized dopants enables a higher dispersion of the catalytically active species, and thus facilitates higher mass transfer reactions. Nanocomposites based on polymer matrix and functionalized carbon nanotubes possess unique microstructure for physisorption of hydrogen atom/molecule on the surface and inside the bulk. This review paper discussed briefly various nanomaterials for hydrogen storage and also presented hydrogen uptake and release characteristics for polyaniline nanospheres at room temperature.

ACKNOWLEDGMENTS

Financial support from US Department of Energy (Contract no. DE-FG36-04GO14224) and QuantumSphere Inc. is gratefully acknowledged. The authors also thank Dr. Rakesh Joshi for his comments and suggestions.

REFERENCES

1. S. Satyapal, J. Petrovic, and G. Thomas, "Gassing up with hydrogen," Scientific American, vol. 296, no. 4, pp. 80–87, 2007.

2. M. S. Dresselhaus and I. L. Thomas, "Alternative energy technologies," Nature, vol. 414, no. 6861, pp. 332–337, 2001. ·

3. B. Sakintuna, F. Lamari-Darkrim, and M. Hirscher, "Metal hydride materials for solid hydrogen storage: a review," International Journal of Hydrogen Energy, vol. 32, no. 9, pp. 1121–1140, 2007. ·

4. E. K. Stefanakos, D. Y. Goswami, S. S. Srinivasan, and J. T. Wolan, "Hydrogen energy," in Environmentally Conscious Alternative Energy Production, M. Kutz, Ed., vol. 4, chapter 7, pp. 165–206, John Wiley & Sons, New York, NY, USA, 2007. ·

5. S. A. Sherif, F. Barbir, T. N. Vieziroglu, M. Mahishi, and S. S. Srinivasan, "Hydrogen energy technologies," in Handbook of Energy Efficiency and Renewable Energy, F. Kreith and D. Y. Goswami, Eds., chapter 27, CRC Press, Boca Raton, Fla, USA, 2007.

6. E. Fontes and E. Nilsson, "Modeling the fuel cell," The Industrial Physicist, vol. 7, no. 4, p. 14, 2001.

7. R. H. Jones and G. J. Thomas, Materials for the Hydrogen Economy, CRC Press, Boca Raton, Fla, USA, 2007, Catalog no. 5024.

8. Report of the Basic Energy Science Workshop on Hydrogen Production, Storage and use prepared by Argonne National Laboratory, May 2003.

9. L. Schlapbach, "Hydrogen as a fuel and its storage for mobility and transport,"MRS Bulletin, vol. 27, no. 9, pp. 675–676, 2002.

10. C. Read, G. Thomas, C. Ordaz, and S. Satyapal, "U.S. Department of Energy's system targets for on-board vehicular hydrogen storage," Material Matters, vol. 2, no. 2, p. 3, 2007.

11. Züttel, "Materials for hydrogen storage," Materials Today, vol. 6, no. 9, pp. 24–33, 2003. ·

12. D. Chandra, J. J. Reilly, and R. Chellappa, "Metal hydrides for vehicular applications: the state of the art," JOM, vol. 58, no. 2, pp. 26–32, 2006. ·

13. M. Seayad and D. M. Antonell, "Recent advances in hydrogen storage in metal-containing inorganic nanostructures and related materials," Advanced Materials, vol. 16, no. 9-10, pp. 765–777, 2004. ·

14. F. E. Pinkerton and B. G. Wicke, "Bottling the hydrogen genie," The Industrial Physicist, vol. 10, no. 1, pp. 20–23, 2004.

15. F. Schüth, "Technology: hydrogen and hydrates," Nature, vol. 434, no. 7034, pp. 712–713, 2005. ·

16. F. Schüth, B. Bogdanović, and M. Felderhoff, "Light metal hydrides and complex hydrides for hydrogen storage," Chemical Communications, vol. 10, no. 20, pp. 2249–2258, 2004. ·

17. N. B. McKeown, S. Makhseed, K. J. Msayib, L.-L. Ooi, M. Helliwell, and J. E. Warren, "A phthalocyanine clathrate of cubic symmetry containing interconnected solvent-filled voids of nanometer dimensions," Angewandte Chemie International Edition, vol. 44, no. 46, pp. 7546–7549, 2005. ·

18. M. Fichtner, "Nanotechnological aspects in materials for hydrogen storage,"Advanced Engineering Materials, vol. 7, no. 6, pp. 443–455, 2005. ·

19. G. Wong-Foy, A. J. Matzger, and O. M. Yaghi, "Exceptional H2 saturation uptake in microporous metal-organic frameworks," Journal of the American Chemical Society, vol. 128, no. 11, pp. 3494–3495, 2006. ·

20. V. Renugopalakrishnan, A. M. Kannan, S. S. Srinivasan, et al., "Nanomaterials for energy conversion applications," Journal of Nanoscience and Nanotechnology. In press.

21. E. G. Baburaj, F. H. Froes, V. Shutthanandan, and S. Thevuthasan, "Low cost synthesis of nanocrystalline titanium aluminides," Interfacial Chemistry and Engineering Annual Report, Pacific Northwest National Laboratory, Oak Ridge, Tenn, USA, 2000.

22. R. Schulz, S. Boily, L. Zaluski, A. Zaluka, P. Tessier, and J. O. Ström-Olsen, "Nanocrystalline materials for hydrogen storage," Innovation in Metallic Materials, pp. 529–535, 1995.

23. L. Schlapbach and A. Züttel, "Hydrogen-storage materials for mobile applications," Nature, vol. 414, no. 6861, pp. 353–358, 2001.

24. W. Grochala and P. P. Edwards, "Thermal decomposition of the non-interstitial hydrides for the storage and production of hydrogen,"Chemical Reviews, vol. 104, no. 3, pp. 1283–1316, 2004. ·

25. B. Bogdanović and M. Schwickardi, "Ti-doped alkali metal aluminium hydrides as potential novel reversible hydrogen storage materials,"Journal of Alloys and Compounds, vol. 253-254, pp. 1–9, 1997. ·

26. C. M. Jensen and R. A. Zidan, "Hydrogen storage materials and method of making by dry homogenation," US patent 6471935, 2002.

27. P. Chen, Z. Xiong, J. Luo, J. Lin, and K. L. Tan, "Interaction of hydrogen with metal nitrides and imides," Nature, vol. 420, no. 6913, pp. 302–304, 2002.·

28. Y. H. Hu and E. Ruckenstein, "H2 storage in Li3N. Temperature-programmed hydrogenation and dehydrogenation," Industrial and Engineering Chemistry Research, vol. 42, no. 21, pp. 5135–5139, 2003. ·

29. J. J. Vajo, S. L. Skeith, and F. Mertens, "Reversible storage of hydrogen in destabilized LiBH4," Journal of Physical Chemistry B, vol. 109, no. 9, pp. 3719–3722, 2005. ·

30. M. Au, "Destabilized and catalyzed alkali metal borohydrides for hydrogen storage with good reversibility," US patent Appl. Publ 0194695 A1, 2006.

31. S. S. Srinivasan, D. Escobar, M. Jurczyk, Y. Goswami, and E. K. Stefanakos, "Nanocatalyst doping of Zn(BH4)2 for on-board hydrogen storage," Journal of Alloys and Compounds, vol. 462, no. 1-2, pp. 294–302, 2008.

32. J. Yang, A. Sudik, D. J. Siegel, et al., "Hydrogen storage properties of 2LiNH2 +LiBH4 + MgH2," Journal of Alloys and Compounds, vol. 446-447, pp. 345–349, 2007. ·

33. G. J. Lewis, J. W. A. Sachtler, J. J. Low, et al., "High throughput screening of the ternary LiNH2-MgH2-LiBH4 phase diagram," Journal of Alloys and Compounds, vol. 446-447, pp. 355–359, 2007. ·

34. Züttel, "Hydrogen storage methods," Naturwissenschaften, vol. 91, no. 4, pp. 157–172, 2004. ·

35. S. H. Joo, S. J. Choi, I. Oh, et al., "Ordered nanoporous arrays of carbon supporting high dispersions of platinum nanoparticles," Nature, vol. 412, no. 6843, pp. 169–172, 2001. ·

36. Zaluska, L. Zaluski, and J. O. Ström-Olsen, "Structure, catalysis and atomic reactions on the nano-scale: a systematic approach to metal hydrides for hydrogen storage," Applied Physics A, vol. 72, no. 2, pp. 157–165, 2001. ·

37. K.-J. Jeon, A. Theodore, C.-Y. Wu, and M. Cai, "Hydrogen absorption/ desorption kinetics of magnesium nano-nickel composites synthesized by dry particle coating technique," International Journal of Hydrogen Energy, vol. 32, no. 12, pp. 1860–1868, 2007. ·

38. X. Yao, C. Z. Wu, H. Wang, H. M. Cheng, and G. Q. Lu, "Effects of carbon nanotubes and metal catalysts on hydrogen storage in magnesium nanocomposites," Journal of Nanoscience and Nanotechnology, vol. 6, no. 2, pp. 494–498, 2006. ·

39. C. Z. Wu, P. Wang, X. Yao, et al., "Hydrogen storage properties of MgH2/SWNT composite prepared by ball milling," Journal of Alloys and Compounds, vol. 420, no. 1-2, pp. 278–282, 2006. ·

40. S. R. Johnson, P. A. Anderson, P. P. Edwards, et al., "Chemical activation ofMgH2; a new route to superior hydrogen storage materials," Chemical Communications, no. 22, pp. 2823–2825, 2005. ·

41. W. Luo, "(LiNH2-MgH2): a viable hydrogen storage system," Journal of Alloys and Compounds, vol. 381, no. 1-2, pp. 284–287, 2004. ·

42. J. Huot, G. Liang, S. Boily, A. Van Neste, and R. Schulz, "Structural study and hydrogen sorption kinetics of ball-milled magnesium hydride," Journal of Alloys and Compounds, vol. 293–295, pp. 495–500, 1999. ·

43. P. Sudan, A. Züttel, Ph. Mauron, Ch. Emmenegger, P. Wenger, and L. Schlapbach, "Physisorption of hydrogen in single-walled carbon nanotubes,"carbon, vol. 41, no. 2, pp. 2377–2383, 2003.

44. B. Viswanathan, M. Sankaran, and M. A. Schibioh, "Carbon nanomaterials: are they appropriate candidates for hydrogen storage?" Bulletin of the Catalysis Society of India, vol. 2, pp. 12–32, 2003.

45. M. G. Nijkamp, J. E. M. J. Raaymakers, A. J. van Dillen, and K. P. de Jong, "Hydrogen storage using physisorption—materials demands," Applied Physics A, vol. 72, no. 5, pp. 619–623, 2001. ·

46. C. Dillon, K. M. Jones, T. A. Bekkedahl, C. H. Kiang, D. S. Bethune, and M. J. Heben, "Storage of hydrogen in single-walled carbon nanotubes," Nature, vol. 386, no. 6623, pp. 377–379, 1997. ·

47. P. M. F. J. Costa, K. S. Coleman, and M. L. H. Green, "Influence of catalyst metal particles on the hydrogen sorption of single-walled carbon nanotube materials," Nanotechnology, vol. 16, no. 4, pp. 512–517, 2005. ·

48. H. W. Kroto, J. R. Heath, S. C. O'Brien, R. F. Curl, and R. E. Smalley, "C60: buckminsterfullerene," Nature, vol. 318, no. 6042, pp. 162–163, 1985. ·

49. Y. Kojima and Y. Kawai, "IR characterizations of lithium imide and amide,"Journal of Alloys and Compounds, vol. 395, no. 1-2, pp. 236–239, 2005. ·

50. Y.-Y. Fan, A. Kaufmann, A. Mukasyan, and A. Varma, "Single- and multi-wall carbon nanotubes produced using the floating catalyst method: synthesis, purification and hydrogen up-take," Carbon, vol. 44, no. 11, pp. 2160–2170, 2006. ·

51. S. J. Cho, K. S. Song, J. W. Kim, T. H. Kim, and K. Choo, "Hydrogen sorption in HCl-treated polyaniline and polypyrrole: new potential hydrogen storage media," Fuel Chemistry Division Preprints, vol. 47, no. 2, pp. 790–791, 2002.

52. B. Panella, L. Kossykh, U. Dettlaff-Weglikowska, M. Hirscher, G. Zerbi, and S. Roth, "Volumetric measurement of hydrogen storage in HCl-treated polyaniline and polypyrrole," Synthetic Metals, vol. 151, no. 3, pp. 208–210, 2005. ·

53. N. B. McKeown, B. Gahnem, K. J. Msayib, et al., "Towards polymer-based hydrogen storage materials: engineering ultramicroporous cavities within polymers of intrinsic microporosity," Angewandte Chemie International Edition, vol. 45, no. 11, pp. 1804–1807, 2006. ·

54. J. L. C. Rowsell, E. C. Spencer, J. Eckert, J. A. K. Howard, and O. M. Yaghi, "Gas adsorption sites in a large-pore metal-organic framework," Science, vol. 309, no. 5739, pp. 1350–1354, 2005. ·

Chapter 9

A REVIEW ON CARBON NANOTUBES IN AN ENVIRONMENTAL PROTECTION AND GREEN ENGINEERING PERSPECTIVE

Yit Thai Ong, Abdul Latif Ahmad, Sharif Hussein Sharif Zein and Soon Huat Tan

School of Chemical Engineering, Engineering Campus,, Universiti Sains Malaysia, Seri Ampangan, 14300, Nibong Tebal, SPS, Pulau Pinang, Malaysia.

ABSTRACT

Recent developments in nanotechnologies have helped to benchmark carbon nanotubes (CNTs) as one of the most studied nanomaterials. By taking advantages of CNTs extraordinary physical, chemical and electronic properties, a wide variety of applications has been proposed in various engineering fields. In this short review, the contribution of CNTs is addressed in terms of sustainable environment and green technologies perspective, such as waste water treatment, air pollution monitoring, biotechnologies, renewable energy technologies, supercapacitors and green nanocomposites. Consideration of CNTs for large scale application from the aspect of cost and potential hazards are also discussed. Based on the literature studied, CNTs pose a great potential as a promising material for application in various environmental fields.

INTRODUCTION

Since the discovery of carbon nanotubes (CNTs), they have eventually revolutionized the future nanotechnologies area. CNTs as reported by Iijima (1991) and Bethune et al. (1993), are seamless cylinder-shaped macromolecules with a radius as small as a few nanometers, and up to several micrometers in length. The walls of these tubes are constructed of a hexagonal lattice of carbon atoms and capped by fullerene-like structures. The unique structure of CNTs can be divided mainly into multi-walled carbon nanotubes (MWCNTs) and single-walled carbon nanotubes (SWCNTs). MWCNTs are composed of two or more concentric cylindrical shells of graphene sheets coaxially arranged around a central hollow area with spacing between the layers. In contrast, SWCNTs are made of a single cylinder graphite sheet held together by van der Waals bonds (Balasubramanian and Burghard, 2005; Daniel et al., 2007). Current synthesis techniques including electric arc discharge (Journet et al., 1997), laser ablation (Guo et al., 1995) and

(Dai et al., 1996) are used commercially to produce large quantities of CNTs.

CNTs mutable hybridization states and sensitivity of the structure to perturbations in synthesis conditions exploit their unique physical, chemical and electronic properties (as stated in Table 1) which inspire innovation in new technologies and applications. Moreover, these unique and tunable properties offer potential advances in environmental systems from proactive (prevention of environmental degradation, optimizing energy efficiency) to retroactive (waste water reuse, pollutant transformation) (Mauter and Elimelech, 2008).

In this short review, current applications of CNTs in waste water treatment, air pollution monitoring, biotechnology, renewable energy and supercapacitors are explored and a proposal for green nanocomposite design that embraces the 3R (reduce, reuse and recycle) concept has also been discussed. At the end, consideration of CNTs in large scale applications is surveyed from the aspect of cost and potential hazards.

Carbon Nanotubes in Waste Water Treatment

Waste water discharge from domestic, industrial or agricultural sources encompasses a wide range of contaminants and has drawn major concern worldwide since they adversely affect the quality of water. The contaminants found in waste water, such as heavy metal ions (Li et al., 2002; Lu and Chiu, 2006; Türker, 2007; Hsieh and Horng, 2007; Rao et al., 2007; Liu et al., 2008; Lu and Chiu, 2008; Lu et al., 2008; Stafiej and Pyrzynska, 2008; Xu et al., 2008; Gao et al., 2009), 1,2-dichlorobenzene (Lin et al., 2002; Peng et al., 2003) and dioxin (Long and Yang, 2001; Fagan et al., 2007) are non-degradable, highly toxic and carcinogenic and can result in accumulative poisoning, cancer and nervous system damage. Removal of these contaminants relies on the sorption behavior of a sorbent. CNTs, with their high surface active site to volume ratio and controlled pore size distribution, have an exceptional sorption capability and high sorption efficiency compared to conventional granular and powder activated carbon, which have intrinsic limitations like surface active sites and the activation energy of sorption. Extensively studies found that the adsorption capacity of CNTs depends on both the surface functional groups and the nature of the sorbate. For instance, the amounts of surface acidity (carboxylic, lactonic and phenolic groups) favor the adsorption of polar compounds (Rao et al., 2007). On the other hand, the unfunctionalized CNTs surface is proved to have higher adsorption capacity towards non-polar compounds such as polycyclic aromatic hydrocarbons (Yang et al., 2006b; Wang et al., 2008). The sorption behaviors of CNTs mainly involve chemical interaction for polar compounds and physical interaction for non-polar compounds. The sorption of both polar and non-polar compounds is normally fitted with Langmuir or Freundlich isotherms (Long and Yang, 2001; Li et al., 2002; Peng et al., 2003; Lu and Chiu, 2006; Yang et al., 2006b; Lu et al., 2008; Xu et al., 2008).

Sorption capacity of CNTs is effective over a broad pH range. Particularly, optimum performance was reported in the pH range of 7 to 10 (Rao et al., 2007). Other than this pH range, ionization and competition between ionic species could occur (Boehm, 2002; Weng and Huang, 2004; Lu and Chiu, 2006; Stafiej and Pyrzynska, 2008). Although CNTs are more expensive compare to conventional activated carbon, their sorption and desorption cycles are more efficient than conventional activated carbon. Sorption/desorption studies have shown the availability performance of CNTs under a numbers of sorption and desorption cycles (Rao et al., 2007; Lu et al., 2008). A regeneration study performed by Lu et al. (2008) reported that the adsorption and desorption of Ni^{2+} in CNTs slightly decreased, but those of granular activated carbon (GAC) sharply decreased after a number of cycles. This phenomenon could be explained by the fact that the porous structure of GAC make desorption of Ni^{2+} more difficult as the ions have to move from the inner surface to the external surface of the pores.

In addition to serving as sorbent for organic and inorganic contaminants, current technology has used CNTs as nanofilters to reduce particle concentrations in waste water (Srivastava et al., 2004; Jin et al., 2007; Tahaikt et al., 2007). Similar to sorbents, specific selectivity on CNTs filters can be manipulated through the attachment of different functionalities at the pore entrances (Fornasiero et al., 2008). Despite their hydrophobic characteristics, CNTs have shown an extraordinary performance in transporting water. Molecular dynamics simulations indicate that the hydrophobic nature of CNTs pores creates weak interactions with water molecules, thus enabling a fast and nearly frictionless flow of water (Noy et al., 2007). Another explanation from Hummer et al. (2001) is that the frictionless flow of water is attributed to the nanoscale confinement that leads to narrowing of the interaction energy distribution, and minimizes the interaction with water. Apart from that, recent filtration studies using CNTs have also revealed the capability of CNT nanofilters to remove pathogenic microorganisms such as protozoa, bacteria and viruses in waste water treatment, with microorganisms being retained on the surface of CNT based on a depth-filtration mechanism (Bohonak and Zydney, 2005; Mostafavi et al., 2009). Brady-Estévez et al. (2008) reported an effective way to remove *E. coli* bacteria at low pressure using SWCNT filters. As covering on a poly(vinylidene fluoride) (PVDF) based microporous membrane, nanotube bundles inside the SWCNTs were able to completely capture and retain *E. coli* cells. A modification through immobilization of SWCNTs on a microporous ceramic filter was also proposed to further enhance robustness, reusability and thermal resistance of the filter without sacrificing its performance. In another study, Mostafavi et al. (2009) synthesized a controllable nanoscale porosity CNT-based filter by using a spray pyrolysis method and observed a maximum efficiency in removal of MS2 virus at a pressure of 8-11 bar.

Application of CNTs in waste water treatment is not limited to filtration and sorbent; several researchers observed strong antimicrobial properties of CNTs. Such behavior allows CNTs to replace chemical disinfectants as a new effective way to control microbial pathogens (Savage and Diallo, 2005; Kang et al., 2007; Li et al., 2008a; Nepal et al., 2008; Cortes et al., 2009). Applying CNTs in water disinfection treatment avoids the formation of harmful disinfection byproducts (DBPs) such as trihalomethanes, haloacetic acids and aldehydes, because they are not strong oxidants and are relatively inert in water. In order to facilitate their dispersion, surfactant or polymers like sodium dodecyl benzenesulfonate, polyvinylpyrolidone or Triton-X are generally used. Highly purified CNTs exhibit strong antimicrobial activity toward Gram positive and Gram negative bacteria, as well as bacterial spores. The activities inflicted by the antimicrobial property can be attributed to impairment of pathogen cellular function by destruction of major constituents (e.g., cell wall), interference with the pathogen cellular metabolic processes, and inhibition of pathogen growth by blockage of the synthesis of key cellular constituents (e.g., DNA, coenzymes and cell wall proteins). Kang et al. (2007) confirmed the strong antimicrobial activities of SWCNT; direct contact of *E. coli* cell with SWCNTs leads to severe membrane damage and subsequent cell inactivation. Some studies have also proposed CNTs as scaffolding for antimicrobial agents like Ag nanoparticles (Morones et al., 2005; Yuan et al., 2008) and antimicrobial lysozyme (Nepal et al., 2008) due to their excellent mechanical properties.

Carbon Nanotubes in Air Pollution

The outstanding electrical, electrochemical and optical properties of CNTs aroused the interest of researchers to explore the potential applications of

CNTs as sensing elements to detect and monitor the concentration of toxic gases released in the environment (Wei et al., 2006; Van Hieu et al., 2008; Bondavalli et al., 2009; Di Francia et al., 2009; Lu et al., 2009; Penza et al., 2009a; Penza et al., 2009b; Zhang and Zhang, 2009). CNTs possess an unique and tunable electronic properties whereby their metallic or semi-conductivity is greatly influenced by their one dimensional cylindrical structure, such as size and chirality. CNT-based gas sensors offer a number of advantages over conventional metal oxide semi-conductor gas sensors including low power consumption, low operating temperature, and high sensitivity (Endo et al., 2008). An example of a CNT-based gas sensor is shown in Fig. 1. A thin film array of CNTs acts as a cathode and is separated from an aluminum anode by a 180 micron-thick glass insulator. The individual CNTs create a high electric field near their ultra-fine tips and increase the overall field to speed up the gas ionization process (Bogue, 2004). The detection by this gas sensor is based on the changes of resistance or conductance in CNTs as a result of direct contact with gas.

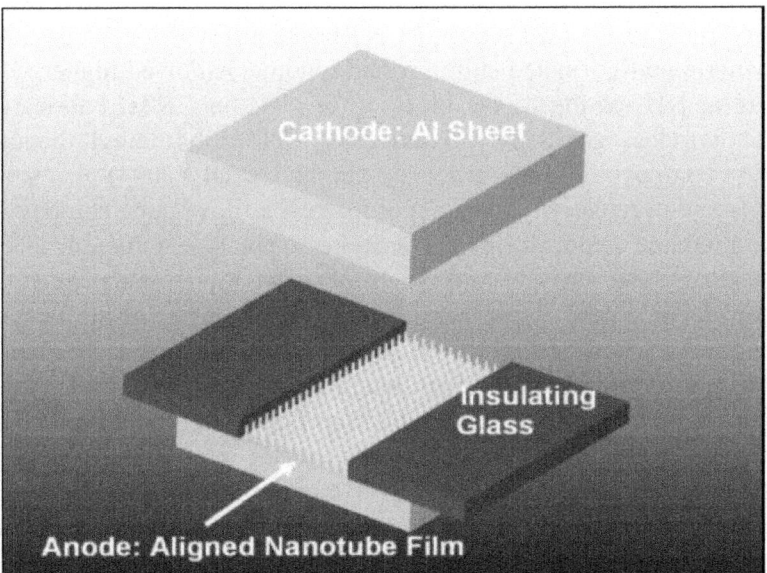

Figure 1: Schematic of a CNT-based gas sensor. Reprinted from Sensor Review, Vol. 24, Bogue, R. W., Nanotechnology: what are the prospects for sensors, Pages No. 253-260, (2004) with permission from © Emerald Group Publishing Limited all rights reserved.

CNT-based gas sensors have been used in several researches for detection of nitrogen oxides (NOx) (Ueda et al., 2008a; Ueda et al., 2008b), nitrogen dioxide (NO_2) (Kong et al., 2000; Cantalini et al., 2003; Valentini et al., 2004; Cho et al., 2006; Moon et al., 2008), ammonia (NH_3) (Nguyen and Huh, 2006; Quang et al., 2006; Nguyen et al., 2007), and sulfur dioxide (SO_2) (Suehiro et al., 2005) at room temperature. Although the results showed high and prompt response from the gas sensors, the time consuming recovery represents a challenge. In order to improve gas desorption from the sensor, several strategies have been proposed. For instance, heating the sensor by using ultraviolet (UV) illumination and increasing the flux rate of purging gas can improve gas desorption from the sensor. Efforts have also been made to improve the sensitivity and affinity of CNT-based gas sensors through polymer functionalization (Wei et al., 2006; Lu et al., 2009). Incorporation of CNTs in conducting polymers such as polyaniline and polypyrrole leads to an increase in sensitivity of the sensor toward selected gases or vapors. A study of polymer coating by Qi et al. (2003) reported that a CNT-based gas sensor coated with polyethyleneimine (PEI) showed high affinity for NO_2 detection at concentrations less than 1 ppb without interference from NH_3 due to the low binding affinity and sticking coefficient of NH_3 on the electron-rich CNTs. Contrary to the PEI coated sensor, a CNT-based gas sensor coated with Nafion (a polymeric perfluorinated sulfonic acid ionomer) allowed higher selectivity for sensing NH_3 owing to the blocking of NO_2 on CNTs. Fabrication of a compact wireless gas sensor based on a CNT/poly(methyl methacrylate) (PMMA) composite chemiresistor by Abraham et al. (2004) also showed an improvement in sensitivity upon exposure to volatile organic compounds likes dichloromethane, chloroform and acetone vapor. Even though CNT-based gas sensors exhibit strong competition for conventional metal oxide sensors, non-stop developments have been carried out to improve the conventional gas sensor. An incorporation of CNTs in metal oxide sensors has been found to overcome the limitation of high operating temperature (Wei et al., 2004; Espinosa et al., 2007; Hoa et al., 2007; Van Duy et al., 2008; Van Hieu et al., 2008; Wang et al., 2008). Moreover, significant improvement in sensitivity and response time was observed in CNT/metal oxide sensors for detection of pollutant gasses such as NO_2, and NH_3 at room temperature. The enhanced performance was attributed to the effective accessing of nanopassages by the gas and the variation in conductance upon adsorption of the gas provided by CNTs (Hoa et al., 2007; Gong et al., 2008).

Carbon Nanotubes in Biotechnology

Owing to the increasing demand for innovative, environmental friendly technologies, there has been a rapid growth in biotechnology in recent years in which living organism are utilized to make products or processes for specific uses. The emergence of biotechnology provides an opportunity for the participation of CNTs, especially in biological fuel cells (biofuel cells). Biofuel cells, as defined by Palmore and Whitesides (1994), are fuel cells that rely on biocatalytic activity to generate electricity. Generally, they can be classified as microbial fuel cells (MFC) or enzymatic biofuel cells (EFC).

MFC exploit microbial catabolic activities to generate electric power. They have been considered as future options in waste water treatment since a variety of materials, including complex organic waste and renewable biomass in waste water, can be used as source materials (Logan et al., 2006; Watanabe, 2008). However, the technology is not feasible practically, primarily due to low performance and lack of technical maturity. Intensive work has been focused on employing and modifying CNTs as electrodes to increase power production in MFC because of their high conductivity and large surface area (Morozan et al., 2007; Qiao et al., 2007; Sharma et al., 2008; Tsai et al., 2009). Tsai et al. (2009) prepared a new type of electrode architecture by coating CNTs onto a carbon cloth to form a highly conductive electrode with high specific surface-area in MFC. The presence of CNTs resulted in an improvement of 250% in the power density compared with a non-CNT coated electrode. Sharma et al. (2008) compared the performance of MFC based on the CNT electrode against a plain graphite electrode; the CNT electrode showed an increment of approximately 6-fold in power density compared to the graphite electrode. Further optimization of the power density can be obtained by utilizing the biocompatibility of CNTs with microorganisms in MFC. Although CNTs were claimed to exhibit antimicrobial capability as discussed previously, the cytotoxicity behavior decreases after their modification and functionalization (Sayes et al., 2006). Investigation of the biocompatibility of *Staphylococcus aureus* with CNTs by Morozan et al. (2007) showed rapid microorganism growth in CNT-modified cell culture media. This points to the potential of reducing the loss of power generation if applied to MFC anodic design.

On the other hand, instead of microbial catalytic activities, EFC utilize enzyme or protein catalysis to convert chemical energy into electrical energy. So far, the application of EFC as power source for low power sensors, communication devices and medical implants has been hampered by their short lifetime, poor enzyme stability and low power density (Kim et al., 2006; Minteer et al., 2007). Nonetheless, the introduction of CNTs as bioelectrodes constitutes a major breakthrough in EFC. Other than mediating charge transfer, CNTs have manifested great efficiency in EFC by supplying a strong platform for enzyme immobilization (Fischback et al., 2006; Asuri et al., 2007; Li et al., 2008b; Zhao et al., 2009; Zhou et al., 2009). CNTs allowed enzyme molecules to be covalently attached onto their surface and promote high enzyme loading by permitting crosslinked enzyme aggregate coating (Fischback et al., 2006). In addition, the nanoscale environment provided by CNTs could enhance enzyme stability. Enzymes covalently attached onto CNTs were found to have a high degree of stability due to their covalent linkage, which afforded significant resistance against denaturation (Govardhan, 1999; Asuri et al., 2007; Sheldon, 2007). In another point of view, the curvature of CNTs increases the distance between enzyme molecules, thereby reducing detrimental interaction between the enzymes and leading to an increased enzyme stability (Asuri et al., 2006). The enzyme stability in EFC is important as it offers an excellent operational stability, which is anticipated to pave the way for increasing the power density and prolonging the lifetime of EFC. For instance, the stabilized activity of glucose oxidase coated on CNTs enabled the continuous operation of an EFC for more than 16 hours (Fischback et al., 2006).

Carbon Nanotubes in Renewable Energy

Worldwide consumption of marketed energy is anticipated to increase by 57% between 2004 and 2030 (International Energy Outlook, 2007). This phenomenon foresees the requirement for advance renewable energy source technologies in order to meet the long term energy demand challenge and protect the environmental balance.

The major breakthrough contributed by CNTs in the solar energy sector lies in their application in photovoltaic devices. Photovoltaic devices generate electricity through conversion of photons absorbed from the sun. To date, several drawbacks, including high cost and low stability under illumination, have been found for commercially available silicon and semiconductor-based photovoltaic devices (Aberle, 2000; Green, 2002). Therefore, CNTs are sought as an alternative material in various solar cell architectures, especially in silicon-based solar cells, organic solar cells, and dye-sensitized solar cells (as illustrated in Fig. 2), due to their affordability and remarkable energy conversion. The nanoscale active surface area of CNTs also allows massive photon absorption for harvesting solar energy, while the presence of a delocalized π-electron system increases the mobility of the charge transfer (Kamat, 2007; Scarselli et al., 2009). Good alignment between CNTs can further enhance their photoconductivity upon illumination. (Liu et al., 2009b).

Silicon-based solar cells utilize the simplest p-n junction to separate electrons/holes and create current upon illumination. CNTs, when incorporated into silicon, serve as a heterojunction component for charge separation, as a highly conductive percolated network for charge transport, and as a transparent electrode for light illumination and charge collection (Khatri et al., 2009; Zhu et al., 2009). Therefore, modest cell efficiency with improvement in stability was observed in CNT/Si heterojunctions (Jia et al., 2008; Liang and Roth, 2008).

As a consequence of being flexible and having low production costs compared with silicon based solar cells, the development of organic solar cells has attracted a great deal of interest from researchers. Organic solar cells depend on a conductive organic polymer like poly(3-octylthiophene) (P3OT), poly(3-hexylthiophene) (P3HT) or [6,6]-phenyl-C_{61}-butyric acid methyl ester (PCBM) for light absorption and charge transfer (Janssen et al., 2005). Current research shows an improvement in efficiency upon incorporation of CNTs in the top electrode (Pasquier et al., 2005; Ulbricht et al., 2006), the photoactive layer (Landi et al.,2005) and the back electrode (Rowell et al., 2006; Ulbricht et al., 2007) of organic solar cells. In the photoactive layer, CNTs serve as photoactive material and optimize the performance of the cells by providing efficient hole or electron transport at the CNT/polymer interface. The photoactive component constructed with P3OT/CNT showed a higher open circuit voltage by taking advantage of the high electron transport capability of CNTs (Landi et al., 2005). In the top and back electrodes, CNTs manage to provide a large surface area for high optical transmittance and low sheet resistance to minimize power loss. Therefore, an increase in photocurrent was observed for the design of a top electrode composed of CNT films and Indium-tin oxide (ITO) (Pasquier et al., 2005; Ulbricht et al., 2006).

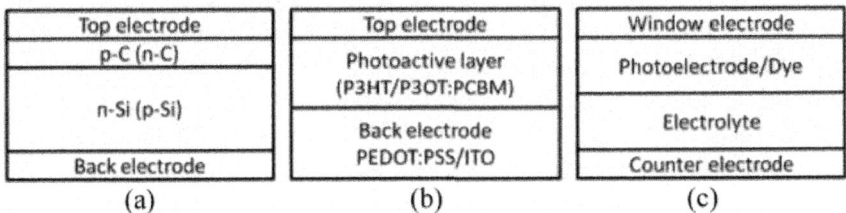

Top electrode	Top electrode	Window electrode
p-C (n-C)	Photoactive layer (P3HT/P3OT:PCBM)	Photoelectrode/Dye
n-Si (p-Si)	Back electrode PEDOT:PSS/ITO	Electrolyte
Back electrode		Counter electrode
(a)	(b)	(c)

Figure 2: Basic solar cell architectures: (a) silicon-based solar cell, (b) organic solar cell and (c) dyesensitized solar cell. Reprinted from Solar Energy Materials and Solar Cells, Vol. 93, Zhu, H., Wei, J., Wang, K. and Wu, D., Applications of carbon materials in photovoltaic solar cells, Pages No. 1461-1470, (2009), with permission from Elsevier.

Dye-sensitized solar cells (DSSC) have been hailed as the promising solar cell for their low cost and simple preparation (O'Regan and Grätzel, 1991; Grätzel, 2003). Electricity is produced in DSSC when semiconducting materials create an electron-hole pair and transfer the charge through a circuit to a counter electrode in contact with a redox couple in the electrolyte solution (Mills and Le Hunte, 1997). Applying CNTs as photoelectrode can help to increase the mobility of carrier transport upon exposure to visible light. However, direct use of CNTs as photoelectrodes could result in modest efficiency due to ultrafast recombination of photogenerated charge carriers. Hence, Brown et al. (2008) suggested CNT/TiO_2 as a strategy to improve charge separation and promote charge flow since TiO_2 particles present on CNTs are capable of injecting electrons from their excited state. Besides photoelectrodes, CNTs are also a popular choice of materials for DSSC counter electrode fabrication. In addition to enhance conversion efficiency, the cells with CNT counter electrode are expected to afford several advantages including nanoscale conducting channels, lightweight, and low cost, as well as improved mechanical properties and thermal stability. The use of a CNT counter electrode in an anthocyanin-sensitized cell showed an efficiency of 1.46%, which is the highest value ever reported compared to a cell using natural dye and platinum counter (Zhu et al., 2008).

Other than solar energy, recent developments in hydrogen storage media have focused on CNTs as one of the ongoing strategic research areas. Hydrogen, a relatively clean fuel compared to conventional fuel, has been considered to be an attractive approach for developing technologies of green energy. The United State Department of Energy (DOE) had targeted minimum hydrogen storage of 6.5wt% in order to meet the demand of commercial storage requirement (U.S. Department of Energy: Washington, 2007). From the investigation on strategies used to achieve hydrogen storage, CNTs have received an exceptional consideration as a potential storage material due to their affordability, recycling characteristics, low density, nanoscale pore size distribution, and reasonable chemical stability (Cheng et al., 2001; Tibbetts et al., 2001; Ritschel et al., 2002). Storage of hydrogen in CNTs mainly involves physiosorption. Theoretical studies conducted by Wang and Johnson (1999) and Dodziuk and Dolgonos (2002) showed that the amount of hydrogen adsorbed depends on the nature of the array and orientation of CNTs, with the adsorption of hydrogen being preferred at the outer surface of CNTs rather than the inner core.

CNTs, particularly SWCNTs, outperform activated carbon for their large bulk density, which enhances volumetric storage. Experimental results (Schur et al., 2002; Furuya et al., 2004; Kayiran et al., 2004; Zhou et al., 2004a; Banerjee and Puri, 2008) indicated that storage of hydrogen in CNTs was lower than 1wt% at ambient temperature and high pressure but reasonable storage of hydrogen could be achieved at higher pressure and lower temperature. In fact, hydrogen physiosorption alone was inadequate to meet the DOE specification; hence, research has focused in the direction of hybrid CNT/metal compounds to promote chemisorption. Transition or alkali metal-doped CNTs, with s-p-d hybridization served to reinforce the notable increase in hydrogen storage via a spill-over mechanism (Yang et al., 2006a; Zacharia et al., 2007). A nearly 30% increase in hydrogen storage capacity was reported for palladium and vanadium doped CNTs at 2 MPa under room temperature (Zacharia et al., 2005). Schaller et al. (2009) reinforced magnesium-nickel (Mg-23.5wt% Ni) with CNT by a powder metallurgy method and reported a hydrogen storage as high as 6.1wt%. Another research performed by Iyakutti et al. (2009) indicated that CNTs coated with aluminum hydride can bind up to four hydrogen molecules, leading to an increase of 8.3 wt% in hydrogen storage capacity. As an alternative to the metal doping methods, structural defects also appear as another potential approach to enhance chemisorptions of hydrogen in CNTs (Gayathri and Geetha, 2007; Chen and Huang, 2008; He and Pan, 2009; Wang et al., 2009). The existence of structural defects in CNTs can be anticipated with an increase in surface area and pore volume. Such structures definitely enhance the interaction between CNTs and hydrogen, which enables an increase in adsorption binding energy of up to 50%.

Unfortunately, until now, there is controversy, both experimentally and theoretically, surrounding claims that CNTs possess abnormal performance as hydrogen storage since the high storage capacity could not be reproduced by other researchers in the same field (Zhou et al., 2004b). So far, the storage capacity of CNTs still remains far from meeting the DOE target. More efforts are needed in order to prepare CNTs as the base materials for hydrogen storage technology.

Carbon Nanotubes in Supercapacitors

Electrochemical capacitors or supercapacitors have been considered as an alternative to replace traditional batteries given their miniature size, high power density, long cycle life and high energy density, with potential for reducing waste disposal to the environment. Supercapacitors, as illustrated in Fig. 3, are composed of high surface area activated capacitors that use a molecule-thin layer of electrolyte as dielectric (Capek, 2009).

Recent advancements in nanotechnology have proposed the application of CNTs as electrode material for the capacitor. Utilization of the large surface area of CNTs in the electrode couple with a thinner layer between the electrode and electrolyte enhances the ability of the capacitor to store higher energy densities. Furthermore, the use of vertically aligned CNTs with several atomic diameters in width can significantly increase the supercapacitor capacity and power density as a result of the dramatically increased surface area of the electrode (Hadjipaschalis et al., 2009). Although they possess high surface area, stability and strong mechanical properties, CNTs are not preferable to use alone as the electrode due to their low capacitance (Chou et al., 2008). Therefore, CNTs have been proposed as a substrate for high specific capacitance transition metal oxides such as manganese oxides (MnO_2) and ruthenium oxide, (RuO_2) (Simon and Gogotsi, 2008; Liu et al., 2009a; Yan et al., 2009). Capacitances of approximately 5000F have been reported with supercapacitors and energy densities up to 5Wh/kg, which is about 10-fold higher than conventional capacitors, with only 0.5Wh/kg (Hadjipaschalis et al., 2009).

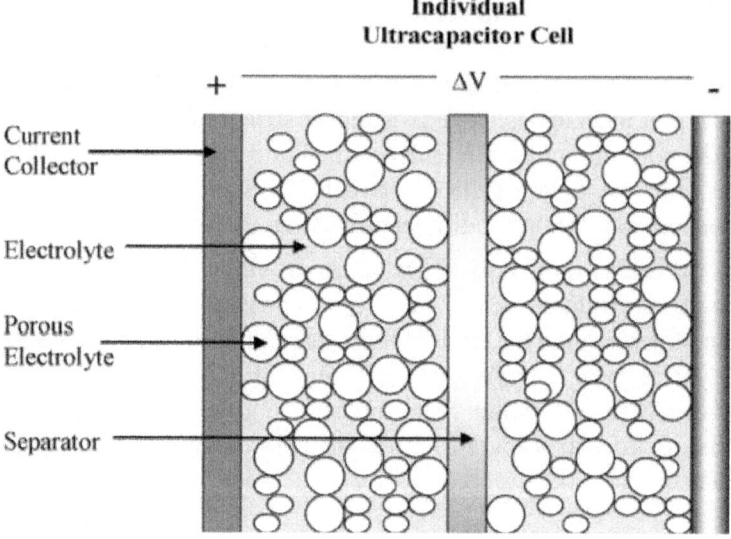

Figure 3: Schematic of a supercapacitor cell. Reprinted from Renewable and Sustainable Energy Reviews, Vol. 13, Hadjipaschalis, I., Poullikkas, A., and Efthimiou, V., Overview of current and future energy storage technologies for electric power applications., Pages No. 1513-1522, (2009), with permission from Elsevier.

Carbon Nanotubes in Green Nanocomposites Design

Waste generation is proportional to the world's economic growth. Wastes, especially synthetic polymer waste, cause negative impacts on the environment. Thus, as a solution to overcome this problem, the European Community has suggested a waste management concept based on two complementary strategies: avoiding waste by improving product design and increasing the recycling and re-use of waste with an emphasis on life-cycle assessment (LCA) to generate transparent and complete assessments of the environmental impact resulting from all stages of the life cycle of the product or activity in question and to use this to evaluate its environmental attributes (Baillie, 2004).

The future perspective point of view for solving waste disposal issues has driven the challenge to synthesize green nanocomposites by using biodegradable polymers as part of the wave of next generation materials (Mojumdar and Raki, 2005; Wang et al., 2005). The green nanocomposite trend employing natural renewable resources embraces the concept of LCA favoring recycling and reuse of waste. Biodegradable polymers have a great potential commercial value and have attracted a great deal of interest from researchers as an alternative to replace non-renewable petroleum-based polymers because of their degradability. However, most of the biodegradable polymers possess poorer mechanical properties and low heat distortion temperatures, which restrict their use in broad applications. Therefore, CNTs could act as nano-reinforcements for the biodegradable polymers in order to provide a suite of composite materials with improvement in mechanical properties, extended durability and better thermal stability. The quality of the biodegradable polymer/CNT nanocomposite is generally determined by the CNT alignment, the CNT-biodegradable polymer adhesion, and the CNT dispersion in the biodegradable polymer matrix (Grossiord et al., 2005; Ray et al., 2006; Vaudreuil et al., 2007).

Song and Qiu (2009) claimed that there was an increase in thermal stability of poly(butylene succinate) (PBSU) by about 10°C upon the incorporation with CNTs. Sitharaman et al. (2008) showed that ultra-short CNTs can significantly reinforce poly(propylene fumarate) (PPF), whose inferior mechanical properties often limit its usage in load bearing applications. Apart from this, CNT/PPF nanocomposites exhibit a comparable *in vivo*biocompatibility that creates the potential for application as a prototype bone tissue engineering scaffold.

Another advantage offered by the green nanocomposites is the ability to recycle the incorporated CNTs due to the degradability of the biodegradable polymer. Degradation of the biodegradable polymer can be achieved through either microbial degradation or enzymatic degradation under specific conditions of pH and temperature. After degradation, the recovered CNTs may act as reinforcement filler for producing new composites. The reuse and recycling process of CNTs could reduce the waste disposal and, at the same time, be cost effective for material processing.

Carbon Nanotubes in Large Scale Applications

CNTs have been hailed as promising materials to be applied in various environmental fields. Upon large scale application, there are several issues that must be overcome in order to make them feasible practically.

The high cost of synthesis is always one of the main issues blocking the application of CNTs on a larger scale. However, recent experimental studies have demonstrated the possibility of mass producing high quality CNTs at lower cost (Colomer et al., 2000; Dalton et al., 2004; Agboola et al., 2007). CNTs with a production rate of 595 kg/h were reported via decomposition of hydrocarbons by a catalytic chemical vapor deposition (CCVD) method in either a plug flow reactor or a fluidized bed reactor containing a solid catalyst, with an average cost between \$25 and \$38/kg (Agboola et al., 2007). Nanothinx, a spin-off company of the Institute of Chemical Engineering and High Temperature Chemical Processes (ICE-HT), was able to reduce the cost of CNT production by using a low cost novel catalyst developed by them. Studies carried out by Zhang et al. (2007) showed that, by using liquefied petroleum gas (\$400/ton) as carbon source material and the ceramic sphere as substrate, high purity CNTs can be easily produced in large scale at low cost. The possibility of mass production of CNTs has catalyzed the vision to commercialize and apply CNT technology in large scale applications.

Despite the high cost of synthesis, concerns regarding the potential implications of CNTs on the environment and on human health have drawn great attention as well. The widespread applications envisioned for CNTs may lead to their disposal into the environmental compartments of air and water. Through inhalation or via the food chain, disposed CNTs may enter into the human body and result in harmful effects. The harmful effects arise as a function of their physical dimension (Magrez et al., 2006; Wick et al., 2007), physical state (Wick et al., 2007), the presence of impurities (Wick et al., 2007) and chemical treatment (Dumortier et al., 2006; Magrez et al., 2006). *In vivo* studies of CNTs found that CNTs could induce pulmonary inflammation (Lam et al., 2004) and lung cellular proliferation (Muller et al., 2005) and inhibited the growth of heart muscle (CBC, 2007) in rats. The toxicity of well dispersed CNTs is less than that of agglomerated CNTs (Wick et al., 2007). Therefore, functionalized CNTs with hydrophilic groups possess less of a toxic impact on mammalian cell viability since their dispersion is improved (Dumortier et al., 2006). This behavior is opposite to what has been observed in bacterial cells, where dispersal increased toxicity (Kang et al., 2008). Up to now, there is still inadequate study regarding the impact of CNTs on the environmental and human health. However, the best way to avoid human and environmental exposure to CNTs is source reduction (Reijnders, 2006).

CONCLUSIONS

In this short review, potential proactive to retroactive applications of CNTs in environment systems have been discussed. As a result of theirs excellent mechanical, electrical, physical and chemical properties, CNTs play a major role in waste water treatment and air pollution monitoring. In waste water treatment, CNTs serve as sorbents, nanofilters and antimicrobial agents to remove organic and inorganic contaminants, as well as pathogenic microorganisms. In air pollution monitoring, development of CNT-based gas sensors results in high sensitivity with prompt sensor response toward pollutant gases. In addition, CNTs are also described as one of the key challenges in producing green energy, which involves clean combustion. The production of electricity by green energy technologies based on biomass catalytic activity (Biofuel cells), renewable sources such as solar (photovoltaic device) and hydrogen (hydrogen fuel cell) prevents the release of toxic gases to the atmosphere and precludes the high demand for fossil fuel. However, these technologies are still not yet been commercially practical because they are still in an emerging stage and more time is required to achieve technical maturation. Apart from this, the superior power density provided by CNT-based supercapacitors has been viewed as an alternative way to replace traditional batteries. Supercapacitors with extended usage time could reduce the amount of waste disposed to the environment over time. Crucial development is still ongoing in order to meet the long term environmental protection challenge. In green materials, the involvement of CNTs in green nanocomposite design embraces the LCA concept, which promotes reduction, reuse and recycle-ability of raw materials. For large scale application, continuous production of CNTs from low-cost sources has taken a closer step to overcoming the problem of the high cost of synthesis. Considering the health implications resulting from widespread use of CNTs, the best way to avoid human and environmental exposure to CNTs is source reduction.

ACKNOWLEDGEMENT

Financial supports from Universiti Sains Malaysia Research University Grant via Golden Goose project, Ministry of Science, Technology and Innovation (MOSTI), Fundamental Research Grant Scheme (FRGS) and USM Fellowship are gratefully acknowledged.

REFERENCES

1. Aberle, A. G., Surface passivation of crystalline silicon solar cells: a review. Progress in Photovoltaics: Research and Applications, 8, No. 5, 473 (2000).

2. Abraham, J. K., Philip, B., Witchurch, A., Varadan, V. K., Reddy, C., A compact wireless gas sensor using a carbon nanotube/PMMA thin film chemiresistor. Smart Materials and Structures, 13, No. 5, 1045 (2004).

3. Agboola, A. E., Pike, R. W., Hertwig, T. A. and Lou, H. H., Conceptual design of carbon nanotube processes. Clean Technologies and Environmental Policy, 9, No. 4, 289 (2007).

4. Asuri, P., Bale, S. S., Pangule, R. C., Shah, D. A., Kane, R. S. and Dordick, J. S., Structure, function, and stability of enzymes covalently attached to single-walled carbon nanotubes. Langmuir, 23, No. 24, 12318 (2007).

5. Asuri, P., Karajanagi, S. S., Yang, H., Yim, T. J., Kane, R. S. and Dordick, J. S., Increasing protein stability through control of the nanoscale environment. Langmuir, 22, No.13, 5833 (2006).

6. Baillie, C., Green Composites - Polymer Composites and the Environment. CRC Press, New York (2004).

7. Balasubramanian, K. and Burghard, M., Chemically functionalized carbon nanotubes. Small, 1, No. 2, 180 (2005).

8. Banerjee, S. and Puri, I. K., Enhancement in hydrogen storage in carbon nanotubes under modified conditions. Nanotechnology, 19, No. 15, 155702 (2008).

9. Bethune, D. S., Kiang, C. H., De Vries, M. S., Gorman, G., Savoy, R., Vazquez, J. and Beyers, R., Cobalt-catalysed growth of carbon nanotubes with single-atomic-layer walls. Nature, 363, No. 6430, 605 (1993).

10. Boehm, H. P., Surface oxides on carbon and their analysis: a critical assessment. Carbon, 40, No. 2, 145 (2002).

11. Bogue, R. W., Nanotechnology: What are the prospects for sensors?. Sensor Review, 24, No. 3, 253 (2004).

12. Bohonak, D. M. and Zydney, A. L., Compaction and permeability effects with virus filtration membranes. Journal of Membrane Science, 254, No. 1-2, 71 (2005).

13. Bondavalli, P., Legagneux, P. and Pribat, D., Carbon nanotubes based transistors as gas sensors: State of the art and critical review. Sensors and Actuators B: Chemical, 140, No. 1, 304 (2009).

14. Brady-Estévez, A. S., Kang, S. and Elimelech, M., A single-walled-carbon-nanotube filter for removal of viral and bacterial pathogens. Small, 4, No. 4, 481 (2008).

15. Brown, P., Takechi, K. and Kamat, P. V., Single-walled carbon nanotube scaffolds for dye-sensitized solar cells. Journal of Physical Chemistry C, 112, No. 12, 4776 (2008).

16. Cantalini, C., Valentini, L., Lozzi, L., Armentano, I., Kenny, J. M. and Santucci, S., NO_2 gas sensitivity of carbon nanotubes obtained by plasma enhanced chemical vapor deposition. Sensors and Actuators B: Chemical, 93, No. 1-3, 333 (2003).

17. Capek, I., Dispersions, novel nanomaterial sensors and nanoconjugates based on carbon nanotubes. Advances in Colloid and Interface Science, 150, No. 2, 63 (2009).

18. CBC, Bacteria thrive amid carbon nanotubes, study find, CBC News, Toronto, Canada (2007).

19. Chen, C. H. and Huang, C. C., Enhancement of hydrogen spillover onto carbon nanotubes with defect feature. Microporous and Mesoporous Materials, 109, No. 1-3, 549 (2008).

20. Cheng, H. M., Yang, Q. H. and Liu, C., Hydrogen storage in carbon nanotubes. Carbon, 39, No. 10, 1447 (2001).

21. Cho, W. S., Moon, S. I., Paek, K. K., Lee, Y. H., Park, J. H. and Ju, B. K., Patterned multiwall carbon nanotube films as materials of NO_2 gas sensors. Sensors and Actuators, B: Chemical, 119, No. 1 180 (2006).

22. Chou, S. L., Wang, J. Z., Chew, S. Y., Liu, H. K. and Dou, S. X., Electrodeposition of MnO_2 nanowires on carbon nanotube paper as free-standing. flexible electrode for supercapacitors. Electrochemistry Communications, 10, No. 11, 1724 (2008).

23. Colomer, J. F., Stephan, C., Lefrant, S., Van Tendeloo, G., Willems, I., Kónya, Z., Fonseca, A., Laurent, C. and Nagy, J. B., Large-scale synthesis of single-wall carbon nanotubes by catalytic chemical vapor deposition (CCVD) method. Chemical Physics Letters, 317, No. 1-2, 83 (2000).

24. Cortes, P., Deng, S. and Smith, G. B., The adsorption properties of bacillus atrophaeus spores on single-wall carbon nanotubes. Journal of Sensors 2009 (2009).

25. Dai, H., Rinzler, A. G., Nikolaev, P., Thess, A., Colbert, D. T. and Smalley, R. E., Single-wall nanotubes produced by metal-catalyzed disproportionation of carbon monoxide. Chemical Physics Letters, 260, No. 3-4, 471 (1996).

26. Dalton, A. B., Collins, S., Razal, J., Munoz, E., Ebron, V. H., Kim, B. G., Coleman, J. N., Ferraris, J. P. and Baughman, R. H., Continuous carbon nanotube composite fibers: Properties. potential applications, and problems, Journal of Materials Chemistry, 14, No. 1, 1 (2004).

27. Daniel, S., Rao, T. P., Rao, K. S., Rani, S. U., Naidu, G. R. K., Lee, H. Y. and Kawai, T., A review of DNA functionalized/grafted carbon nanotubes and their characterization. Sensors and Actuators, B: Chemical, 122, No. 2, 672 (2007).

28. Di Francia, G., Alfano, B. and La Ferrara, V., Conductometric gas nanosensors. Journal of Sensors 2009 (2009).

29. Dodziuk, H. and Dolgonos, G., Molecular modeling study of hydrogen storage in carbon nanotubes. Chemical Physics Letters, 356, No. 1-2, 79 (2002).

30. Dumortier, H., Lacotte, S., Pastorin, G., Marega, R., Wu, W., Bonifazi, D., Briand, J. P., Prato, M., Muller, S. and Bianco, A., Functionalized carbon nanotubes are non-cytotoxic and preserve the functionality of primary immune cells. Nano Letters, 6, No. 7, 1522 (2006).

31. Endo, M., Strano, M. and Ajayan, P., Potential Applications of Carbon Nanotubes. Carbon Nanotubes, p. 13 (2008).

32. Espinosa, E. H., Ionescu, R., Chambon, B., Bedis, G., Sotter, E., Bittencourt, C., Felten, A., Pireaux, J. J., Correig, X. and Llobet, E., Hybrid metal oxide and multiwall carbon nanotube films for low temperature gas sensing. Sensors and Actuators, B: Chemical, 127, No. 1, 137 (2007).

33. Fagan, S. B., Santos, E. J. G., Souza Filho, A. G., Mendes Filho, J. and Fazzio, A., Ab initio study of 2,3,7,8-tetrachlorinated dibenzo-p-dioxin adsorption on single wall carbon nanotubes. Chemical Physics Letters, 437, No. 1-3, 79 (2007).

34. Fischback, M. B., Jong, K. Y., Zhao, X., Wang, P., Hyun, G. P., Ho, N. C., Kim, J. and Ha, S., Miniature biofuel cells with improved stability under continuous operation. Electroanalysis, 18, No.19-20, 2016 (2006).

35. Fornasiero, F., Hyung, G. P., Holt, J. K., Stadermann, M., Grigoropoulos, C. P., Noy, A. and Bakajin, O., Ion exclusion by sub-2-nm carbon nanotube pores. Proceedings of the National Academy of Sciences of the United States of America, 105, No. 45, 17250 (2008).

36. Furuya, Y., Hashishin, T., Iwanaga, H., Motojima, S. and Hishikawa, Y., Interaction of hydrogen with carbon coils at low temperature. Carbon, 42, No. 2, 331 (2004).

37. Gao, Z., Bandosz, T. J., Zhao, Z., Han, M. and Qiu, J., Investigation of factors affecting adsorption of transition metals on oxidized carbon nanotubes. Journal of Hazardous Materials, 167, No. 1-3, 357 (2009).

38. Gayathri, V. and Geetha, R., Hydrogen adsorption in defected carbon nanotubes. Adsorption, 13, No. 1, 53 (2007).

39. Gong, J., Sun, J. and Chen, Q., Micromachined sol-gel carbon nanotube/ SnO_2 nanocomposite hydrogen sensor. Sensors and Actuators, B: Chemical, 130, No. 2, 829 (2008).

40. Govardhan, C. P., Crosslinking of enzymes for improved stability and performance. Current Opinion in Biotechnology, 10, No. 4, 331 (1999).

41. Grätzel, M., Dye-sensitized solar cells. Journal of Photochemistry and

Photobiology C: Photochemistry Reviews, 4, No. 2, 145 (2003).

42. Green, M. A., Third generation photovoltaics: Solar cells for 2020 and beyond. Physica E: Low-dimensional Systems and Nanostructures, 14, No. 1-2, 65 (2002).

43. Grossiord, N., Loos, J. and Koning, C. E., Strategies for dispersing carbon nanotubes in highly viscous polymers. Journal of Materials Chemistry, 15, No. 24, 2349 (2005).

44. Guo, T., Nikolaev, P., Thess, A., Colbert, D. T. and Smalley, R. E., Catalytic growth of single-walled manotubes by laser vaporization. Chemical Physics Letters, 243, No. 1-2, 49 (1995).

45. Hadjipaschalis, I., Poullikkas, A. and Efthimiou, V., Overview of current and future energy storage technologies for electric power applications. Renewable and Sustainable Energy Reviews, 13, No. 6-7, 1513 (2009).

46. He, H. Y. and Pan, B. C., Studies on structural defects in carbon nanotubes. Frontiers of Physics in China, 4, No. 3, 297 (2009).

47. Hoa, N. D., Quy, N. V., Cho, Y. S. and Kim, D., Nanocomposite of SWNTs and SnO_2 fabricated by soldering process for ammonia gas sensor application. Physica Status Solidi (A) Applications and Materials, 204, No. 6, 1820 (2007).

48. Hsieh, S. H. and Horng, J. J., Adsorption behavior of heavy metal ions by carbon nanotubes grown on microsized Al_2O_3 particles. Journal of University of Science and Technology Beijing, Mineral, Metallurgy, Material, 14, No. 1, 77 (2007).

49. Hummer, G., Rasaiah, J. C. and Noworyta, J. P., Water conduction through the hydrophobic channel of a carbon nanotube. Nature, 414, No. 6860, 188 (2001).

50. Iijima, S., Helical microtubules of graphitic carbon. Nature, 354, No. 6348, 56 (1991).

51. International Energy Outlook, U. S., DOE/EIA-0484: Department of Energy Washington, DC (2007).

52. Iyakutti, K., Kawazoe, Y., Rajarajeswari, M. and Surya, V. J., Aluminum hydride coated single-walled carbon nanotube as a hydrogen storage medium. International Journal of Hydrogen Energy, 34, No. 1, 370 (2009).

53. Janssen, R. A. J., Hummelen, J. C. and Sariciftci, N. S., Polymer-fullerene bulk heterojunction solar cells. MRS Bulletin, 30, No. 1, 33 (2005).

54. Jia, Y., Wei, J., Wang, K., Cao, A., Shu, Q., Gui, X., Zhu, Y., Zhuang, D., Zhang, G., Ma, B., Wang, L., Liu, W., Wang, Z., Luo, J. and Wu, D.,

Nanotube-silicon heterojunction solar cells. Advanced Materials, 20, No. 23, 4594 (2008).

55. Jin, S., Fallgren, P. H., Morris, J. M. and Chen, Q., Removal of bacteria and viruses from waters using layered double hydroxide nanocomposites. Science and Technology of Advanced Materials, 8, No. 1-2, 67 (2007).

56. Journet, C., Maser, W. K., Bernier, P., Loiseau, A., Lamy de la Chapelle, M., Lefrant, S., Deniard, P., Lee, R. and Fischer, J. E., Large-scale production of single-walled carbon nanotubes by the electric-arc technique. Nature, 388, No. 6644, 756 (1997).

57. Kamat, P. V., Meeting the clean energy demand: Nanostructure architectures for solar energy conversion. Journal of Physical Chemistry C, 111, No. 7, 2834 (2007).

58. Kang, S., Herzberg, M., Rodrigues, D. F. and Elimelech, M., Antibacterial effects of carbon nanotubes: Size does matter. Langmuir, 24, No. 13, 6409 (2008).

59. Kang, S., Pinault, M., Pfefferle, L. D. and Elimelech M., Single-walled carbon nanotubes exhibit strong antimicrobial activity. Langmuir, 23, No. 17, 8670 (2007).

60. Kayiran, S. B., Lamari, F. D. and Levesque, D., Adsorption properties and structural characterization of activated carbons and nanocarbons. Journal of Physical Chemistry B, 108, No. 39, 15211 (2004).

61. Khatri, I., Adhikari, S., Aryal, H. R., Soga, T., Jimbo, T. and Umeno, M., Improving photovoltaic properties by incorporating both single walled carbon nanotubes and functionalized multiwalled carbon nanotubes. Applied Physics Letters, 94, No. 9 (2009).

62. Kim, J., Jia, H. and Wang, P., Challenges in biocatalysis for enzyme-based biofuel cells. Biotechnology Advances, 24, No. 3, 296 (2006).

63. Kong, J., Franklin, N. R., Zhou, C., Chapline, M. G., Peng, S., Cho, K. and Dai, H., Nanotube molecular wires as chemical sensors. Science, 287, No. 5453, 622 (2000).

64. Lam, C. W., James, J. T., McCluskey, R. and Hunter, R. L., Pulmonary toxicity of single-wall carbon nanotubes in mice 7 and 90 days after intractracheal instillation. Toxicological Sciences 77(1): 126-134.(2004)

65. Landi, B. J., Raffaelle, R. P., Castro, S. L. and Bailey, S. G., Single-wall carbon nanotube-polymer solar cells. Prog. Photovoltaics: Res. Appl, 13, 165 (2005).

66. Li, Q., Mahendra, S., Lyon, D. Y., Brunet, L., Liga, M. V., Li, D. and Alvarez, P. J. J., Antimicrobial nanomaterials for water disinfection

and microbial control: Potential applications and implications. Water Research, 42, No. 18, 4591 (2008a).

67. Li, X., Zhou, H., Yu, P., Su, L., Ohsaka, T. and Mao, L., A Miniature glucose/O$_2$ biofuel cell with single-walled carbon nanotubes-modified carbon fiber microelectrodes as the substrate. Electrochemistry Communications, 10, No. 6, 851 (2008b).

68. Li, Y. H., Wang, S., Wei, J., Zhang, X., Xu, C., Luan, Z., Wu, D. and Wei, B., Lead adsorption on carbon nanotubes. Chemical Physics Letters, 357, No. 3-4, 263 (2002).

69. Liang, C. W. and Roth, S., Electrical and optical transport of GaAs/ carbon nanotube heterojunctions. Nano Letters, 8, No. 7, 1809 (2008).

70. Lin, H. F., Ravikrishna, R. and Valsaraj, K. T., Reusable adsorbents for dilute solution separation. 6. Batch and continuous reactors for the adsorption and degradation of 1,2-dichlorobenzene from dilute wastewater streams using titania as a photocatalyst. Separation and Purification Technology, 28, No. 2, 87 (2002).

71. Liu, X., Huber, T. A., Kopac, M. C. and Pickup, P. G., Ru oxide/carbon nanotube composites for supercapacitors prepared by spontaneous reduction of Ru(VI) and Ru(VII). Electrochimica Acta, 54, No. 27, 7141 (2009a).

72. Liu, Y., Li, Y. and Yan, X. P., Preparation, Characterization, and Application of L-Cysteine Functionalized Multiwalled Carbon Nanotubes as a Selective Sorbent for Separation and Preconcentration of Heavy Metals. Advanced Functional Materials, 18, No. 10, 1536 (2008).

73. Liu, Y., Lu, S. and Panchapakesan, B., Alignment enhanced photoconductivity in single wall carbon nanotube films. Nanotechnology, 20, No. 3 (2009b).

74. Logan, B. E., Hamelers, B., Rozendal, R., Schröder, U., Keller, J., Freguia, S., Aelterman, P., Verstraete, W. and Rabaey, K., Microbial fuel cells: Methodology and technology. Environmental Science and Technology, 40, No. 17, 5181 (2006).

75. Long, R. Q. and Yang, R. T., Carbon Nanotubes as Superior Sorbent for Dioxin Removal. J. Am. Chem. Soc, 123, No. 9, 2058 (2001).

76. Lu, C. and Chiu, H., Adsorption of zinc(II) from water with purified carbon nanotubes. Chemical Engineering Science, 61, No. 4, 1138 (2006).

77. Lu, C. and Chiu, H., Chemical modification of multiwalled carbon nanotubes for sorption of Zn^{2+} from aqueous solution. Chemical Engineering Journal, 139, No. 3, 462 (2008).

78. Lu, C., Liu, C. and Rao, G. P., Comparisons of sorbent cost for the removal of Ni^{2+} from aqueous solution by carbon nanotubes and granular activated carbon. Journal of Hazardous Materials, 151, No. 1, 239 (2008).

79. Lu, J., Kumar, B., Castro, M. and Feller, J. F., Vapour sensing with conductive polymer nanocomposites (CPC): Polycarbonate-carbon nanotubes transducers with hierarchical structure processed by spray layer by layer. Sensors and Actuators, B: Chemical, 140, No. 2, 451 (2009).

80. Magrez, A., Kasas, S., Salicio, V., Pasquier, N., Seo, J. W., Celio, M., Catsicas, S., Schwaller, B. and Forró, L., Cellular toxicity of carbon-based nanomaterials. Nano Letters, 6, No. 6, 1121 (2006).

81. Mauter, M. S. and Elimelech, M., Environmental applications of carbon-based nanomaterials. Environmental Science and Technology, 42, No. 16, 5843 (2008).

82. Mills, A. and Le Hunte, S., An overview of semiconductor photocatalysis. Journal of Photochemistry and Photobiology A: Chemistry, 108, No. 1, 1 (1997).

83. Minteer, S. D., Liaw, B. Y. and Cooney, M. J., Enzyme-based biofuel cells. Current Opinion in Biotechnology, 18, No. 3, 228 (2007).

84. Mojumdar, S. C. and Raki, L., Preparation and properties of calcium silicate hydrate-poly(vinyl alcohol) nanocomposite materials. Journal of Thermal Analysis and Calorimetry, 82, No. 1, 89 (2005).

85. Moon, S. I., Paek, K. K., Lee, Y. H., Park, H. K., Kim, J. K., Kim, S. W. and Ju, B. K., Bias-heating recovery of MWCNT gas sensor. Materials Letters, 62, No. 16, 2422 (2008).

86. Morones, J. R., Elechiguerra, J. L., Camacho, A., Holt, K., Kouri, J. B., Ramírez, J. T. and Yacaman, M. J., The bactericidal effect of silver nanoparticles. Nanotechnology, 16, No. 10, 2346 (2005).

87. Morozan, A., Stamatin, L., Nastase, F., Dumitru, A., Vulpe, S., Nastase, C., Stamatin, I. and Scott, K., The biocompatibility microorganisms-carbon nanostructures for applications in microbial fuel cells. Physica Status Solidi (A) Applications and Materials, 204, No. 6, 1797 (2007).

88. Mostafavi, S. T., Mehrnia, M. R. and Rashidi, A. M., Preparation of nanofilter from carbon nanotubes for application in virus removal from water. Desalination, 238, No. 1-3, 271 (2009).

89. Muller, J., Huaux, F., Moreau, N., Misson, P., Heilier, J. F., Delos, M., Arras, M., Fonseca, A., Nagy, J. B. and Lison, D., Respiratory toxicity of multi-wall carbon nanotubes. Toxicology and Applied Pharmacology,

207, No. 3, 221 (2005).

90. Nepal, D., Balasubramanian, S., Simonian, A. L. and Davis, V. A., Strong antimicrobial coatings: Single-walled carbon nanotubes armored with biopolymers. Nano Letters, 8, No. 7, 1896 (2008).

91. Nguyen, H. Q. and Huh, J. S., Behavior of single-walled carbon nanotube-based gas sensors at various temperatures of treatment and operation. Sensors and Actuators, B: Chemical, 117, No. 2, 426 (2006).

92. Nguyen, L. H., Phi, T. V., Phan, P. Q., Vu, H. N., Nguyen-Duc, C. and Fossard, F., Synthesis of multi-walled carbon nanotubes for NH_3 gas detection. Physica E: Low-dimensional Systems and Nanostructures, 37, No. 1-2, 54 (2007).

93. Noy, A., Park, H. G., Fornasiero, F., Holt, J. K., Grigoropoulos, C. P. and Bakajin, O., Nanofluidics in carbon nanotubes. Nano Today, 2, No. 6, 22 (2007).

94. O›Regan, B. and Grätzel, M., A low-cost, high-efficiency solar cell based on dye-sensitized colloidal TiO_2 films. Nature, 353, No. 6346, 737 (1991).

95. Palmore, G. T. R. and Whitesides, G. M., Enzymatic conversion of biomass for fuels cells. American Chemical Society, 566, 271 (1994).

96. Pasquier, A. D., Unalan, H. E., Kanwal, A., Miller, S. and Chhowalla, M., Conducting and transparent single-wall carbon nanotube electrodes for polymer–fullerene solar cells. Appl. Phys. Lett, 87, No. 20, 1 (2005).

97. Peng, X., Li, Y., Luan, Z., Di, Z., Wang, H., Tian, B. and Jia, Z., Adsorption of 1,2-dichlorobenzene from water to carbon nanotubes. Chemical Physics Letters, 376, No. 1-2, 154 (2003).

98. Penza, M., Rossi, R., Alvisi, M., Cassano, G. and Serra, E., Functional characterization of carbon nanotube networked films functionalized with tuned loading of Au nanoclusters for gas sensing applications. Sensors and Actuators, B: Chemical, 140, No. 1, 176 (2009a).

99. Penza, M., Rossi, R., Alvisi, M., Signore, M. A., Cassano, G., Dimaio, D., Pentassuglia, R., Piscopiello, E., Serra, E. and Falconieri, M., Characterization of metal-modified and vertically-aligned carbon nanotube films for functionally enhanced gas sensor applications. Thin Solid Films, 517, No. 22, 6211 (2009b).

100. Qi, P., Vermesh, O., Grecu, M., Javey, A., Wang, Q., Dai, H., Peng, S. and Cho, K. J., Toward Large Arrays of Multiplex Functionalized Carbon Nanotube Sensors for Highly Sensitive and Selective Molecular Detection. Nano Lett, 3, No. 3, 347 (2003).

101. Qiao, Y., Li, C. M., Bao, S. J. and Bao, Q. L., Carbon nanotube/polyaniline

composite as anode material for microbial fuel cells. Journal of Power Sources, 170, No. 1, 79 (2007).

102. Quang, N. H., Van Trinh, M., Lee, B. H. and Huh, J. S., Effect of NH_3 gas on the electrical properties of single-walled carbon nanotube bundles. Sensors and Actuators, B: Chemical, 113, No. 1, 341 (2006).

103. Rao, G. P., Lu, C. and Su, F., Sorption of divalent metal ions from aqueous solution by carbon nanotubes: A review. Separation and Purification Technology, 58, No. 1, 224 (2007).

104. Ray, S. S., Vaudreuil, S., Maazouz, A. and Bousmina, M., Dispersion of multi-walled carbon nanotubes in biodegradable poly(butylene succinate) matrix. Journal of Nanoscience and Nanotechnology, 6, No. 7, 2191 (2006).

105. Reijnders, L., Cleaner nanotechnology and hazard reduction of manufactured nanoparticles. Journal of Cleaner Production, 14, No. 2, 124 (2006).

106. Ritschel, M., Uhlemann, M., Gutfleisch, O., Leonhardt, A., Graff, A., Täschner, C. and Fink, J., Hydrogen storage in different carbon nanostructure. Applied Physics Letters, 80, No. 16, 2985 (2002).

107. Rowell, M. W., Topinka, M. A., McGehee, M. D., Prall, H. J., Dennler, G., Sariciftci, N. S., Hu, L. B. and Gruner, G., Organic solar cells with carbon nanotube network electrodes. Appl. Phys. Lett, 88, No. 23 (2006).

108. Savage, N. and Diallo, M. S., Nanomaterials and water purification: Opportunities and challenges. Journal of Nanoparticle Research, 7, No. 4-5, 331 (2005).

109. Sayes, C. M., Liang, F., Hudson, J. L., Mendez, J., Guo, W., Beach, J. M., Moore, V. C., Doyle, C. D., West, J. L., Billups, W. E., Anusman, K. D. and Colvin, V. L., Functionalization density dependence of single-walled carbon nanotubes cytotoxicity in vitro. Toxicology Letters, 161, No. 2, 135 (2006).

110. Scarselli, M., Scilletta, C., Tombolini, F., Castrucci, P., De Crescenzi, M., Diociaiuti, M., Casciardi, S., Gatto, E. and Venanzi, M., Photon harvesting with multi wall carbon nanotubes. Superlattices and Microstructures, 46, No. 1-2, 340 (2009).

111. Schaller, R., Mari, D., dos Santos, S. M., Tkalcec, I. and Carreño-Morelli, E., Investigation of hydrogen storage in carbon nanotube-magnesium matrix composites. Materials Science and Engineering A, 521-522, 147 (2009).

112. Schur, D. V., Tarasov, B. P., Zaginaichenko, S. Y., Pishuk, V. K., Veziroglu,

T. N., Shul)ga, Y. M., Dubovoi, A. G., Anikina, N. S., Pomytkin, A. P. and Zolotarenko, A. D., The prospects for using of carbon nanomaterials as hydrogen storage systems. International Journal of Hydrogen Energy, 27, No. 10, 1063 (2002).

113. Sharma, T., Reddy, A. L. M., Chandra, T. S. and Ramaprabhu, S., Development of carbon nanotubes and nanofluids based microbial fuel cell. International Journal of Hydrogen Energy, 33, No. 22, 6749 (2008).

114. Sheldon, R. A., Enzyme immobilization: The quest for optimum performance. Advanced Synthesis and Catalysis, 349, No. 8-9, 1289 (2007).

115. Simon, P. and Gogotsi, Y., Materials for electrochemical capacitors. Nature Materials, 7, No. 11, 845 (2008).

116. Sitharaman, B., Shi, X., Walboomers, X. F., Liao, H., Cuijpers, V., Wilson, L. J., Mikos, A. G. and Jansen, J. A., In vivo biocompatibility of ultra-short single-walled carbon nanotube/biodegradable polymer nanocomposites for bone tissue engineering. Bone, 43, No. 2, 362 (2008).

117. Song, L. and Qiu, Z., Crystallization behavior and thermal property of biodegradable poly(butylene succinate)/functional multi-walled carbon nanotubes nanocomposite. Polymer Degradation and Stability, 94, No. 4, 632 (2009).

118. Srivastava, A., Srivastava, O. N., Talapatra, S., Vajtai, R. and Ajayan, P. M., Carbon nanotubes filters. Nature Materials, 3, No. 9, 610 (2004).

119. Stafiej, A. and Pyrzynska, K., Solid phase extraction of metal ions using carbon nanotubes. Microchemical Journal, 89, No. 1, 29 (2008).

120. Suehiro, J., Zhou, G. and Hara, M., Detection of partial discharge in SF6 gas using a carbon nanotube-based gas sensor. Sensors and Actuators, B: Chemical, 105, No. 2, 164 (2005).

121. Tahaikt, M., El Habbani, R., Ait Haddou, A., Achary, I., Amor, Z., Taky, M., Alami, A., Boughriba, A., Hafsi, M. and Elmidaoui, A., Fluoride removal from groundwater by nanofiltration. Desalination, 212, No. 1-3, 46 (2007).

122. Tibbetts, G. G., Meisner, G. P. and Olk, C. H., Hydrogen storage capacity of carbon nanotubes, filaments, and vapor-grown fibers. Carbon, 39, No. 15, 2291 (2001).

123. Tsai, H. Y., Wu, C. C., Lee, C. Y. and Shih, E. P., Microbial fuel cell performance of multiwall carbon nanotubes on carbon cloth as electrodes. Journal of Power Sources, 194, No. 1, 199 (2009).

124. Türker, A. R., New Sorbents for Solid-Phase Extraction for Metal

Enrichment. CLEAN - Soil, Air, Water, 35, No. 6, 548 (2007).

125. U.S. Department of Energy: Washington, DC. Multi-Year Research, Development, and Demonstration Plan: Planned Program Activities for 2003-2010 (2007).

126. Ueda, T., Bhuiyan, M. M. H., Norimatsu, H., Katsuki, S., Ikegami, T. and Mitsugi, F., Development of carbon nanotube-based gas sensors for NOx gas detection working at low temperature. Physica E: Low-dimensional Systems and Nanostructures, 40, No. 7, 2272 (2008a).

127. Ueda, T., Katsuki, S., Takahashi, K., Narges, H. A., Ikegami, T. and Mitsugi, F., Fabrication and characterization of carbon nanotube based high sensitive gas sensors operable at room temperature. Diamond and Related Materials, 17, No. 7-10, 1586 (2008b).

128. Ulbricht, R., Lee, S. B., Jiang, X. M., Inoue, K., Zhang, M., Fang, S. L., Baughman, R. H. and Zakhidov, A. A., Transparent carbon nanotube sheets as 3-D charge collectors in organic solar cells. Sol. Energy Mater. Sol. Cells, 91, No. 5, 416 (2007).

129. Ulbricht, R., Jiang, X., Lee, S.B., Zhang, M., Fang, S., Baughman, R. H., Zakhidov, A., Polymeric solar cells with oriented and strong transparent carbon nanotube anode. Phys. Stat. Sol. B, 243, 3528 (2006).

130. Valentini, L., Cantalini, C., Armentano, I., Kenny, J. M., Lozzi, L. and Santucci, S., Highly sensitive and selective sensors based on carbon nanotubes thin films for molecular detection. Diamond and Related Materials, 13, No. 4-8, 1301 (2004).

131. Van Duy, N., Van Hieu, N., Huy, P. T., Chien, N. D., Thamilselvan, M. and Yi, J., Mixed SnO_2/TiO_2 included with carbon nanotubes for gas-sensing application. Physica E: Low-dimensional Systems and Nanostructures, 41, No. 2, 258 (2008).

132. Van Hieu, N., Thuy, L. T. B. and Chien, N. D., Highly sensitive thin film NH_3 gas sensor operating at room temperature based on SnO_2/MWCNTs composite. Sensors and Actuators, B: Chemical, 129, No. 2, 888 (2008).

133. Vaudreuil, S., Labzour, A., Sinha-Ray, S., Mabrouk, K. E. and Bousmina, M., Dispersion characteristics and properties of poly(methyl methacrylate)/multi-walled carbon nanotubes nanocomposites. Journal of Nanoscience and Nanotechnology, 7, No. 7, 2349 (2007).

134. Wang, H., Xu, P., Zhong, W., Shen, L. and Du, Q., Transparent poly(methyl methacrylate)/silica/ zirconia nanocomposites with excellent thermal stabilities. Polymer Degradation and Stability, 87, No. 2, 319 (2005).

135. Wang, J., Liu, L., Cong, S. Y., Qi, J. Q. and Xu, B. K., An enrichment

method to detect low concentration formaldehyde. Sensors and Actuators, B: Chemical, 134, No. 2, 1010 (2008).

136. Wang, Q. and Johnson, J. K., Optimization of carbon nanotube arrays for hydrogen adsorption. Journal of Physical Chemistry B, 103, No. 23, 4809 (1999).

137. Wang, X., Jialong, L. U. and Xing, B., Sorption of organic contaminants by carbon nanotubes: Influence of adsorbed organic matter. Environmental Science and Technology, 42, No. 9, 3207 (2008).

138. Wang, Y., Deng, W., Liu, X. and Wang, X., Electrochemical hydrogen storage properties of ball-milled multi-wall carbon nanotubes. International Journal of Hydrogen Energy, 34, No. 3, 1437 (2009).

139. Watanabe, K., Recent Developments in Microbial Fuel Cell Technologies for Sustainable Bioenergy. Journal of Bioscience and Bioengineering, 106, No. 6, 528 (2008).

140. Wei, B.Y., Hsu, M. C., Su, P. G., Lin, H. M., Wu, R. J. and Lai, H. J., A novel SnO_2 gas sensor doped with carbon nanotubes operating at room temperature. Sensors and Actuators, B: Chemical, 101, No. 1-2, 81 (2004).

141. Wei, C., Dai, L., Roy, A. and Tolle, T. B., Multifunctional chemical vapor sensors of aligned carbon nanotube and polymer composites. Journal of the American Chemical Society, 128, No. 5, 1412 (2006).

142. Weng, C. H. and Huang, C. P., Adsorption characteristics of Zn(II) from dilute aqueous solution by fly ash. Colloids and Surfaces A: Physicochemical and Engineering Aspects, 247, No. 1-3, 137 (2004).

143. Wick, P., Manser, P., Limbach, L. K., Dettlaff-Weglikowska, U., Krumeich, F., Roth, S., Stark, W. J. and Bruinink, A., The degree and kind of agglomeration affect carbon nanotube cytotoxicity. Toxicology Letters, 168, No. 2, 121 (2007).

144. Xie, X. L., Mai, Y. W. and Zhou, X. P., Dispersion and alignment of carbon nanotubes in polymer matrix: A review. Materials Science and Engineering R: Reports, 49, No. 4 (2005).

145. Xu, D., Tan, X., Chen, C. and Wang, X., Removal of Pb(II) from aqueous solution by oxidized multiwalled carbon nanotubes. Journal of Hazardous Materials, 154, No. 1-3, 407 (2008).

146. Yan, J., Fan, Z., Wei, T., Cheng, J., Shao, B., Wang, K., Song, L. and Zhang, M., Carbon nanotube/MnO_2 composites synthesized by microwave-assisted method for supercapacitors with high power and energy densities. Journal of Power Sources, 194, No. 2, 1202 (2009).

147. Yang, F. H., Lachawiec, Jr. A. J. and Yang, R. T., Adsorption of spillover hydrogen atoms on single-wall carbon nanotubes. Journal of Physical Chemistry B, 110, No. 12, 6236 (2006a).

148. Yang, K., Zhu, L. and Xing, B., Adsorption of polycyclic aromatic hydrocarbons by carbon nanomaterials. Environmental Science and Technology, 40, No. 6, 1855 (2006b).

149. Yuan, W., Jiang, G., Che, J., Qi, X., Xu, R., Chang, M. W., Chen, Y., Lim, S. Y., Dai, J. and Chan-Park, M. B., Deposition of Silver Nanoparticles on Multiwalled Carbon Nanotubes Grafted with Hyperbranched Poly(amidoamine) and their Antimicrobial Effects. The Journal of Physical Chemistry C, 112, No. 48, 18754 (2008).

150. Zacharia, R., Kim, K. Y., Fazle Kibria, A. K. M. and Nahm, K. S., Enhancement of hydrogen storage capacity of carbon nanotubes via spillover from vanadium and palladium nanoparticles. Chemical Physics Letters, 412, No. 4-6, 369 (2005).

151. Zacharia, R., Rather, S. U., Hwang, S. W. and Nahm, K. S., Spillover of physisorbed hydrogen from sputter-deposited arrays of platinum nanoparticles to multi-walled carbon nanotubes. Chemical Physics Letters, 434, No. 4-6, 286 (2007).

152. Zhang, Q., Huang, J., Wei, F., Xu, G., Wang, Y., Qian, W. and Wang, D., Large scale production of carbon nanotube arrays on the sphere surface from liquefied petroleum gas at low cost. Chinese Science Bulletin, 52, No. 21, 2896 (2007).

153. Zhang, W. D. and Zhang, W. H., Carbon nanotubes as active components for gas sensors. Journal of Sensors 2009 (2009).

154. Zhao, H. Y., Zhou, H. M., Zhang, J. X., Zheng, W. and Zheng, Y. F., Carbon nanotube-hydroxyapatite nanocomposite: A novel platform for glucose/O_2 biofuel cell. Biosensors and Bioelectronics, 25,No. 2, 463 (2009).

155. Zhou, L., Zhou, Y. and Sun, Y., Enhanced storage of hydrogen at the temperature of liquid nitrogen. International Journal of Hydrogen Energy. 29, No. 3, 319 (2004a).

156. Zhou, L. J., Tan, J. G., Hu, B. H. and Feng, H. L., Ultrastructural study of sclerotic dentin in non-carious cervical lesions disposed by a total-etching dentin adhesive. Journal of Peking University, 36, No. 3, 319 (2004b).

157. Zhou, M., Deng, L., Wen, D., Shang, L., Jin, L. and Dong, S., Highly ordered mesoporous carbons-based glucose/O_2 biofuel cell. Biosensors and Bioelectronics, 24, No. 9, 2904 (2009).

158. Zhu, H., Wei, J., Wang, K. and Wu, D., Applications of carbon materials in photovoltaic solar cells. Solar Energy Materials and Solar Cells, 93, No. 9, 1461 (2009).

159. Zhu, H. W., Zeng, H. F., Subramanian, V., Masarapu, C., Hung, K. H. and Wei, B. Q., Anthocyanin-sensitized solar cell using carbon nanotube films as counter electrodes. Nanotechnology, 19, No. 46 (2008).

Chapter 10

MECHANICAL PROPERTIES OF METAL DIHYDRIDES

Peter A Schultz[1] and Clark S Snow[2]

[1] Multiscale Science Department, Sandia National Laboratories, Albuquerque, NM 87185, USA

[2] Applied Science and Technology Maturation Department, Sandia National Laboratories, Albuquerque, NM 87185, USA

ABSTRACT

First-principles calculations are used to characterize the bulk elastic properties of cubic and tetragonal phase metal dihydrides, MM_2{M = Sc, Y, Ti, Zr, Hf, lanthanides} to gain insight into the mechanical properties that govern the aging behavior of rare-earth di-tritides as the constituent ^3H, tritium, decays into ^3He. As tritium decays, helium is inserted in the lattice, the helium migrates and collects into bubbles, that then can ultimately create sufficient internal pressure to rupture the material. The elastic properties of the materials are needed to construct effective mesoscale models of the process of bubble growth and fracture. Dihydrides of the scandium column and most of the rare-earths crystalize into a cubic phase, while dihydrides from the next column, Ti, Zr, and Hf, distort instead into the tetragonal phase, indicating incipient instabilities in the phase and potentially significant changes in elastic properties. We report the computed elastic properties of these dihydrides, and also investigate the off-stoichiometric phases as He or vacancies accumulate. As helium builds up in the cubic phase, the shear moduli greatly soften, converting to the tetragonal phase. Conversely, the tetragonal phases convert very quickly to cubic with the removal of H from the lattice, while the cubic phases show little change with removal of H. The source and magnitude of the numerical and physical uncertainties in the modeling are analyzed and quantified to establish the level of confidence that can be placed in the computational results, and this quantified confidence is used to justify using the results to augment and even supplant experimental measurements.

INTRODUCTION

Rare earth metals (lanthanides) and many transition metals readily form hydrides, making them useful for storage of hydrogen and its isotopes. The rare-earth metal hydrides are particularly suitable materials for long term storage of tritium. Erbium tritide has been frequently studied [1–4] as a prominent example of such a material, with the goal to gain insight into the aging process as the tritium decays. The decay of tritium into helium introduces a complication not present in conventional studies of hydride chemistry, as the composition of the material changes via radioactive decay of the constituent tritium, ^3H. The consequent build-up of helium, ^3He, in the material ultimately leads to formation of high-pressure He bubbles [2]. This pressure can eventually rupture the material, the release of helium leading to failure of devices dependent upon the integrity of the material.

Fundamental understanding is lacking in each step of this degradation process—nucleation of bubbles, migration of He through the lattice to the bubbles, bubble growth, coalescence of bubbles, and finally fracture and release—and integrating these individual physical scales into an useful and actionable comprehensive understanding is an ongoing challenge for effective modeling and simulation to quantify aging effects in tritides. Some progress has been made in microstructural-scale models of tritide aging and helium bubble growth and resultant fracture [5–9], which examine the evolution of bubbles and quantify the structural integrity of the material before fracture. These models are dependent upon empirical parameters such as the initial bubble nucleation density and, the focus of this study, the elastic properties of the target tritide [10–12]. It is not the purpose of this paper to assess the relative merits of these models, but rather speak to the requirements of quantified mechanical properties of the materials that are common to all of them. These data are difficult to obtain experimentally. Rare-earth hydrides samples are exceedingly brittle, partly because of a 10–15% lattice expansion upon hydriding. While thin film samples can be manufactured that lack the brittleness of their bulk counterparts, these films are not amenable to standard techniques used to measure bulk elastic properties. First-principles calculations are particularly well-suited for this purpose and, furthermore, also can provide insight into the causes of the changes in mechanical behavior. While the energetics and electronic properties of these hydrides have been frequently studied [13–15], these crucial elastic properties have received much less attention.

Our investigation centers around ErH_2. ErH_2 has been the subject of several first-principles studies looking at the defects and impurities, seeking to identify and quantify the atomic processes that govern the material chemical evolution [16–18]. Stoichiometric ErH_2, just as most of its rare earth (RE) brethren, crystallizes into a CaF_2 structure (Fm$\bar{3}$m, #225), a cubic (fcc) form illustrated in figure 1. Almost all the rare earth dihydrides (including the dihydrides of all the Group IIIB metals, Sc, Y, and La,), also form a bulk ground state in the CaF_2 structure. Shifting one column to the right in the Periodic table, the Group-IVB TiH_2, ZrH_2, and HfH_2 instead form a tetragonally distorted fluorite structure (*I4/mmm*, #139), where the cube in figure 1 is compressed along one axis.

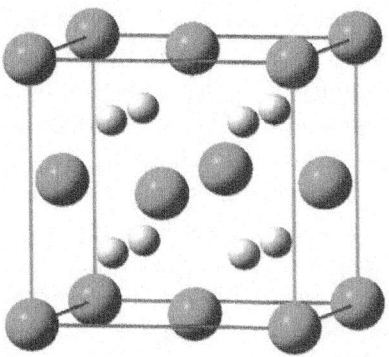

Figure 1: CaF_2 fcc crystal structure of MM_2. The large grey spheres represent the metal M atoms and the small white spheres represent the hydrogen atoms.

The analysis of the energies of this structural distortion has been often studied [15, 19–22]. The consensus for the cause of the distortion of the crystal into the tetragonal form has coalesced around the high density of states (DOS) at the Fermi level for the Group IVB dihydrides. This high DOS has been associated with a John-Teller-like instability, that drives this distortion although that interpretation has recently been disputed [22].

The band structure of these materials bear great similarity to one another, demonstrated in many of these studies [15, 19–22]. A rigid band picture of occupation has been proposed to rationalize the properties [15], with the changes being determined by the position of the Fermi level—the number of valence electrons—in a mostly similar band structure. This indicates that there will be a strong dependence of the elastic properties upon the stoichiometry. This dependence might need to be incorporated in any microstructural model to obtain accurate description of the material aging.

The stoichiometry of the material is initially determined by the manufacturing, and then modified by entropic effects, which can create vacancies in the H lattice and create H interstitials. In addition, in the tritide, the aging process also modifies the stoichiometry over time. As the ^{3}H decays, ^{3}He is generated in the lattice. The modest nuclear recoil energy in this decay, ~1 eV, is insufficient to cause atomic displacement, leaving the daughter helium atom initially near the same location as its tritium parent. To dislodge a ^{3}He from a tetrahedral site into the lattice requires an activation energy of 1.31 eV [16] in Er_2.

In a rigid band picture, this transmutation of hydrogen into helium fills more of the band structure, shifting it toward a band filling and Fermi level that favors a tetragonal distortion. In the aging material, however, the He is observed not to stay in place long enough for sufficient buildup of He to cause a distortion. Instead, the He migrates away to join bubbles, leaving behind a H vacancy and substoichiometric hydride. We investigate the importance of these stoichiometry changes on the elastic properties of the material through an atom-potential averaging method. Changing the nuclear charge on the H atom, increasing its charge toward He to model an average population of He substituting for H, decreasing its charge to model partial vacancy population on the H site, we probe the elastic properties of the resultant off-stoichiometric material. First-principles results of the energetics of these dihydrides, particularly in the tetragonal distortions of the Group-IVB dihydrides, have not given consistent results [15, 19, 22], and calculations of elastic constants in metals are particularly prone to numerical problems [23, 24]. In this work, we apply particularly careful attention to the numerical aspects of the modeling that inject uncertainties into the computation of structure and elastic constants. Examination of the convergence behavior provides numerical uncertainties that quantify the level of confidence in the simulations, and key results are verified through comparison of different means of computation and through comparisons to experiments.

Computational Details

We used the generalized gradient approximation with the Perdew, Burke, Ernzerhof (PBE) functional [25] for all the density functional theory (DFT) calculations. The rationale for using PBE and not the local density approximation (LDA) [26] had been discussed previously [16]: the LDA is unable to describe the ground state structure of the Er bulk metal correctly, disqualifying it for the longer-term goals of the study. Additional details for the Er-H calculations can also be found in this previous work [16], salient details are summarized here.

The DFT calculations for Er, Group-IIIB and Group-IVB hydrides were done with SeqQuest[27], using Hamann-type generalized norm-conserving pseudopotentials [28], and well-optimized, valence double-ζ plus polarization, contracted Gaussian basis sets. The metal atoms all included their respective semi-core p-shell in the valence, and then added a non-linear core correction [29] to treat the non-linear exchange-correlation effects of the valence electrons interacting with the core, using a cutoff radius that was tested to be converged to an asymptotic full-core limit. The hydrogen atom pseudopotential was generated using a new method by Hamann [30] (verified to give equivalent results to the standard Hamann pseudopotentials [31]). To represent non-integral charge off-hydrogen atoms Hz, $Z = \frac{3}{4}, \frac{5}{4},$ and $\frac{6}{4}$ atoms, standard Hamann pseudopotentials [28] were developed. All the hydrogen and off-hydrogen atoms used triple-ζ- s plus single-ζ- p polarization basis sets. The triple-ζ- $s$results are almost identical to results using the double-ζ- s results for the metal hydrides.

All the SEQQUEST DFT calculations were done in the 3-atom primitive cells using a 24^3 real space grid: the fcc rhombohedral unit cell for the cubic phase and the corresponding distorted cell for the tetragonal phase.

The transition metal dihydrides considered here are non-magnetic. For the Er and other rare-earth dihydrides, any magnetism is confined to the f-electrons. The lanthanide series fills the first shell of f-electrons, electrons localized very close to the metal core and only weakly interacting with valence electrons. These only very minimally affect the bonding in the hydrides. For computational convenience, the f-electrons (and, hence, all magnetism) have been place in the core of the pseudopotential. Spin effects in the hydrides are ignored throughout this work.

These dihydrides are all metallic, requiring an artificial electronic temperature to converge. We use a simple Fermi function with an effective electronic temperature of ~41 meV. While artificial electron temperatures can skew results, particularly causing discrepancies between stress- and energy-derived quantities [24], the consistency checks below indicate that this modest electronic temperature is small enough to be converged to the 0 K limit, well within the other numerical uncertainties in the calculation.

Calculations of elastic constants, particularly shear moduli, can be very sensitive to k-point sampling [23, 24] and other computational model parameters, so extra diligence was exercised to converge and verify these results. The calculations (see below) sampled the Brillouin Zone with regular grids ranging from 8^3 through 20^3 (in some cases, to 24^3) centered at Γ. Zone samples less than 12^3 were typically poorly converged, those denser than 12^3 exhibited random scatter about a single value. We adopt the (unweighted) average of results from samples with k-point samples 13^3 and larger as the 'k-limit' converged bulk value, and the range of scatter about that k-limit value to represent the uncertainty. The optimal lattice constant converges relatively quickly with k-point sampling, hence, all elastic constants (for every k-sample) were computed at the k-limit lattice constant.

The elastic constants were computed from linear fits to the strain derivatives of the appropriate component of the computed stress tensor, obtained as finite differences between stress calculations for strains of 0.001 to 0.008, in increments of 0.001 (0.1%). For the cubic crystals, two strains, a compression along the z-axis and the xz-shear, were sufficient to obtain the three unique elastic constants. For the tetragonal crystal, an x-axis compression and yz-axis shear were added to obtain the additional unique elastic constants. The real-space integral cutoffs were extended to converge the stress tensor calculations sufficiently to extract numerically meaningful stress derivatives from the small incremental strain calculations. In the sheared crystals, it is necessary to relax the (hydrogen) atoms in order to get the ground state of the relaxed crystals, the need to resolve accurate stress derivatives from stress differences for such small strains required that atomic configurations be much more finely converged than for typical configuration relaxations, to forces less than 0.26 meV $\overset{\circ}{\text{A}}^{-1}$.

As a verification of the stress-derived results, the bulk modulus was also computed from fits to the potential energy surface. For the cubic crystal, fixed point DFT calculations were performed for 13–15 regularly spaced lattice parameters (0.02 Bohr increments, to ~$\pm$1.5%) about the equilibrium lattice constant, and fit with a Birch–Murnaghan equation of state, using a third order polynomial [32–34]. Previous experience indicates that the uncertainties in the computed bulk modulus—from the number and selection of the sample points, form of the equation of state, etc—are about 1 GPa, and this experience is reflected in the results presented later. For the tetragonal crystals, the sampling points for the equation of state calculations were obtained from a series of cell-optimization calculations with applied pressures. Cells were optimized with applied pressures ranging from +12 GPa to −12 GPa, in increments of 1 GPa. The resulting cell volumes and energies, like the cubic crystals, were fit with a Birch–Murnaghan equation of state, again using a third order polynomial. As seen later, these potential energy fits for the tetragonal crystal also have an apparent accuracy of ~1 GPa.

Additional density functional theory calculations, to compute the elastic constants of cubic rare earth hydrides, were performed with the Vienna *ab initio* simulation package (VASP) [35], using the projector augmented wave (PAW) method. The PAW pseudopotentials were taken from the VASP database [36]. These calculations were all performed in the conventional 12-atom cubic cell (four formula units), rather than the 3-atom fcc primitive unit cells used for the SEQQUEST calculations. First, we performed systematic tests on ErH_2 to determine parameters sufficient to numerical converge the computed properties. The convergence testing and computation of elastic constants were done in the integrated MedeA computational environment [37]. The total energy was converged to less than 0.0001 eV atom^{-1} with a plane wave energy cutoff of 562.5 eV and with a $21 \times 21 \times 21$ sampling of the Brillouin Zone (centered at the Γ point), and these converged settings were subsequently used for all VASP calculations for cubic rare earth dihydrides. The elastic constants computed with VASP were calculated using a 'stress-strain' approach [38], using a strain of 0.5% from the equilibrium lattice structure.

Results

A. Numerical uncertainties

The dihydrides considered in this study are all metals, observed in the current calculations, and illustrated in the band structures presented in many previous computational studies of metal dihydrides[13–15, 22]. In metallic systems, the calculations of elastic constants can be profoundly sensitive to the Brillouin Zone sampling and the effective electronic temperature [23, 24]. In bcc-Ta, the equilibrium lattice parameter and total energy converged relatively quickly with k-point density, but the elastic constants, and particularly the shear moduli, C_{44} and Cs, defined as

$$C_s = \frac{(C_{11} - C_{12})}{2}$$

(1)

showed large dependence with k-sampling (5 GPa variability) out to k-points samples as large as 40^3 [24]. Hence, while converged calculations of the ground state lattice constant, total energy, and bulk modulus are relatively straightforward to obtain, computation of elastic constants might have much greater uncertainties and need to be approached more skeptically.

Figure 2 illustrates the k-sampling dependence of the bulk modulus and the elastic constants computed for cubic ErH_2, from a 8^3 through a 24^{3} k-sampling of the full Brillouin Zone for the 3-atom primitive cell. In all cases we consider, the lattice constant and total energy exhibit almost no k-point dependence, i.e. they have converged to within 0.001 Å and 2 meV/formula-unit (f.u.) over the range of k-point samples investigated. The bulk modulus also converges very well (to within 0.1–0.2 GPa) and relatively quickly. The other elastic constants exhibit much greater variability, as seen in earlier analyses of elastic constant computations in metals [23, 24]. Once the sampling gets denser than 12^3, the computed constants appear to converge around an average value, and there is little apparent diminution in the scatter of the computed values about this average out to the largest k-point sampling we consider. This variability with k-sampling dominates the uncertainties associated with computation of elastic constants and is the principal factor in the numerical precision with which the elastic constants can be predicted. The typical numerical uncertainties due to k-sampling for computation of shear moduli in the cubic dihydrides are 2–3 GPa, but are as large as 5–6 GPa in some cases.

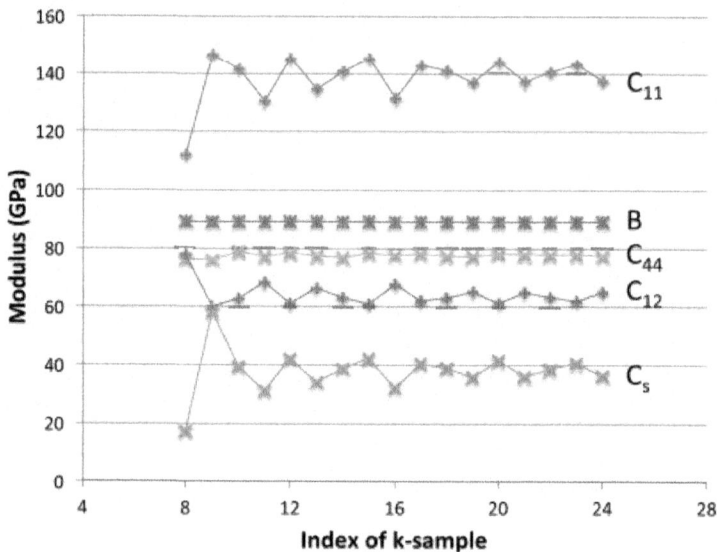

Figure 2: The computed elastic constants of cubic ErH_2 are plotted as a function of the k-point sampling. The stress-derived bulk modulus converges very quickly (to within 0.1 GPa), but the other elastic constants converge much more slowly, becoming unreliable at a k-sampling smaller than 12^3. The true DFT value would be the k-converged value, which is estimated as a k-limit average with an uncertainty due to the large-k scatter (see text), e.g. the shear constant C_{44} varies by ~3 GPa around an average value of 39 GPa.

It is pointless to attempt to numerically refine these elastic constants any further. First, the apparent rate of convergence, narrowing of the width of the scatter, is very slow with number of k-points but, more importantly, this k-dependent scatter has become smaller than the expected accuracy of the DFT. The B computed with semi-local functionals such as PBE can typically only predict experimental results for a bulk modulus to within 10–20%. The rapidly increasing computational cost of using denser k-sampling does not improve the total quality of the numerical predictions, i.e. the increased computational cost is not recouped in increased fidelity when these physical uncertainties are also considered.

This variability does not indict the quality of the calculation of C_{11} and C_{12} at each k-sampling density (from which the elastic modulus Cs is constructed). The bulk modulus in a cubic system, is also constructed from these same elastic constants:

$$B\,[\text{cubic}] = \frac{C_{11} + 2C_{12}}{3}.$$

$$(2)$$

The stress-derived $B(\{Cij\})$ show very small variability (typically ~0.1 GPa) with k-point sampling. As seen in figure 2, C_{11} and C_{12} suffer significant variations with k-point sample out to the largest sampling we consider, but these values are very closely correlated, leading to an almost constant bulk modulus. An independent calculation of the bulk modulus, $B(E)$ from a fit to a third-order Birch–Murnaghan equation of state of the total energies at different lattice parameters, yields the same result for the bulk modulus as the stress-derived calculation, to within ~1 GPa, verifying this evaluation of $B(\{Cij\})$ from the elastic constants. Moreover, the lattice constant and B from the stress-derived calculation and from the total energy Birch–Murnaghan equation of state closely match, indicating the (artificial) electronic temperature used to facilitate self-consistency in these metallic calculations gives a close approximation to a 0 K limit. This isolates the k-dependence as responsible for the variation in evaluated Cij, and not numerical artifacts from other aspects of the computational model.

Figure 3 illustrates the convergence of the elastic constants with k-sampling for the tetragonal structure HfH_2. In tetragonal dihydrides, the k-dependence of bulk properties becomes more severe, and spreads to the assessment of the lattice parameters. This can be attributed to a sensitivity of the structure and energy to occupation of states at the Fermi level. The associated computational challenge is to resolve the larger DOS at the Fermi level in the tetragonal dihydrides [15]. As in the cubic structure, the shear elastic constants are particularly sensitive. While the lattice constants a_0 and c_0 (not plotted) show greater variability, the equilibrium volume converges better with k-point sampling, converging as well as the volume (and lattice constant) had for the cubic crystal.

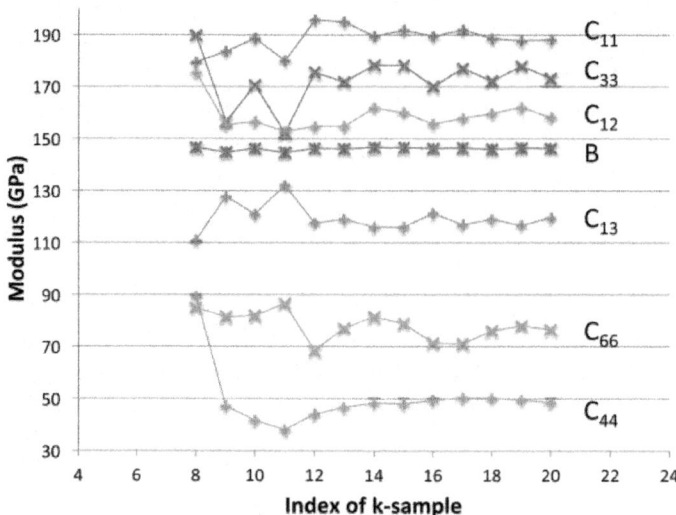

Figure 3: The computed elastic constants of tetragonal HfH$_2$ are plotted as a function of the k-point sampling index (sampling is a regular k^3 grid). The stress-derived bulk modulus converges quickly, only slightly less well than for a cubic structure (to within 0.2 GPa), but the other elastic constants converge much more slowly, becoming unreliable at a k-sampling smaller than 12^3, and only becoming converged to within ~2–3 GPa after k-sampling larger than 16^3 (the C_{66} shear constant only to within 5 GPa).

The bulk modulus can be evaluated in terms of the elastic constants as in a tetragonal crystal [39].

$$B \text{ [tetragonal]} = \frac{(C_{11} + C_{12})C_{33} - 2C_{13}^2}{C_{11} + C_{12} + 2C_{33} - 4C_{13}} \tag{3}$$

This stress-derived $B(\{Cij\})$ also shows more rapid convergence with k-sampling, although not quite as fast nor quite as tight as for the cubic structure.

The bulk modulus can alternatively be computed from the equation of state. The $B(E)$ can be obtained from analytic derivatives of a Birch–Murnaghan fit to the total energies of a sequence of pressure-optimized tetragonal cells. The energy-fit $B(E)$ match very well the stress-derived $B(\{Cij\})$, to within the same ~1 GPa that was observed for the cubic crystal, verifying that the calculation of elastic constants using strain-derivatives of the stress is consistent with the total energy calculations for the tetragonal structure. This verifies the form and evaluation of equation (3), and verifies the computation of the elastic constants within the formula given in equation (3).

B. Cubic structure metal hydrides

The results for the structure and elastic constants for the cubic structure of ErH_2 and the Group-IIIB and Group-IVB dihydrides are presented in table 1. The k-limit values are the full converged prediction of the DFT simulations. The quoted uncertainties in the k-limit values reflect numerical uncertainty in the converged average value, taken as the magnitude of the scatter from the average as the k-point sampling is increased. The value using a 12^3 k-sampling is also presented, a representative of the value one obtains with just a single k-point sample, at the threshold at which the simulations appear to converge to a k-limit asymptotic average. The 12^3 k-sample results in table 1 are at the boundary of the asymptotic range, i.e. converged within the uncertainties of the k-limit values. The formation energy and lattice constants are fully k-sampling converged, to the precision quoted in table 1.

Table 1. Structural properties and elastic constants computed for cubic metal dihydrides

	$-\Delta E_f$ (eV)	a_0 (Å)	C_{11} (GPa)	C_{12} (GPa)	C_{44} (GPa)	C_s (GPa)	$B(\{C_{ij}\})$ (GPa)	$B(E)$ (GPa)
ErH_2	2.281[a]	5.126[b]						
k-limit	—	a_0(expt.)	137(5)	61(3)	76.8(0.5)	38(5)	86.5(.1)	—
k-limit	2.071	5.113	140(4)	63(2)	77.8(0.5)	39(3)	89.1(.1)	88.3(.1)
$k = 12^3$	2.075	5.113	145.4	61.1	78.5	42.2	89.2	88.4
ScH_2	2.073[a]	4.783[c]						
k-limit	1.990	4.766	166(5)	63(2)	80(1)	51(4)	97.4(.2)	97.3(.1)
$k = 12^3$	1.992	4.766	168.8	61.7	76.9	53.6	97.5	97.4
YH_2	2.346[a]	5.195[c]						
k-limit	2.091	5.206	132(3)	58(2)	73(1)	37(3)	82.8(.1)	82.9(.1)
$k = 12^3$	2.095	5.205	132.6	56.2	73.6	38.2	82.8	82.9
LaH_2	2.151[a]	5.663[d]						
k-limit	1.697	5.710	95(2)	48(1)	53(1)	23(1)	63.5(.1)	63.8(.1)
$k = 12^3$	1.698	5.710	95.7	47.5	54.1	24.1	63.6	63.9
TiH_2	1.283[a]	4.475[e],4.463[f]						
k-limit	1.434	4.417	119(1)	155(1)	−21(4)	−18(1)	143.1(.1)	140.2(.1)
$k = 12^3$	1.435	4.417	120.5	154.3	−21	−17	143.0	140.3
ZrH_2	1.687[a]	4.794[g]						
k-limit	1.614	4.811	88(2)	157(1)	−30(4)	−34(2)	133.9(.1)	134.4(.1)
$k = 12^3$	1.615	4.810	89.0	156.3	−30	−33.7	133.9	134.5
HfH_2	1.375[a]	4.713[h]						
k-limit	1.376	4.713	88(3)	179(2)	−42(3)	−45(3)	148.5(.1)	149.7(.1)
$k = 12^3$	1.376	4.713	81.5	182.1	−49	−50.3	148.6	149.6

[a] Experimental result from Beavis [40].

[b] [41]

[c] [42]

[d] At 0 K, [43]

[e] At 300 K [44] (N.B., ground state is tetragonal below 300 K [45]).

[f]Extrapolated to TiT2, from sub-stoichiometric data using linear relationship in [46]

[g]Extrapolated to ZrH2 from sub-stoichiometric data using linear relationship in [47]

[h]For HfH1.7, from [48].

Note: The value in the parentheses for the k-limit values reflects its k-sampling uncertainty, variation with respect to

k-sampling in the large k-sampling limit.

The elastic properties are not much changed for ErH_2 if computed at the experimental lattice parameter rather than the computed a_0. Usually the PBE functional significantly overestimates the lattice constant and therefore strongly underestimates the resulting computed bulk modulus, and elastic properties computed at the experimental lattice parameter improve the accuracy of the calculations. For the rare earth dihydrides, the bare PBE lattice constant, absent zero-point effects and magnetism, yields exceptionally good agreement with experiment, PBE only <0.3% different and smaller. This fortuitously good agreement in a_0 means the predictions lack a bias from a mistaken volume. Nonetheless, that the small 0.3% change in a_0 leads to almost a 3 GPa change in the computed bulk modulus serves as a cautionary result, slight changes in the structure could result in significant changes in dependent properties such as the bulk modulus, injecting another source of error into the analysis.

The formation energy ΔE_f for a dihydride of metal M is obtained from the computed total energy of MH_2, and of the ground state of the bulk metal and hydrogen molecule, E_M and E_{H_2}:

$$\Delta E_f = E_{MH_2} - E_M - E_{H_2}. \tag{4}$$

Thermal effects are neglected in equation (4), they are minor on the scale of errors in the physical approximation of the DFT. Zero-point effects are also ignored. The largest zero-point correction to the formation energy reduces the formation energy by the zero-point vibration of the hydrogen molecule [49]. This molecular constant is partially counterbalanced by the zero-point vibrational energy of the hydrogen atom in each metal lattice. The H_2 ZPE gives a sense of the magnitude of the remaining errors in E_f and could be trivially added to the computed values for improved comparison to experiment. The ZPE had been computed in the PBE simulation context to be 0.224 eV [49] for this ZPE, but we note that it would be a more accurate and appropriate physical comparison to use the 'exact' ZPE for H_2, 0.270 eV/H_2 [50] in any such adjustment. The agreement of our computed formation energy, ΔE_f in table 1, with experimental heats of formation, as given by Beavis [40], is surprisingly good for ErH_2 and ScH_2 and still reasonable for the other hydrides, despite neglecting thermal and quantum nuclear effects. The ΔEf calculated here for ErH_2 matches a VASP-calculated value to within 3 meV [49].

The computed lattice constant for β-ErH_2, 5.113 Å, agrees well with experiment [41], 5.126 Å. Our VASP calculation gives $a_0 = 5.129$ Å, consistent with this result. This agreement between the two codes, with very different pseudopotentials and basis sets, provides additional verification of the computational methods. Our results also agree with previous VASP calculations [49], who quote $a_0 = 5.128$ Å. The computed lattice parameters for the other dihydrides presented table 1 agree well with experiment [42–48] and with previous calculations [18, 21, 22, 51–55]. The stoichiometric group-IVB dihydrides are known to be tetragonal, and adopt the cubic phase for a small range of substoichiometric H. The TiH_2 transforms into a fcc-CaF_2 phase above 310 K [45], but adopts the tetragonal structure at lower temperatures. In table 1, the first quoted experimental lattice parameter for TiH_2 is for this observed high-temperature cubic phase [44]. The substoichiometric hydrides show only a small dependence of lattice parameter on the H content over the range of H content where the cubic phase is stable, and the relationship between H content and lattice constant is roughly linear. The second quoted 'experimental' lattice parameter for TiT_2 is obtained by extrapolating to the stoichiometric ditritide using the relationship developed [46] over the range of substoichiometric tritide where the fcc phase is stable ($1.5 < T/Ti < 1.8$). This extrapolated a_0 in TiT_2 is slightly smaller, partially attributable to an isotope effect, the hydrogen vibrations causing a larger expansion in the lattice than the tritium.

Unlike for TiH_2, no cubic form of the stoichiometric ZrH_2 has been observed, but, similarly, a linear relationship was developed for the substoichiometric ZrH_{2-x} system [47], which we use to extrapolate to obtain the 'experimental' value of $a_0 = 4.794$ Å quoted in table 1 for the dihydride. The lattice constant measured for $HfH_{1.7}$ is used as the experimental value here for the dihydride; we were unable to find the experimental data to perform a comparable extrapolation from substoichiometric data for HfH_2. The agreement of the calculated a_0 with these inferred 'experimental' lattice constants for the cubic structure of group-IVB dihydrides is very good.

The C_{11}, C_{12}, and C_{44} elastic constants are computed from the strain-derivatives of the stress, the shear modulus Cs and $B(\{Cij\})$ are evaluated from these using the analytical relationships in equations (1) and (2). The alternate energy-fit calculation of $B(E)$ agrees with $B(\{Cij\})$ across this set of dihydrides, verifying the computation of B. The computed bulk modulus for

TiH_2 is less good, the difference is slightly larger, 3 GPa. This signals a potential numerical issue in the calculation, and upon investigation we discovered that this larger discrepancy was due to inaccuracy in the real space integration in the total energy evaluation. A finer real space grid, increased from 24^3 to 32^3 (or more) brought the $B(E)$ verification into the same agreement as for the other dihydrides, and had minimal effect on the elastic constants computed from the stress-strain relations. This exercise illustrates how seemingly redundant verification checks, with meaningful estimates of expected and actual uncertainties, can identify potential inaccuracies in the calculation, and thereby add greater confidence to the computational results.

Quijano, *et al* [22] had done a notably careful study of the structural properties of the group-IVB dihydrides, examining stability of the fcc versus the fct phase, additionally examining the importance of relativistic effects. Their results for fcc phase TiH_2, ZrH_2, and HfH_2 lattice parameters of cubic (4.428, 4.817, and 4.727 Å, respectively) and the bulk modulus (144, 136, 148 GPa) are closely mimicked in the current results. We find that these group-IVB cubic dihydrides all have negative shear moduli, both the C_{44} and the tetragonal shear Cs, indicating that the cubic structure is unstable to a distortion. Given that these are all known to take tetragonal ground states, this is not surprising, but the result indicates that the cubic phase does not even have a locally stable region, and, further, suggests that these shear elastic constants might be used computationally to detect the instability of the cubic phase, a notion we examine later. This structural instability leads to numerical instabilities in the assessment of the stresses in the calculation of the C_{44}. The computed stresses are negative, and not cleanly linear, so the numerical evaluations of C_{44} from the stress-strain results are much more uncertain at each k-point.

C. Rare earth cubic dihydrides

We now shift to erbium's immediate neighbors, to examine the mechanical properties of cubic dihydrides in the rare earths (RE). A search was performed on the Pearson database [56] of crystals to identify all the RE dihydrides that formed a cubic ($Fm\bar{3}m$) dihydride ground state. The cubic structure is the stable ground state across almost the entire lanthanide series. The ground state crystal structure of PmH_2 is not known. Promethium has no stable isotope and only trace amounts are found in naturally occurring ores (it has never been made). The dihydrides of Eu and Yb instead crystallize into an orthorhombic structure (Pnma, #62), rather than the cubic structure of ErH_2 and the other RE dihydrides. This could be anticipated, as Eu and Yb do not always exhibit the same [+3] oxidation state that characterizes the other rare earths, but instead can also adopt a [+2] oxidation state, i.e. with respect to the other RE atoms, they will have one fewer valence electron.

Table 2 presents the DFT results for the lattice parameter and elastic constants computed for the RE dihydrides in the fcc CaF_2 structure. Results are presented both from SEQQUESTcalculations using a linear combination of atomic orbitals (lcao) basis set (using the 3-atom primitive cell and a $20^{3 k}$-point sampling), and also from calculations using VASP with a plane wave (pw) basis. In these results, to examine the trends across the RE series, the lcao calculations for EuH_2 and YbH_2 use pseudopotentials with a [+3] oxidation state (rather than the [+2] oxidation state pseudopotential), and in the cubic (fcc) structure rather than the Pnma orthorhombic structure observed in nature.

Table 2: Lattice parameters a_0 (Å), elastic constants (GPa), and computed Debye temperature (K) for cubic RE dihydrides

Hydride		a_0(Exp.)[a]	a_0(DFT)	C_{11}	C_{12}	C_{44}	$B(\{C_{ij}\})$	C_s	Θ_D (DFT)[d]
LaH_2	lcao	5.667	5.710	93.5	48.5	53.7	63.5	22.5	364
	pw		5.658	89.3	48.5	47.3	62.1	20.4	343
CeH_2	lcao	5.581	5.622	98.6	50.6	56.2	66.6	24.0	369
	pw		5.437	90.7	55.7	46.7	67.4	17.5	324
PrH_2	lcao	5.518	5.552	102.2	51.6	58.4	68.5	25.3	374
	pw		5.555	102.0	52.1	53.1	68.7	25.0	363
NdH_2	lcao	5.464	5.487	106.3	52.8	60.6	70.7	26.7	376
	pw		5.490	107.4	54.0	55.5	71.8	26.7	366
PmH_2	lcao	[b]	5.428	111.8	54.5	63.1	73.6	28.7	382
	pw		5.431	111.6	55.6	58.3	74.3	28.0	372
SmH_2	lcao	5.374	5.375	116.8	55.5	65.5	75.9	30.6	383
	pw		5.385	115.8	56.3	60.2	76.1	29.8	372
EuH_2	lcao	[c]	5.325	121.8	56.4	67.8	78.2	32.7	388
GdH_2	lcao	5.303	5.279	126.8	57.5	70.4	80.6	34.7	388
	pw		5.285	122.3	58.7	65.4	79.9	31.8	374
TbH_2	lcao	5.246	5.234	130.9	58.2	72.4	82.4	36.3	391
	pw		5.237	127.0	59.5	68.1	82.0	33.8	379
DyH_2	lcao	5.201	5.191	135.4	59.5	74.3	84.8	37.9	392
	pw		5.200	131.7	62.0	68.7	85.2	34.8	377
HoH_2	lcao	5.165	5.151	139.5	60.2	76.8	86.6	39.7	395
	pw		5.160	133.7	62.6	73.5	86.3	35.6	382
ErH_2	lcao	5.123	5.113	144.4	61.3	78.2	89.0	41.6	396
	pw		5.122	137.5	63.7	74.6	88.3	36.9	382
TmH_2	lcao	5.090	5.074	147.5	62.4	81.2	90.8	42.5	399
	pw		5.090	140.1	66.0	75.4	90.7	37.1	381
YbH_2	lcao	[c]	5.039	150.3	63.3	82.7	92.3	43.5	397
LuH_2	lcao	5.033	5.004	153.7	64.5	84.4	94.2	44.6	398
	pw		5.020	145.7	66.6	78.5	93.0	40.0	381

[a] Experimental lattice parameter for LaH_2 and CeH_2 from [57]; PrH_2 through LuH_2 from [58], except GdH_2 from [59].

[b] PmH2 has never been made.

[c] Computed result shown is for a pseudopotential with the [3+] oxidation state.

[d] Computed from elastic constants using relation from [60].

The agreement between the lcao and pw calculations for each RE dihydride is very good across the series. The differences in elastic constants between the pw and lcao results are consistent with the numerical uncertainties with respect to k-sampling reported in table 1. The bulk modulus, as illustrated in figure 2 for ErH_2 and analyzed in table 1 for several dihydrides, exhibits much finer convergence with k-point sampling, so that the total uncertainty in this quantity is dominated by the form of the equation of state, estimated above to be ~1 GPa, rather than k-sampling dependence. The differences between the lcao and pw results for B in the RE series are typically ~1 GPa, consistent with the estimated uncertainty in the model form used to extract the elastic properties. Because the k-point sampling used in computing the elastic constants for the RE series are well within the asymptotic range used for computing the k-limit result, the uncertainties (with respect to the k-limit value) due to k-sampling in the values quoted in table 2 will be the same as those presented for the stable cubic dibidrides in table 1, i.e. ~4 GPa for C_{11} and Cs, ~2 GPa for C_{12}, ~3 GPa for C_{44}. The values in table 2 are quoted to tenths of GPa only to facilitate finer comparisons between the lcao and pw calculations, and also to analyze trends across the series, this finer precision is otherwise physically meaningless. That the results using very different pseudopotentials and basis sets, using different methods for extracting elastic constants for the results, and using different cells and k-point sampling, are in good agreement lends greater confidence to the numerical predictions. The different approaches give chemically and numerically consistent results.

The computed properties of the RE dihydrides are similar to one another and follow a monotonic trend from one end of the lanthanide series to the other. The small exception is the plane wave result for C_{12} (and the Cs derived from it) for CeH_2. Ce is another rare earth that often adopts an oxidation state different from [+3]: it can promote its f-electron into the valence electrons to access a [+4] oxidation state. The lcao result uses a pseudopotential leaving exactly one f-electron in the core, imposing the [+3] oxidation state that is consistent with its neighbors, while the PAW pseudopotential of the PW calculation does not make this constraint on the oxidation state.

The lattice parameter steadily shrinks across the series, the effective size of the RE atom getting smaller as the nuclear charge increases. The computed a_0 are in very good agreement with experiment [43, 58, 59]. The PBE calculation gives a_0 slightly larger than experiment early in the series and then trends to become slightly smaller than experiment at the end of the series.

All the computed shear moduli are positive, indicating that the RE dihydrides are (at least locally) stable in the cubic phase, consistent with the experimental observations [56]. The elastic constants all steadily become stiffer, as would be expected as the lattice parameter decreases for a given structure. This supports the notion that the bonding and electronic structure are very similar across the RE dihydrides, the dominant difference being the shrinking in size in the RE atom across the series. Overall, DFT with the PBE functional appears to give a very good description of the mechanical properties of cubic dihydrides. Using the relations developed by Deligoz *et al* [60], an estimate for the Debye temperature can be obtained directly from the elastic constants. The computed results are presented in the last entry in table 2. Unremarkably, the variation across the lanthanides is small.

D. Group IVB tetragonal dihydrides

The negative shear elastic constants for the group-IVB dihydrides presented in table 1 signal that the cubic phase is unstable to a distortion for these materials. Table 3 summarizes the structural properties of the tetragonal (fct) ground state group-IVB dihydrides. In agreement with experiment[61–63], and recent theory [21, 22], the tetragonal form with c/a <1 is the ground state for all of these, favored over the c/a > 1 form. The preference for the c/a < 1 is weak, only 4 meV f.u.$^{-1}$ in TiH$_2$ and 12 meV f.u.$^{-1}$ in both ZrH$_2$ and HfH$_2$, but meaningful within the numerical uncertainties present in these calculations. These lattice parameters and energy preferences match well with the relative structural distortion energies computed by Quijano, *et al* [22].

Table 3: Structural properties for tetragonal metal dihydrides

		$-\Delta E_f$ (eV)	a_0 (Å)	c_0 (Å)	c/a	V(fct) (Å^3)	V(fcc) (Å^3)	ΔE(fcc) (meV f.u.$^{-1}$)
TiH$_2$	Exp.[a]		4.528	4.279	.945			
c/a < 1	k-limit	1.445	4.531(2)	4.184(4)	0.924(1)	21.47	21.54	−11
	$k = 12^3$	1.445	4.524	4.198	0.928	21.48	21.54	−10
c/a > 1	k-limit	1.441	4.329(1)	4.591(2)	1.061(1)	21.51	21.54	−7
	$k = 12^3$	1.440	4.331	4.587	1.059	21.51	21.54	−5
ZrH$_2$	Exp.[b]		4.9825(9)	4.4488(9)	0.8929(8)			
	Exp.[c]		4.975(3)	4.447(3)	0.894(1)			
c/a < 1	k-limit	1.646	5.011(3)	4.400(6)	0.878(1)	27.62	27.82	−32
	$k = 12^3$	1.646	5.018	4.385	0.874	27.61	27.83	−31
c/a > 1	k-limit	1.634	4.647(1)	5.135(3)	1.105(1)	27.73	27.82	−20
	$k = 12^3$	1.634	4.644	5.142	1.107	27.73	27.83	−19
HfH$_2$	Exp.[d]		4.919	4.363	0.886			
c/a < 1	k-limit	1.418	4.935(3)	4.267(5)	0.865(2)	25.98	26.17	−42
	$k = 12^3$	1.418	4.942	4.254	0.861	25.98	26.17	−42
c/a > 1	k-limit	1.406	4.525(2)	5.094(3)	1.126(2)	26.08	26.17	−30
	$k = 12^3$	1.405	4.520	5.107	1.130	26.08	26.17	−32

[a] [45].
[b] At 293 K, [61].
[c] At 294 K, [62].
[d] At 300 k, [63].

[a] [45].

[b] At 293 K, [61].

[c] At 294 K, [62].

[d] At 300 k, [63].

We performed additional calculations in the LDA, and obtained the same structural preference for TiH_2 and ZrH_2, of roughly the same magnitudes as with the PBE functional. The result by Ackland favoring $c/a > 1$ for ZrH_2 [19] is probably attributable to an inadequate k-point sampling (the quoted k-sampling [19], a sampling less than the 12^3 asymptotic threshold for reliable results determined in this study). The result favoring $c/a > 1$ for TiH_2 by Wolf and Herzig [15], using an all-electron method, is more mysterious. But the energy preference in TiH_2 involved is so small, 3–4 meV f.u.$^{-1}$, and may not be physically significant; this is perhaps within the numerical accuracy that can reasonably be expected of older computational DFT methods.

Negative shear elastic constants signal the instability of the cubic structure to a distortion, but do not predict the form that this distortion will take. Given that both the planar shear and tetragonal shear constant presented in table 1 for the group IVB fcc dihydrides are negative, any candidate distortion is plausible. A negative Cs alone would be suggestive of a tetragonal distortion, but that a negative C_{44} accompanies it indicates other distortions would also be energy-lowering. We investigated rhombohedral crystal distortions, compressing and expanding the cubic structure along the (1 1 1)-axis, for both TiH_2 and ZrH_2. Both compression and expansion lead to energy-lowering distortions, and these hexagonally strained structures are favored over the fcc structure just as both tetragonally-strained fct structures are favored over the fcc structure.

The ZrH_2 favors the (1 1 1)-compressed rhombohedral structure over the fcc structure by 7 meV, and the (1 1 1)-expanded structure is favored by 5 meV, somewhat smaller than energy-lowering obtained with the tetragonally distorted structures. The TiH_2 also favors the (1 1 1)-compressed hexagonal structure over (1 1 1)-expanded structure, at 6 and 1 meV f.u.$^{-1}$ below the cubic structure. We note that the compressed hexagonal structure is actually more stable than the $c/a > 1$ tetragonal distortion. These results suggest that the cubic fcc structure is a local maximum in the structure, surrounded by a valley of lower energy distortions in all directions. This would more readily facilitate the fct $\rightarrow$ fcc transition observed at 310 K [45] for TiH_2, the barrier to a homogeneous distortion would be lower than a barrier represented by a transition through the cubic structure.

The lattice parameters for the fct structures show slightly more variability than for the cubic structure with k-point sampling. Whereas the cubic structures converged to less than 0.001 Å at a 12^3 k-sample, the k-sampling variability in the optimized a_0 and c_0 for the fct structure can be as much as 0.2%, a numerical uncertainty that is as large or larger than the uncertainties in the experimental measurements. However, this variation in the computed lattice parameters a_0 and c_0 is tightly correlated. The total cell volume is converged to better than 0.01 Å^3 already at 12^3, indicating that the details of the sampling of the electronic states at the Fermi level have a more significant effect on the shape of the structure, the c/a ratio, than its volume.

Table 4 presents the computed elastic constants of the ground state ($c/a < 1$) tetragonal metal dihydrides. The uncertainties in the evaluated elastic constants due to k-sampling are comparable, if slightly larger than for the cubic structure presented in table 1 (leaving aside the pathologies that manifest in the calculation of C_{44} due to the instability of the cubic structure). The tetragonal crystal has a lowered symmetry from the cubic crystal, there are more independent elastic constants to compute than for the fcc structure, and the calculations are more computationally demanding, making it more challenging to verify the results. A couple of the consistency checks we performed are presented in table 4.

Table 4: Computed elastic properties (in GPa) for tetragonal metal dihydrides

	C_{11}	C_{12}	C_{13}	C_{31}	C_{33}	C_{44}	C_{66}	$B(\{C_{ij}\})$	$B(E)$
TiH$_2$									
k-limit	178(2)	149(2)	119(2)	120(2)	159(2)	17(1)	59(2)	140.2(.1)	—
$k = 12^3$	178.1	146.2	120.9	122.1	155.5	18.7	57.7	140.0	139.8
ZrH$_2$									
k-limit	177(2)	143(2)	106(2)	107(1)	151(3)	40(2)	74(4)	130.7(.2)	—
$k = 12^3$	181.0	138.7	106.9	106.9	149.4	35.4	65.6	130.5	131.1
PBE($k = 12^3$)[a]	165.6	140.9	106.8	—	145.5	30.5	60.6	130	—
PBE($k = 20^3$)[b]	166	149	109	—	149	26.5	55.8	133	—
HfH$_2$									
k-limit	190(2)	159(2)	118(2)	118(2)	175(4)	49(2)	76(6)	146.3(.2)	—
$k = 12^3$	195.6	154.8	117.4	117.4	175.3	44.0	66(3)	146.3	147.0

[a][54].

[b][52].

First, we computed C_{13} and C_{31} independently. To obtain C_{11}, an axial strain along the c-axis is used, and the C_{13} can be obtained from the derivative of the off-axis stress within the same strain. To obtain C_{33}, it is necessary to perform a strain along the a-axis. The C_{31} can be obtained from a stress-derivative along a different axis than this strain. By symmetry, C_{13} and C_{31} should be identical, but numerically will differ due to different real-space grid spacing and reciprocal space grid spacing along the direction of the strain. This difference can be used as a consistency check, to assess how well the elastic constant calculation is numerically converged with respect to the integration grids. The difference between C_{13} and C_{31} is ~1 GPa or less for all the tetragonal dihydride calculations, less than the expected k-sampling uncertainty for any individual elastic constant.

A second consistency check involved independent calculations of the bulk modulus. The bulk modulus for these tetragonal crystals was first obtained from the computed elastic constants using the analytic relationship of equation (3). For the 12^3 k-sample, we also obtained the energy-fit bulk modulus $B(E)$ from an analytic derivative of a Birch–Murnaghan equation of state, fit to a series of optimized cells under a series of compressive and tensile pressures. This more holistic test assesses both the consistency of the energy and stress calculations and also the efficacy and proper application of the relationship between the elastic constants and the bulk modulus for a tetragonal crystal given in equation (3). The agreement between the stress-derived and the energy-fit bulk modulus is better than 1 GPa, once more within the expected accuracy of the elastic constant calculations, and also matches the known uncertainty in the numerical fit to the equation of state.

For ZrH$_2$, there are two earlier sets of results for the elastic constants using PBE [52, 54], that are reasonably well-converged and compatible with the current study. Our computed elastic constants match the previous results of Zhang *et al* [54], and Olsson, *et al* [52] reasonably well. Our calculations were derived from stress-strain relationships, and these authors instead used a series of energy-fits to obtain their elastic constants, which might explain the modest differences between the different sets of results. All the IVB dihydrides pass the stability tests for the tetragonal structure, and, despite differences in an individual elastic constant as large as 10 GPa (in C_{11}), the resultant bulk modulus still agrees well between these different DFT calculations.

E. Off-stoichiometry stability

These metals tend to form hydrides over a wide range of H stoichiometry; for instance, erbium and the other RE elements can be readily hydrided to form up to a trihydride [40]. The stability of the different erbium hydrides, ErH, ErH$_2$, and ErH$_3$, was investigated with first-principles calculations [49]. The interest in erbium for tritium storage is in the cubic structure. The single phase cubic form is stable for Er:H ratios ranging from 1.9 to 2.1 [40]. In the tritide, the stoichiometry of the initial manufactured sample will change with aging, as tritium decays into helium and modify the mechanical properties of the material. Helium replaces hydrogen in the lattice, and then the helium diffuses away and coalesces into bubbles, leaving behind a substoichiometric tritide. These compositional changes can potentially affect the elastic properties (and the stability of the cubic phase) and therefore are important for modeling the mechanical behavior of aging tritides.

To examine the trends of the elastic properties with a change in stoichiometry, we adopt an averaged-atom approach, where every hydrogen H in the lattice is replaced with a hydrogen atom Hz with a modified nuclear charge z. The advantage of this averaged-atom approach is that it preserves the full symmetry of the crystal in the small primitive cell while exploring off-stoichiometry effects. For $z < 1$, this approach mimics the average population of $x = 2(1 - z)$ vacancies on the H site in the structure, i.e. a random ErH_{2-x} structure, while for $z > 1$, this approach mimics a random $ErH_{2-x}Hex$ structure where $x = 2(z - 1)$. The results for the elastic properties of ErHz are presented in table 5 for $z = 0.75$ to $z = 1.50$, using the theoretical lattice constant appropriate to each Hz.

Table 5: Computed structural and elastic properties (in GPa) for cubic ErH$_2^z$ ($k = 20^3$)

	a_0(Å)	C_{11}	C_{12}	C_{44}	C_s	$B(\{C_{ij}\})$
$ErH_2^{0.75}$	5.226	124.1	49.0	58.8	37.5	74.0
$ErH_2^{1.00}$	5.113	144.4	61.3	78.2	41.6	89.0
$ErH_2^{1.25}$	5.060	117.9	80.3	73.1	18.8	92.9
$ErH_2^{1.50}$	5.088	82.6	105.5	−39.1	−34.3	82.6
ErHHe	5.238					

The averaged-atom results indicate that C_{11} decreases and C_{12} increases with replacement of H with He in the lattice, culminating with $C_{11} < C_{12}$ for $ErH_2^{1.5}$, indicating a structural instability. The shear moduli C_{44} and Cs decrease and then sharply drop at $z = 1.5$. The optimized tetragonal structure for $ErH_2^{1.5}$ is more stable by 57 meV f.u.$^{-1}$ than the cubic crystal. Given that $ErH^{1.5}$ is isovalent—the same number of valence electrons—with HfH_2, this is not surprising. The lattice parameter first decreases and then turns slightly larger with increasing He concentration. The bulk modulus does not change much across this composition change; only the shear moduli exhibit strong changes.

The numerical results for the averaged-atom structures should not be taken too literally. An ErHHe fcc structure (where every second H atom in ErH_2 is replaced by a He atom), isovalent with the $ErH_2^{1.5}$, has a significantly larger lattice constant than $ErH_2^{1.5}$. This cubic ErHHe model, nonetheless, exhibits the same structural instability as $\text{ErH}_2^{1.5}$, although with a larger margin, 157 meV, favoring the tetragonal distortion.

The effects of transmutation of H into He in the lattice have been studied previously [64] using an explicit-atom approach. Using a larger supercell (96 atoms), between one and six H atoms in a larger 96 atom ErH_2 supercell were replaced with He atoms, probing a gradation in He composition much finer than considered here. For small x, those results show C_{12} decreases rather than increases, and the bulk modulus decreases significantly, and over a much smaller gradation of He composition. This indicates that effects for small concentrations of clustered He on H lattice sites the mechanical properties will be different than for dispersed concentrations of He.

Results for dispersed He should be taken as illustrative of trends rather than as physically predictive. The tritides are not observed to sustain even small concentrations of He located on H sites, but rather the He generated from decayed tritium diffuses away and segregates into bubbles almost immediately [9]. The macroscopically observed changes in mechanical properties of the aging material are dominated by microstructural effects: the bubble density, shape, and size, the pressures within the He bubble interacting with the elastic properties of the remaining tritide matrix, and the dislocations and mesoscale features this interaction generates [3]. The components that go into such an analysis include the elastic properties of the matrix, but that matrix will retain minimal residual He concentrations. The computational results are more useful, indicative of incipient local structural instabilities, if a few He atoms, before they diffuse into bubbles, are found within close proximity of each other. This instability to local distortions due to heterogeneity of local He concentrations could perhaps facilitate increased diffusion of He, and, more speculatively, the statistical likelihood of a near approach of several He might be a contributor to the initial nucleation of bubbles. Tuning ErH_2^z to $z = 1.5$ creates a system isovalent with HfH_2. If several H are replaced by He in a local region, this might favor local distortions and strains.

Simulations of the substoichiometric tritide, on the other hand, should provide more realistic models of reality, either for a tritide that was less than fully loaded, or because He segregation in the aging tritide leaves behind a substoichiometric matrix. The results in table5 indicate that with an average of 1/4 H vacancies, the lattice becomes slightly (2%) larger and all the elastic constants get slightly softer (~20%)—relatively small changes considering the large composition change. XRD studies of aging ErT_2 show a linear anisotropic increase in lattice constant of ~1% when $1/8 \, {}^3\text{H}$ have converted to ${}^3\text{He}$ [65], consistent with this result. For use in a macroscopic model of the mechanical response of the aging material, this kind of variability in the elastic properties in the matrix could be readily incorporated, or even, given the modest and isotropic change in the elastic properties with (sub)stoichiometry, might be safely ignored.

The examination of the structural phase stability with this stoichiometry yields insight into the nature of the bonding that determines the structure. Consider again the group IVB metals: all crystallize in the tetragonal phase for the dihydride, but all adopt the cubic phase over a range of substoichiometric hydrogen. The single-phase cubic structure, the δ-phase, is present for a M:H ratio of 1.5 to 1.8 for Ti [46], ranging from 1.5 to 1.7 for Zr [47], and appears below ~1.7 for Hf [48]. Above this δ-phase, the lattice converts to the ε-phase (fct structure), below this range it becomes a mixed α(hexagonal)-δ phase, essentially hydrogen dissolved in the pure-metal lattice. This trend of conversion to the cubic phases is consistent with a rigid band picture [19], where the band structure of these different hydrides are very similar to one another, and the principle difference is the number of valence electrons that fill that band structure. The MH_1 (or $MH_2^{0.50}$) would be isoelectronic with the RE cubic dihydrides (the RE atoms having three valence electrons, the IVB atoms four). Within an average-atom picture, the effective substoichiometric $HfH_2^{0.75}$ already favors the cubic structure, even before the $HfH_2^{0.5}$ that is isovalent. As one adds electrons to the ErH_2 lattice, transmuting to $ErH_2^{1.5}$, the crystal distorts to the tetragonal form of the isovalent HfH_2.

Figure 4 plots the computed shear elastic constants, C_{44} and Cs, as one varies Hz, to fill the valence band structure with more ($H \rightarrow H^{z=1.25}, H^{z=1.50}$) and fewer ($H \rightarrow H^{z=0.75}, H^{z=0.50}$) electrons. The elastic constants for cubic LaH_2^z, and ErH_2^z are shown, along with the HfH_2^z versus the number of electrons filling the valence band structure. Plotting against valence electron count, all three show similar behavior, with negative shear moduli at six valence electrons f.u.$^{-1}$ signaling the instability of the cubic phase to a distortion. This supports a rigid band picture, where the filling of the band structure rather than the chemical nature of the metal determines the structure. Whether one attributes this to a Jahn–Teller-like distortion due to a high density of states at the Fermi level for the group IVB dibydrides, or disputes that interpretation otherwise through a detailed analysis of the HfH_2 band structure [22], it is unambiguous from comparing these disparate dihydrides that it is the filling within the band structure with six valence electrons that triggers the instability.

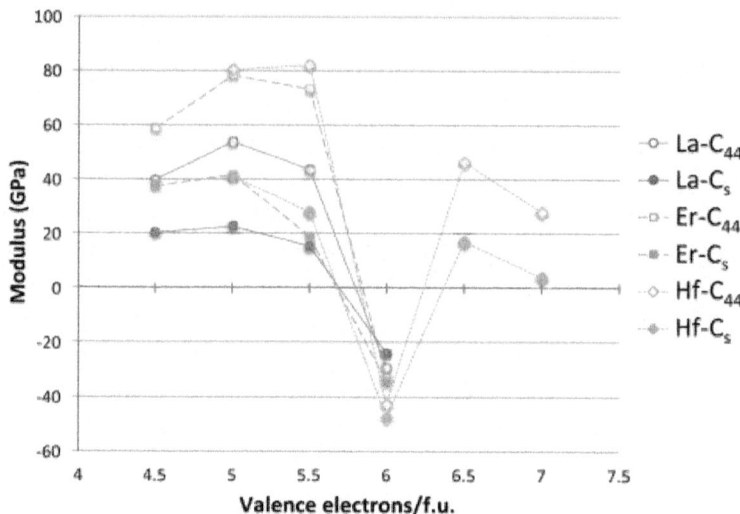

Figure 4: The dependence of the shear elastic constants in the cubic structure MH_2^z on the number of valence electrons, for M = La, Er, Hf, obtained by modifying the core charge on the H to give a non-integer total charge for Hz. The open(filled) symbols present the computed C_{44} (Cs) values, the (red) circles connected by solid lines are for LaH$_2^z$, the (blue) squares by dashed lines for ErH$_2^z$, and the (orange) diamonds by dotted line for HfH$_2^z$. The LaH$_2$ and ErH$_2$, at five valence electrons per f.u., have a positive plateau in the shear constants, but these drop precipitously at $z = 1.5$ and become negative, the negative shear constants indicating an instability to distortion at the electron count (6 E^- f.u.$^{-1}$) that makes this isovalent with fct-HfH$_2$. Conversely, reducing the electron count in HfH$_2^z$ stabilizes the fcc structure, and the HfH$_2^{0.5}$ exhibits the same positive plateau in the shear elastic constants as the isovalent LaH$_2$ and ErH$_2$ with the same number of valence electrons.

F. Analysis and relation to experiment

The parameters crucial for modeling bubble growth and fracture in aging tritides like ErT$_2$, such as surface energies and mechanical properties [11], are exceedingly difficult to obtain with accuracy from experimental measurements, indicating a first-principles approach as a means to obtain the materials properties [4] is needed to model the bubble evolution, growth and ultimate fracture of the material. The first-principles calculations in this paper provide detailed analysis of the mechanical properties, such as the elastic constants, paying careful attention to the numerical uncertainties that attend the DFT calculations.

The thin film materials are polycrystalline, so the quantities of interest for microstructural modeling of the materials are averaged quantities, such as the bulk shear modulus μ, and the Young's modulus E. For single crystals, these quantities can be obtained straightforwardly from the computed elastic constants [33, 39], but assessment of these quantities in a polycrystalline material adds further numerical uncertainties into the analysis. Analyses by Reuss and Voigt for the shear modulus, G_R and G_V, place bounds on the bulk shear modulus in a polycrystalline material. Both of these bounds incorporate a numerical uncertainty roughly equivalent to the numerical uncertainty in computing the crystalline shear elastic constants. Conventionally, the 'predicted' bulk shear modulus G_H is defined as the average of these [33], but in many cases this convention injects additional numerical uncertainties into the 'predicted' shear modulus, as the difference in the G_R and G_V bounds can be significant, larger than 10 GPa, for the hydrides.

This shear modulus G, with the bulk modulus B, is used to estimate Young's modulus E:

$$E = \frac{9BG_H}{3B + G_H}$$

(5)

which, being proportional to G_H, carries roughly the same proportion of numerical uncertainty as G. The consequence is that the numerical calculations for these essential quantities, even obtained from precise DFT calculations, embody numerical uncertainties as large as the rather substantial uncertainties derived from the experimental analysis of these quantities, even discounting the physical uncertainties associated with the accuracy of the DFT approximation.

Table 6 presents the results of the DFT analysis for these bulk mechanical properties, along with their experimental counterparts. The Debye temperature can also be used as an experimental comparison as it can be computed directly from shear moduli [60]. The values in parenthesis are the numerical uncertainties in the DFT predictions. These incorporate uncertainties inherent within the Reuss G_R and Voigt G_H bounds from the numerical k-limit uncertainties of the underlying shear elastic constants from which these are computed, and then adds an uncertainty given estimating the shear modulus G_H as an average midway between these bounds. It is this total uncertainty in G_H that can be compared meaningfully to the appropriate experimental quantity. Even if the DFT approximation were physically perfectly accurate, these numerical uncertainties from the averaging and k-sampling would remain. The shear modulus across these materials can only be predicted to within 10–20%, roughly half of this uncertainty due to the k-convergence in the computation of the elastic constants, and the other half in the estimation of G_H within the bounds given by G_R and G_V.

Table 6: Experimental and computed moduli (in GPa) and Debye temperatures (K)

		Young's modulus	G_R—reuss shear bound	G_V—voigt shear bound	G_H—shear Modulus	Bulk modulus	Debye[a] temperature (K)
ErH$_2$	Exp.	148 ± 20			60 ± 10	97 ± 4	381
	GGA-PBE	144 (17)	55.2	62.1	59 (7)	89 (1)	389.8 (6.5)
YH$_2$	Exp.	135 ± 20			55 ± 10	90 ± 7	537
	GGA-PBE	136 (15)	52.5	58.6	56 (6)	83 (1)	521 (8)
LaH$_2$	Exp.	36 ± 6			14 ± 3	24 ± 3	243
	GGA-PBE	95 (10)	34.6	41.2	38 (4)	89 (1)	365.5 (6.5)
TiH$_2$	Exp.	100 ± 15			40 ± 7	66 ± 5	
(fct)	GGA-PBE	68 (14)	20.8	27.1	24 (5)	143 (1)	
ZrH$_2$	Exp.	175 ± 20			70 ± 10	115 ± 7	
(fct)	GGA-PBE	113 (24)	35.4	47.7	42 (9)	131 (1)	

[a]Experimental Debye temperature from [66].

Young's modulus E is, to lowest order, proportional to G_H, see equation (5), and thereby inherits the same proportion of uncertainty as G_H, which amounts to 10–24 GPa, uncertainties comparable to those seen in the experimental measurements. Once more, in the face of such numerical uncertainties, it is pointless to quote G_H or E to finer than 1 GPa.

The experimental and first-principles-based results for ErH$_2$ and YH$_2$ exhibit very good agreement, the difference in G_H being ~1 GPa. We reiterate that this close agreement, while gratifying, is not very meaningful; both the experimentally derived and first-priniciples computed G_H carry 10 GPa uncertainties. The bulk modulus comparison, where the PBE results are ~10% smaller than the experimentally inferred value, is more typical of the kind of agreement that can typically be expected between DFT and experiment. These results both confirm (validate) the accuracy of the DFT for the elastic properties of the dihydrides, and lend greater confidence to the computed values of these mechanical properties, greater confidence than in the rather involved analysis to obtain these values from experiment. The 10% differences in B are expected, and the computations offer refined guidance for microstructural models.

The comparisons between experimental and computational values for LaH$_2$, TiH$_2$ and ZrH$_2$are much worse. The results for ZrH$_2$ are almost within agreement, strictly speaking, invoking the full combined uncertainties in both the experimentally-derived values and the DFT-derived values. However, the results for LaH$_2$ and TiH$_2$ are clearly well outside of the range of the uncertainties of the experimental analysis and the DFT analysis. These comparisons illustrate the value of using the experimental and computational approaches in concert for such a study. The physical uncertainties in DFT, particularly for B are known to be modest. The PBE is typically prone to be too soft, by 10–20%; this accuracy of DFT for the bulk modulus is confirmed for ErH_2 and YH_2. The likely explanation of the difference between the modeled cubic materials and the experimental results is the difficulty in manufacturing samples of controlled known phase. Both LaH_2 and TiH_2 have complex phase diagrams, and multiple phases in close proximity. Slight variations in manufacturing procedure will produce a variety of phases, possibly even mixed phases, corrupting the analysis of elastic properties of the ideal dihydride material. The problem in ZrH_2 is similar but less extreme: most likely the cubic phase was also in the manufactured material. No x-ray diffraction results were taken, the specific phases or their proportion present in the LaH_2 and TiH_2 samples are not known. What this comparison exposes is the tremendous challenges in obtaining accurate assessment of the mechanical properties from nanoindentation experiments for dihydride films. For these materials, the DFT-derived results can be used with greater confidence in microstructural models than analyses descended from the nanoindentation experiments.

The Debye temperature estimated for ErH_2 and for YH_2 is in extraordinary good agreement with the measured experimental values, as is the measured Θ_D for LuH_2 (361 K) [66] to the computed value in table 2. This would appear to validate the relationship between shear moduli and Θ_D proposed by Deligoz, et al [60] The discrepancy is larger for LaH_2, experiment measuring 243 K [66] while the DFT result predicts 366 K. At this point, it is not possible to state whether this is a failing in the model, or a failing in the PBE functional, or whether the experimental analysis is in some manner flawed. It would be useful to revisit the experimental measurements of Θ_D in light of the sample preparation difficulties encountered in attempting to measure elastic properties in LaH_2, to resolve this discrepancy.

IV. Summary and conclusions

The mechanical properties of cubic and tetragonal dihydrides, focusing on the rare earths and nearby transition metals, have been analyzed using first-principles DFT methods. Specific attention was invested in elucidating the numerical accuracy of the methods used in the calculations. Directly computed quantities such as the elastic shear constants can be evaluated to no better than 3–4 GPa, and derived quantities such as the shear modulus or Young's modulus can only be predicted to no better than 10–20%, entirely distinct from the uncertainties due to the physical accuracy of the DFT approximation. The results as a function of composition explored the underlying physics in the phase change from a cubic form to a tetragonal form. The stability of the cubic structure could be probed using the stability criteria of the shear elastic constants, and is attributed to the number of electrons in the valence band structure. This supports the notion of a rigid band picture dictating the stability of the cubic and the tetragonal structure. The results for the mechanical properties of the various dihydrides considered, such as formation energies, lattice constants, and elastic constants, are shown to be in reasonably good agreement with what limited experimental data is available, and, furthermore, is able to discriminate experimental results that are suspect because of difficulties in manufacturing the materials. This accuracy suggests that the computed DFT data can be used effectively and confidently in microstructural models of tritide aging where experimental data is unavailable, with a realistic estimate of the level of quantitative confidence in the results.

ACKNOWLEDGMENTS

Sandia is a multiprogram laboratory managed and operated by Sandia Corporation, a wholly owned subsidiary of Lockheed Martin Company, for the United States Department of Energy's National Nuclear Security Administration under contract DE-AC04-94AL85000.

REFERENCES

1. Spulak R G Jr 1987 On helium release from metal tritides *J. Less-Common Met.* **132** *L17*

2. Snow C S, Brewer L N, Gelles D S, Rodriguez M A, Kotula P G, Banks J C, Mangan M A and Browning J F 2008 Helium release and micro structural changes in Er(D,T)2−xHex films J. Nucl. Mater. **374** *147*

3. Knapp J A, Browning J F and Bond G M 2009 Evolution of mechanical properties in ErT2 thin films J. Appl. Phys. **105** *053501*

4. Snow C S, Browning J F, Bond G M, Rodriguez M A and Knapp J A 2014 3He bubble evolution in ErT2: a survey of experimental results J. Nucl. Mater. **453** *296*

5. Evans J H 1977 An inter bubble fracture mechanism of blister formation on heium-irradiation metals J. Nucl. Mater. **68** *129*

6. Evans J H 1978 The role of implanted gas and lateral stress in blister formation mechanismsJ. Nucl. Mater. **76–7** *228*

7. Wolfer W G and Drugan W J 1988 Elastic interaction energy between a prismatic dislocation loop and a spherical cavity Phil. Mag. A **57** *923*

8.   Wolfer W G 1988 The pressure for dislocation loop punching by a single bubble Phil. Mag. A**58** *285*

9.   Wolfer W G 1989 Dislocation loop punching in bubble arrays Phil. Mag. A **59** *87*

10. Cowgill D F 2005 Helium nano-bubble evolution in aging metal tritides *Fusion Sci. Tech.* **48**539

11. Cowgill D F, Somerday *B et al (*ed) 2009 Proc. of the 2008 Int. Hydrogen Conf. Jackson, WY(ASM International, Materials Park, OH, 7–10 september 2008) Effects of Hydrogen on MaterialsD F Cowgill, B Somerday *et al (*ed)

12. Cowgill D F 2004 Helium nano-bubble evolution in aging metal tritides *Sandia National Laboratories Report SAND2004-1739*

13. Cowgill D F 2006 Helium bubble linkage and the transition to rapid He release in aging Pd tridide *Sandia National Laboratories Report SAND2006-7779*

14. Switendick A C 1984 Electronic structure of Group IV hydrides and Their alloys J. Less-Common Met. **101** *191*

15. Switendick A C 1987 Electronic structure and total energy calculations for transition metal hydrides J. Less-Common Met. **130** *249*

16. Wolf W and Herzig P 2000 First principles investigations of transition metal dihydrides, TH2: T = Sc, Ti, V, Y, Zr, Nb; energetics and chemical bonding J. Phys.: Condens. Matter **12** *4535*

17. Wixom R R, Browning J F, Snow C S, Schultz P A and Jennison D R 2008 First principles site occupation and migration of hydrogen, helium, and oxygen in β-phase erbium hydride J. Appl. Phys. **103** *123708*

18. Peng S M, Yang L, Long X G, Shen H H, Sun Q Q, Xu X T and Gao F 2011 Bond-order potential for erbium-hydride systems J. Phys. Chem. C **115** *25097*

19. Chen R-C, Li Y, Dai Y-Y, Zhu Z-Q, Peng S-M, Long X-G, Gao F and Zu X-T 2012 Ab initio study of H and He migrations in β-phase Sc, Y, Er hydrides Chin. Phys. B **21** *056601*

20. G J 1988 Embrittlement and the bistable crystal structure of zirconium hydride Phys. Rev. Lett. **80** *2233*

21. Kul'kova S E, Muryzhnikova O N and Naumov I I 1999 Electronic structure and lattice stability in the dihydrides of titanium, zirconium, and hafnium Phys. Solid State **41** *1763*

22. Xu Q and Van der Ven A 2007 First-principles investigation of metal-hydride phase stability: the Ti-H system Phys. Rev. B **76** *064207*

23. Quijano R, de Coss R and Singh D J 2009 Electronic structure and energetics of the tetragonal distortion for TiH2, ZrH2, and HfH2: a first principles study Phys. Rev. B **80** 184103

24. Mehl M J 2000 Occupation number broadening schemes: choice of 'temperature' Phys. Rev.B **61** *1654*

25. Mattsson A E, Schultz P A, Desjarlais M P, Mattsson T R and Leung K 2005 Designing meaningful density functional theory calculations in materials science: a primer Modelling Simul. Mater. Sci. Eng. **13** *R1*

26. Perdew J P, Burke K and Ernzerhof M 1996 Generalized gradient approximation made simple Phys. Rev. Lett. **77** *3865*

27. Perdew J P and Zunger A 1981 Self-interaction correction for density-functional approximations for many-electron systems Phys. Rev. B **23** *5048*

28. Schultz P A SeqQuest program unpublished see: http://dft.sandia.gov/Quest/

29. Hamann D R 1989 Generalized norm-conserving pseudopotentials Phys. Rev. B **40** *2980*

30. Louie S G, Froyen S and Cohen M L 1982 Nonlinear ionic pseudopotentials in spin-density-functional calculations Phys. Rev. B **26** *1738*

31. Hamann D R PUNSLDX program unpublished

32. Mehl M J 1993 Pressure dependence of the elastic moduli in aluminum-rich Al-Li compounds Phys. Rev. B **47** *2493*

33. Birch F 1978 Finite strain isotherm and velocities for single-crystal and polycrystalline NaCl at high pressure and 300 K J. Geophys. Res. **83** *1257*

34. Murnaghan F D 1937 Finite deformations of an elastic solid Am. J. Math. **49** *235*

35. Kresse G and Furthmüller J 1996 Efficient iterative schemes for ab initio total-energy calculations using a plane-wave basis set Phys. Rev. B **54** *11169*

36. Kresse G and Joubert J 1999 From ultrasoft pseudopotentials to the projector augmented-wave method Phys. Rev. B **59** *1758*

37. France-Lanord A, Rigby D, Mavromaras A, Eyert V, Saxe P, Freeman C and Wimmer E 2009 MedeA: atomistic simulations for designing and testing materials for micro/nano electronics systems 15th Int. Conf. on Thermal, Mechanical and Multi-Physics Simulation and Experiments in Microelectronics and Microsystems (EuroSimE), 2014; Materials Design Inc., Angel Fire, NM USA

38. Le Page Y and Saxe P 2002 Least squares extraction of elastic data

for strained materials from ab initio calculations of stress Phys. Rev. B **65** *104104*

39. Mehl M J, Osburn J E, Papaconstantopoulos D A and Klein B M 1990 Structural properties of ordered high-melting-temperature inter metallic alloys from first-principles total-energy calculations Phys. Rev. B **41** *10311*

40. L C 1969 Charactericstics of some binary transition metal hydrides J. Less-Common Met. **19** *315*

41. Grenshaw J A, Spooner F J, Wilson C G and McQuillan A D 1981 The growth of crystals of erbium hydride J. Mater. Sci. **16** *2855*

42. Weaver J H, Rosei R and Peterson D R 1966 Electronic structure of metal hydrides. I. Optical studies of ScH2, YH2, and LuH2 Phys. Rev. B **19** *4855*

43. Goon E J 1959 The non-stoichiometry of lanthanum hydride J. Phys. Chem. **63** *2018*

44. Setoyama D, Matsunage J, Muta H, Uno M and Yamanaka S 2004 Mechanical properties of titanium hydride J. Alloys Compd. **381** *215*

45. Yakel H L Jr 1958 Thermocrystallography of higher hydrides of titanium and zirconium Acta Crystallogr. **11** *46*

46. Zhou X S, Liu Q, Zhang L, Peng S M, Long X G, Ding W, Cheng G J, Wang W D, Liang J H and Fu Y Q 2014 Effects of tritium content on lattice parameter, 3He retention, and structural evolution during aging of titanium tritide *Int. J.* Hydrog. Energy **39** *20062*

47. Yamanaka S, Yoshioka K, Uno M, Katsura M, Anada H, Matsuda T and Kobayashi S 1999 Thermal and mechanical properties of zirconium hydride J. Alloys Compd. **293–5** *23*

48. Azhazha R V, Kovtun K V, Malykhin S V, Merisov B A, Pugachev A T, Reshetnyak E N and Khadzhai G Ya 2008 Accumulation of hydrogen in hafnium: structure and electrical resistivity Phys. Met. Metallogr. **105** *188*

49. Joubert J-M and Crivello J-C 2012 Stability of erbium hydrides studied by DFT calculationsInt. J. Hydrog. Energy **37** *4246*

50. Kołos W and Wolniewicz L 1968 Improved theoretical ground state energy of the hydrogen molecule J. Chem. Phys. **49** *404*

51. Yang J-W, Gao T and Gong Y-R 2014 The disproportionation reactino phase transition, mechanical, and lattice dynamical properties of the lanthanum dihydrides under high pressure: a first principles study Solid State Sci. **32** *76*

52. Olsson P A T, Massih A R, Blomqvist J, Alvarez Holston A-M and

Bjerkén C 2014 Ab initio thermodynamics of zirconium hydrides and deuterides Comput. Mater. Sci. **86** *211*

53. Wang F and Gong H R 2012 Mechanical and structural stability of zirconium dihydride *Int. J.* Hydrog. Chem. **37** *9688*

54. Zhang P, Wang B-T, He C-H and Zhang P 2011 First-principles study of ground state properties of ZrH2 Comput. Mater. Sci. **50** *3297*

55. Ivashchenko V I, Ivashchenko L A, Srynsckyy P L, Grishnova L A and Stegnyy A I 2008 Ab initio study of the electron structure and phonon dispersions for TiH2 and ZrH2 Carbon Nanomaterials in Clean Energy Hydrogen Systems ed B Baranowski *et al (Berlin: Springer) 705*

56. Villars P and Cenzual K 2015 Pearson's crystal data: crystal structure database for Inorganic Compounds, Release 2014/15 ASM Int. (Materials Park, OH, USA)

57. Holley C E Jr, Mulford R N R, Ellinger F H, Koehler W C and Zahariasen W H 1955 The crystal structure of some rare earth hydrides J. Phys. Chem. **59** *1226*

58. Pebler A and Wallace W W 1962 Crystal structures of some lanthanide hydrides J. Phys. Chem. **66** *148*

59. Sturdy G E and Mulford R N R 1956 The gadolinium-hydrogen system J. Am. Chem. Soc. **78** 1083

60. Deligoz E, Colakoglu K and Ciftci Y O 2007 A first principles study of cubic IrO2 polymorphEur. Phys. J. B **60** *477–81*

61. Bowman R C Jr, Craft B D, Cantrell J S and Venturini E L 1985 Effects of thermal treatments on the lattice properties and electronic structure of ZrH2 Phys. Rev. B **31** *5604*

62. Niedźwiedź K, Nowak N and *ogał* O J 1993 91Zr NMR in non-stoichiometric zirconium hydrides, ZrHx () J. Alloys Compd. **194** *47*

63. Sidhu S S 1954 Deuterium effect on hydrogen bond distance in hafnium dihydride *J. Chem. Phys.* **22** *1062*

64. Fan K, Yang L, Zhang Z, Peng S, Long X, Zhou X, Zu X and Gao F 2014 Ab initio calculations of mechanical properties in β-MH2−xHex (M = Er, Sc) Eur. Phys. J. B **87** *295*

65. Rodriguez M A, Ferrizz R M, Snow C S and Browning J F 2008 X-Ray powder diffraction data for ErH2−xDx Powder Diffr. **23** *259*

66. Jacob I, Wolf A and Mintz M H 1981 On the Debye temperatures of metal hydrides Solid State Commun. **40** *877–9*

Chapter 11

REVIEW OF SOLID STATE HYDROGEN STORAGE METHODS ADOPTING DIFFERENT KINDS OF NOVEL MATERIALS

Renju Zacharia[1,2] and Sami ullah Rather[3]

[1]Institut de Recherche sur l'Hydrogène, Université du Québec à Trois-Rivières, P Trois-Rivieres, QC, Canada G9A 5H7

[2]Gas Processing Center, College of Engineering, Qatar University, Doha, Qatar

[3]Chemical and Materials Engineering Department, King Abdulaziz University, Jeddah 21589, Saudi Arabia

ABSTRACT

Overview of advances in the technology of solid state hydrogen storage methods applying different kinds of novel materials is provided. Metallic and intermetallic hydrides, complex chemical hydride, nanostructured carbon materials, metal-doped carbon nanotubes, metal-organic frameworks (MOFs), metal-doped metal organic frameworks, covalent organic frameworks (COFs), and clathrates solid state hydrogen storage techniques are discussed. The studies on their hydrogen storage properties are in progress towards positive direction. Nevertheless, it is believed that these novel materials will offer far-reaching solutions to the onboard hydrogen storage problems in near future. The review begins with the deficiencies of current energy economy and discusses the various aspects of implementation of hydrogen energy based economy.

INTRODUCTION

Currently mankind depends completely on nonrenewable resources such as natural gas, coal, and petroleum to fulfill energy needs. This exquisite dependence on nonrenewable energy sources has twofold consequences: the continuous depletion of energy sources at an alarming rate and the adverse health and environmental impacts [1, 2]. These repercussions have compelled scientists, technologists, economists, and policy-makers to search for alternate, sustainable, and less-polluting energy sources [3–5]. Hydrogen is considered as a clean and sustainable energy carrier, which ultimately can replace nonrenewable fossil fuels and therefore can resolve the availability, environmental, and health concerns of the latter [6, 7]. However, the implementation of an energy economy based on this sustainable and clean fuel is not straight forward but suffers severe hurdles in the production, storage, delivery, and utilization of hydrogen [8–10]. Amongst the various problems that exist in the successful materialization of hydrogen fuel based economy, the formulation of a safe, economical, and efficient hydrogen storage method poses the most confronting challenge [11–13]. This is particularly true, if the utilization of hydrogen fuel in the transportation sector is considered [6, 14]. The transportation sector presently relies exclusively on refined petroleum products that are increasingly unaffordable [6]. This dependence can be eliminated by employing hydrogen as the transportation fuel, which requires high-density hydrogen storage medium [6, 15]. Therefore, there is considerable enthusiasm in devising novel hydrogen media that can be utilized in the transportation sector. Carbon nanotubes (CNTs), metal-doped carbon nanotubes (M/CNT), metal-organic frameworks (MOFs), metal-doped metal-organic frameworks (M/MOFs), covalent-organic frameworks (COFs), zeolites, and clathrates are novel nanoporous materials which can store large quantities of hydrogen [16–23]. Likewise, complex chemical and metallic hydrides represent compounds in which large quantities of hydrogen are stored via chemical bonding [16, 22]. Though studies on their hydrogen storage and release properties of these materials are in the rudimentary stages, they are envisaged to offer long-term onboard hydrogen storage solutions.

The central focus of this review is the recent advances in the solid state hydrogen storage techniques using aforementioned materials. The review is organized as follows: in the introductory section, we present the challenges and problems of the present fossil-fuel based energy economy. The fundamental reasons for adopting a hydrogen fuel based economy are presented subsequently. In the second section, the characteristics and benefits of hydrogen energy economy are detailed. Further, various elements of hydrogen energy infrastructure and requirements of these elements are dealt in this section. The final section of the review presents the problems specific to onboard hydrogen storage. This is followed by the quantitative explanations of the design targets for a successful hydrogen storage medium. The hydrogen storage behavior of nanostructured carbon nanotubes (CNTs), metal-doped carbon nanotubes (M/CNT), metal-organic frameworks (MOFs), metal-doped metal-organic frameworks (M/MOFs), covalent-organic frameworks (COFs), zeolites, complex chemical and metallic hydrides, and clathrates is presented subsequently.

Challenges and Problems in the Present Energy Economy. The present energy economy is based on fossil fuels which comprise mainly three components: petroleum, natural gas, and coal [24]. These nonrenewable forms of energy cannot indefinitely serve as the principal energy sources owing to their continuous depletion and the tremendous rise in their demand. The disproportionate rate of the global production and consumption of fossil fuels is shown in Figure 1 [9, 24]. In Figure 1, the projected trend between the years 2003 and 2040 is expected to cover the world energy needs only if the population remains constant [9, 25, 26]. The massive increase in the energy demand towards the middle of this century is closely associated with the predicted drastic world population growth, technological advances, and increased living standards [9]. Further, on a global scale, the fuel supply demand will be increasingly higher when highly populated countries expand their economies and become more energy intensive.

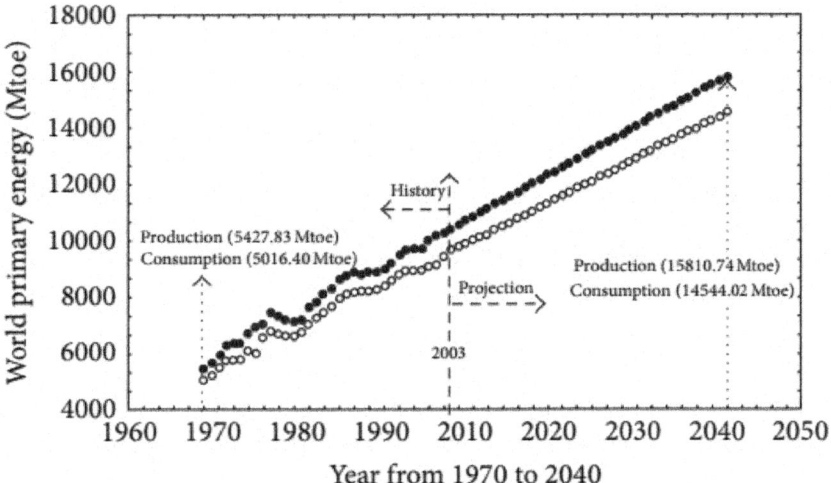

- Production (correlation coefficient (r) = 0.99919)
  (determination coefficient (r^2) = 0.99838)

- Consumption (correlation coefficient (r) = 0.99922)
  (determination coefficient (r^2) = 0.99844)

Figure 1: The global variation in the production and consumption of fossil fuels. All fuels are converted into the units of million tons of oil equivalent (Mtoe). The production and consumption between 2003 and 2040 are predicted based on past data [9].

According to various energy specialists, the fossil-fuel reserves that are presently available can support a maximum of 40 years for petroleum, 60 years for natural gas, and 156 years for coal [9, 24]. It can be noted that less polluting fossil fuels, such as natural gas and petroleum, are in higher demand and have low reserves while fuels, such as coal, are available for longer duration but pose severe environmental threats [24, 27]. Due to the disparity in the utilization pattern and future availability of these fuels, it is expected that they are likely to become unaffordable and unavailable in the near future. This predicted scarcity of fossil-fuel reserves together with the foreseen increase in energy consumption threatens the energy and economic security problems worldwide. Thus, attaining a greater energy security by reducing the dependence on depleting nonrenewable energy sources serves as the primary motivation for the implementation of a sustainable energy economy.

The second confronting problem associated with the indiscriminate use of fossil fuels is the diminishing air quality and subsequent air pollution, while utilizing the fossil fuels [28]. The release of tar, dust, and harmful gases such as CO_2, SO_2, and NO_2 volatile organic components (VOCs) during the combustion of the fossil fuels results in local health hazards [28, 29]. On the other hand, the increased presence of greenhouse gases (GHGs), such as CO_2, NO_2, induces fast changes in the global climate [30]. With the consumption of nonrenewable fuels at a predicted pace of 1000 ($1 = 10^{18}$ J), the increase in annual average temperature of earth is predicted to be around 2°C by 2050 [31, 32]. This increase in temperature is sufficient enough to significantly affect the various life forms across the whole world.

There are three sectors that significantly contribute to the emission of GHGs via fossil fuel consumption: transportation, industry, and electric utilities [33, 34]. Their contributions are graphically displayed in Figure 2[35]. As indicated in Figure 2, major part of GHG-emissions has resulted from the fuel consumption in the transportation sector. Further, within the transportation sector, the light trucks and the cars that are used for private conveyance use significant proportion of the fossil fuels as shown in Figure 3 [36]. Also, it is evident from Figure 3 that disparity in the fuel consumption between private and heavy vehicles becomes substantial by the first quarter of this millennium. These data imply that considerable reduction of GHG-emission can be achieved only if the present energy sources used in the transportation sector are replaced by a cleaner and sustainable energy source, initially focusing the medium-sized vehicles.

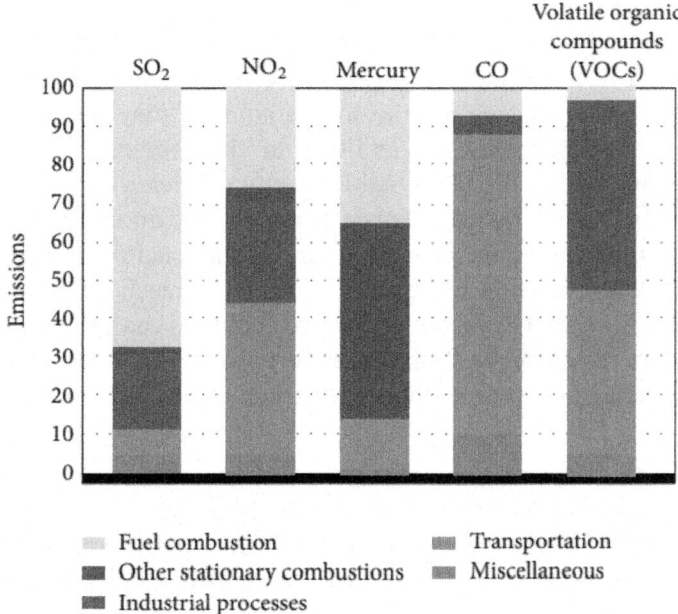

Figure 2: The contribution of various sectors to the emission of greenhouse gases in USA in million metric tons (MMT) and million metric tons of carbon equivalent (MMTCE) per year [35]. It can be seen that the transportation sector contributes significantly to increased emission of oxides of carbon and nitrogen and VOCs.

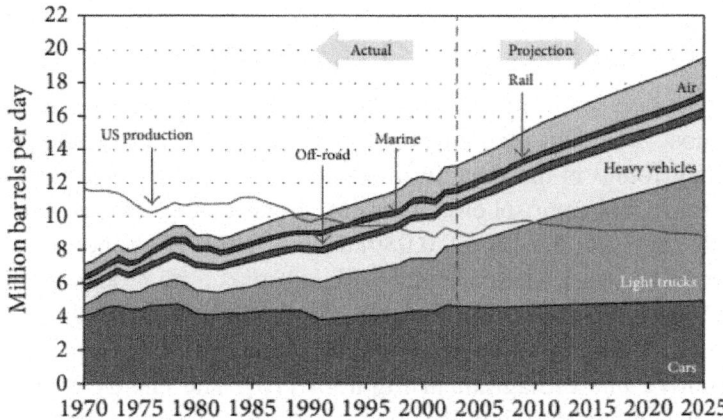

Figure 3: The relative proportions of fossil-fuel use in USA by different transportation modes projected to year 2025 [36]. It is evident that light trucks and cars that are predominantly used for private conveyance use considerable proportion of the fuels.

A formal agreement to mitigate the GHG-emissions and safeguard the environment and human health is reflected in the Kyoto Protocol to the United Nations Framework Convention on Climate Change (FCCC) [37–39]. Under this agreement, the countries which ratify it are committed to reduce their GHG-emission or engage in emissions trading if they maintain or increase emissions of these gases. In addition to the aforementioned limitations, nonrenewable fuels have unfavorable energy efficiency when compared with the renewable energy sources. For instance, a conventional combustion-based power plant typically generates electricity at efficiencies up to 35–42%, while fuel plants that use renewable energy sources can reach efficiencies up to 70–90% [40]. Likewise, the gasoline powered fuel cells can convert less than 30% of the energy into the power, while the fuel cells that are powered by renewable fuels such as hydrogen or methanol can utilize up to 60% of the fuel's energy [41, 42].

In short, considering the above three arguments, it can be seen that the transition from the present fossil-fuel based energy economy to a cleaner and sustainable energy economy will be driven by both economical and environmental reasons. The future of this implementation depends crucially on the energy storage technologies in combination with the generation of renewable energy sources. Hydrogen represents a primary renewable energy source which cannot be used directly but require intermediate conversion steps to maintain its improved attributes in terms of availability, supply, and safety. The challenges pertinent to the implementation of a hydrogen fuel based economy, in particular those associated with the high-density solid state hydrogen storage, are discussed in the following sections.

HYDROGEN FUEL BASED ECONOMY

Hydrogen is a colorless, odorless, tasteless, and nonpoisonous gas. It is the most abundant element in the universe (75% by mass), but it is not commonly found in the pure form, owing to the high reactivity. At the room temperature and atmospheric pressure, hydrogen exists as van der Waals gas with very low-density of $0.08988 \, kg/m^3$ [43]. Hydrogen is a liquid in a small zone between triple and the critical points with a density $70.8 \, kg/m^3$ at $-253°C$. At temperatures below $-262°C$, hydrogen exists as a solid with a density of $70.6 \, kg/m^3$. Various physical phases of hydrogen are depicted in the primitive phase diagram in Figure 4 [16]. The energy characteristics of hydrogen as renewable fuel depend critically on its physical state of existence. Two important parameters that are extremely crucial for the transportation applications of hydrogen are the specific energy and the energy density. The former is the energy per unit mass of the fuel and is the measure of the net energy content of the fuel (measured in kWh/kg). This parameter decides the minimum refueling distance between two stoppages. The energy density of the fuel is the net usable energy/unit volume of the fuel (measured in kWh/m^3). The energy density of hydrogen decides the net system volume of the fuel storage system. At ambient pressure and temperature conditions, the energy density of hydrogen is nearly ten times lower than that of conventional fuels [44, 45]. This low-energy density of hydrogen is a serious obstacle for the implementation of hydrogen fuel for automotive applications. In order to use hydrogen as a successful energy source in the transportation applications, its physical state has to be altered to improve its energy characteristics.

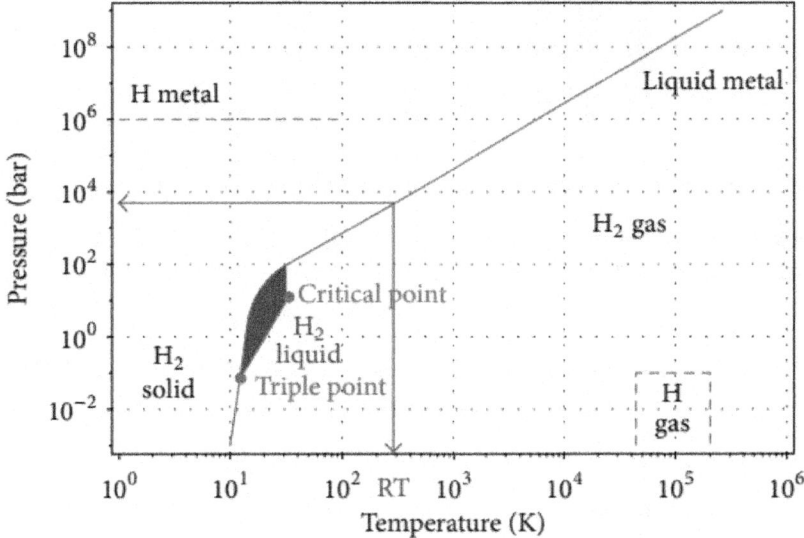

Figure 4: The primitive phase diagram of hydrogen. Figure adapted from [16].

Comparison of Energy Characteristics of Hydrogen with Conventional Fuels

For automotive applications, hydrogen has to meet and supercede the conventional fossil fuels in its energy characteristics. As mentioned previously, two most important parameters are the energy density and specific energy of hydrogen, which are, respectively, known as volume- and mass-based energy density [44, 45]. In Figures 5 and 6, a comparison of mass- and volume-based energy density of hydrogen with that of conventional fuels is made. It is evident from Figure 5 that even in its liquefied form hydrogen has nearly two times lesser energy density when compared with the gasoline.

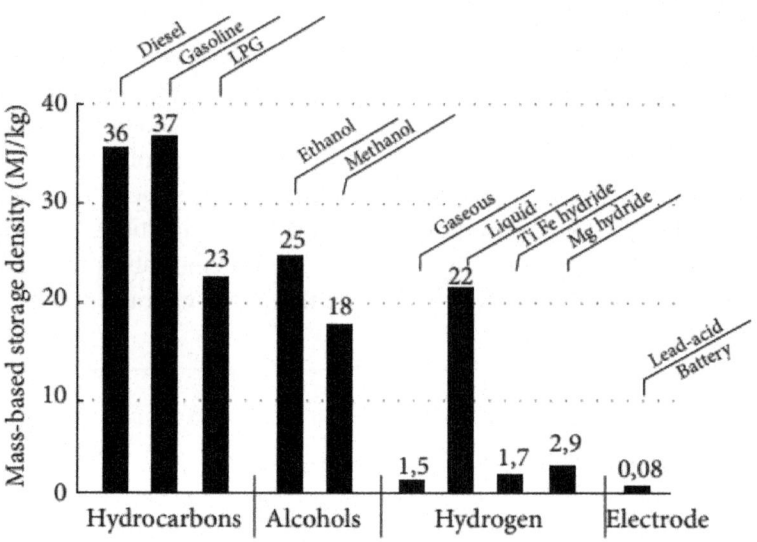

Figure 5: Specific energy of hydrogen compared with that of conventional fuels, based on the data from [44].

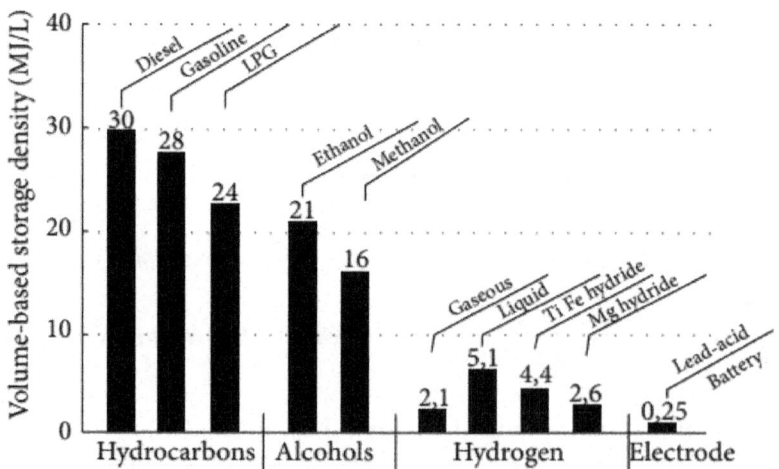

Figure 6: Energy density of hydrogen compared with that of conventional fuels, based on the data from [44].

A comparison of various properties of hydrogen and other conventional fuels relevant in the transportation sector is provided in Table 1 [46].

Table 1: Properties of hydrogen compared to other fuels. Data adapted from [46]

Property	Units	Hydrogen	Methane	Propane	Methanol	Ethanol	Gasoline
Chemical formula		H_2	CH_4	C_3H_8	CH_3OH	C_2H_5OH	C_xH_y ($x = 4$–12)
Molecular weight (a, b)		2.02	16.04	44.1	32.04	46.07	100–105
Density, NTP (3, a, c)	kg/m³	0.0838	0.668	1.87	791	789	751
	lb/ft³	0.00523	0.0417	0.116	49.4	49.3	46.9
Viscosity, NTP (3, a, b)	g/cm-sec	$8.81E-5$	$1.10E-4$	$8.012E-5$	$9.18E-3$	0.0119	0.0037–0.0044
	lb/ft-sec	$5.92E-6$	$7.41E-6$	$5.384E-6$	$6.17E-4$	$7.99E-4$	$2.486E-4$–$2.957E-4$
Normal boiling point (a, b)	°C	−253	−162	−42.1	64.5	78.5	27–225
	°F	−423	−259	−43.8	148	173.3	80–437
Vapor specific gravity, NTP (3, a, d)	Air = 1	0.0696	0.555	1.55	N/A	N/A	3.66
Flash point (b, d)	°C	<−253	−188	−104	11	13	−43
	°F	<−423	−306	−155	52	55	−45
Flammability range in air (c, b, d)	vol%	4.0–75.0	5.0–15.0	2.1–10.1	6.7–36.0	4.3–19.0	1.4–7.6
Autoignition temperature (b, d)	°C	585	540	490	385	423	230–480
	°F	1085	1003	914	723	793	450–900
Maximum flame velocity in air (2, c)	m/s	2.83	0.45	0.46	N/A	N/A	N/A
	ft/s	9.28	1.48	1.52			

Benefits of Hydrogen Fuel Based Energy

A compilation of the salient positive attributes of hydrogen is provided in the following:(a)Hydrogen is a nontoxic, clean energy carrier which does not produce carbon dioxide (CO_2), particulate, or sulfur emissions. However, it can produce oxides of nitrogen under some conditions. Its combinations with oxygen produces water and energy both are essential for the existence of life on earth. (b)Hydrogen has a high specific energy on a mass basis when compared with any conventional fuels. Quantitatively, this can be understood by considering that the energy content of 9.5 kg of hydrogen is equivalent to that of 25 kg of gasoline [44].(c)The energy density of hydrogen under ambient pressure and temperature conditions are lower as compared to the conventional fuels [44]. However, the volumetric energy density of hydrogen can be increased by storing hydrogen at lower temperatures or at higher pressures. Energy density of hydrogen can be improved also by adsorbing it into highly porous solid state materials.(d)Hydrogen is highly inflammable. The energy that is required to ignite and burn hydrogen is nearly 10 times lesser when compared with that of conventional fossil fuels. Therefore, the automobiles that are equipped with hydrogen fuel do not have ignition problems even in most severe winter [47]. (e)Hydrogen has relatively high value for important transport properties such as kinematic viscosity, thermal conductivity, and diffusion coefficient, when compared with the conventional fuels. These together with its extremely low density and luminosity give it unique diffusive and heat transfer characteristics [47].(f)Hydrogen can be produced via a multitude of processes. These include electrolysis of water, direct and indirect thermochemical decompositions, and processes driven directly by the sunlight. Additionally hydrogen can be produced via sequestration of hydrocarbon fossil fuels [48–50].(g)Hydrogen can be safely transported in pipelines due to its high utilization safety.(h) Hydrogen can also be used as a chemical feedstock in the petrochemical, food, microelectronics, ferrous and nonferrous metal, chemical and polymer synthesis, and metallurgical process industries.(i)Compared to electricity, hydrogen can be stored over relatively long periods of time.

Transition of Present Energy Economy to a Hydrogen Fuel Based Economy

The cost of transition in the any energy delivering infrastructure involves investment of huge amounts of money [51]. Such a transition is considered irreversible and permanent step. It is estimated that the investment of the complete transition from the conventional petroleum based economy to the hydrogen based economy in USA alone will be millions of dollars [52]. The complete transition of the present day energy economy to a hydrogen-based energy economy implies the use of hydrogen as the main chemical energy carrier and the electricity as the main nonchemical form of energy. This transition is being made gradually and is likely to continue to the middle or the end of the 21st century.

A pragmatic and quantitative assessment of this transition can be made by following the production trend of hydrogen. The current world production of hydrogen is nearly 50 million tons/year, which is equivalent to approximately 2% of the world energy demand [53, 54]. In order to expand the role of hydrogen in the near future, several approaches are being proposed. One of them is to use hydrogen for transportation by mixing it with the natural gas as fuel for internal combustion engines (ICEs). This increases the engine performance and decreases the pollution [55, 56]. Another approach involves producing hydrogen at central locations and distributing it to refilling stations, where it can be used in the liquid form in the fuel cell powered light motor vehicles. The gradual takeover of the hydrogen fuel based economy will involve several phases and is likely to be introduced over a long period of time. The following time-bound phases can be identified in the process of this takeover [57]:(a)In the near term, the hydrogen will be produced primarily via the advanced steam reforming of natural gas. This process is a well-understood one and cost of the process depends mainly on the feedstock investment [48]. Steam reforming can be performed at the central locations or at the distributed facilities. The steam reforming of natural gas decreases the amount of CO_2 released into the atmosphere, since the by-product of steam reforming is high-purity CO_2 which can be collected and used in many ways.(b)In the medium term, the restructuring of the electric utility industry can be performed. This will give opportunities for the distribution generation of hydrogen fuel. In this term, the on-site generation of electricity will be carried out by hydrogen powered fuel cells. In addition to electricity, the fuel cells also produce thermal energy for hot water and space heating. In this phase, hydrogen required can be increasingly produced from coal and from the gasification of dedicated biomass.(c)In the final phase of establishing the hydrogen fuel based economy, strong hydrogen markets and growing hydrogen infrastructure can establish opportunities for complete renewable hydrogen systems. The hydrogen required for the fuel cells can be produced via the electrolysis of water using intermittent energy technologies such as wind turbines and photovoltaic systems [57]. Fuel cells will use hydrogen to generate electricity during high-demand periods or to supplement the intermittent energy sources. The emergence and growth of advance technologies that produce hydrogen from water and sunlight and the technologies that store hydrogen in high-energy density systems will likely to take place during this phase. Commercialization and market penetration of advanced technologies to produce, store, and use hydrogen in the final phase will mark the establishment of hydrogen energy economy.

The establishment of hydrogen based economy, however, suffers from some uncertainties, mostly associated with the technologies that deal with its infrastructure [57]. For instance, the hydrogen economy is envisaged as the end state based on hydrogen produced from renewable power sources such as solar energy or wind energy. However, it is not yet economic to produce hydrogen in these ways. Secondly, hydrogen powered fuels provide clean and efficient energy for future vehicles and stationary power generation which is meaningful only if the hydrogen is produced cleanly. Further, the cost and technical hurdles associated with mass adoption of fuel cells need to be addressed. Also, there are uncertainties associated with the fuel cell cost and viability that can challenge its market penetration [58].

Elements of a Hydrogen Energy Infrastructure

In spite of various compelling benefits of hydrogen fuel based economy, its realization faces multiple challenges. This is because, unlike the conventional fuels, hydrogen has no existing large-scale supporting infrastructures. Consequently, implementing an economy completely based on hydrogen requires development of various hydrogen energy infrastructures. Further, it is necessary to establish close coordination and integration of these elements. The main elements of a hydrogen energy infrastructure are the ones related to its production, delivery, storage, conversion, and applications. A schematic representation of these elements and their interrelationship is provided in Figure 7 [57].

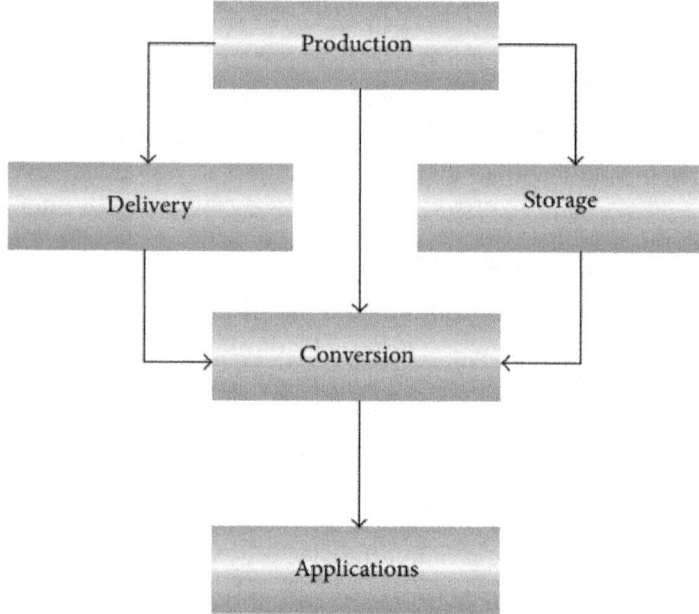

Figure 7: Schematic representation of various elements involved in the establishment of hydrogen fuel infrastructure.

In the following, these elements, their present status, and the technical challenges are discussed.

- Production of Hydrogen. For the establishment of hydrogen as the primary source of energy, it needs to be produced centrally in large refineries, energy complexes, or renewable or nuclear power stations. It can also be produced locally in power plants, fueling stations, communities, rural areas, and on-site customer's premises. Bulk of the present hydrogen production is performed via the catalyzed steam reformation of natural gas. Though this process is considered to be the cheapest option presently available, the cost factors are not favorable for large-scale commercialization for onboard hydrogen supply [59, 60]. This is particularly true when the costs of hydrogen production technologies are compared with that of conventional fuels [14]. Another factor that hinders the advancement of hydrogen production technologies is the low demand for the hydrogen. Also, present sequestration techniques produce large quantities of greenhouse gases and the processes are not optimized for the reduction or environmentally sound carbon capture [61]. Three other major technologies for the hydrogen production currently under consideration are the electrolytic hydrogen production from alkaline, polymer membrane, and ceramic oxide electrolytes [62]. The key challenges associated with the production of hydrogen are the alternative low-cost production techniques for hydrogen from renewable fuels, dedicated biomass and nuclear sources, and environmentally sound carbon capture techniques for existing sequestration methods [63].

- Delivery. The second component of the hydrogen energy infrastructure is the delivery of hydrogen, which involves the transport of hydrogen from its productions site to the end-user device. Most of the hydrogen produced is currently transported by road via cylinders or cryogenic tankers. Delivery via pipelines is also a prevailing method of hydrogen delivery [64, 65]. In high-demand areas, pipelines can be used to distribute hydrogen. To distribute hydrogen to rural and other lower-demand areas, trucks and other transportation means need to be used. This hydrogen can be gaseous, liquid, or solid hydrogen carriers. In order to advance the delivery of hydrogen, low-cost hydrogen transport technology must be identified and developed. Further, an advanced supply network should accommodate both centralized and decentralized hydrogen production facilities.

- Storage of Hydrogen. One of the most important challenges in the implementation of hydrogen based energy is finding an appropriate storage medium for hydrogen [11]. This is particularly true for mobile applications of hydrogen fuel. This can be illustrated by considering the large volume occupied by hydrogen at room temperature and moderate pressures. For realistic driving distances, typically 4 kg of hydrogen is required, which occupies nearly 50 m^3 at ambient pressures and temperature conditions [11]. This large container size adversely affects the vehicle size. Thus, to store hydrogen in any useful form, it must be altered to achieve a higher energy density. The energy density target for a successful hydrogen storage medium, specified by the partnership between US DOE and FreedomCAR, is 1.2 kWh/L [66]. The energy density of hydrogen in its various forms is depicted in Figure 8 [36]. It is evident from the figure that only liquefied hydrogen has the above energy density 1.2 kWh/L, which presently qualifies as a motor fuel [67]. However, it is not economical to store hydrogen in the liquid form because it is an energy intensive process [68]. In addition, liquid hydrogen storage suffers from fuel-loss due to the boil-off [69, 70]. Due to extremely high kinematic transport parameters and low ignition threshold, minor leakage of hydrogen can be huge safety threats [71]. Additionally, this form of storage suffers problem related to safety [47, 71]. The problems in the onboard hydrogen storage can be circumvented only if relatively lightweight, low-cost solid state hydrogen storage devices are identified and developed. Additionally, they should have high storage density as required by the FreedomCAR targets. Ongoing research to identify novel materials for hydrogen storage suggests that nanoporous materials, such as carbon nanotubes, metal-doped carbon nanotubes, metal-organic frameworks (MOFs), covalent-organic frameworks (COFs), complex chemical hydrides, clathrates, and intermetallic alloys, are the most promising materials for future hydrogen storage [16–19].

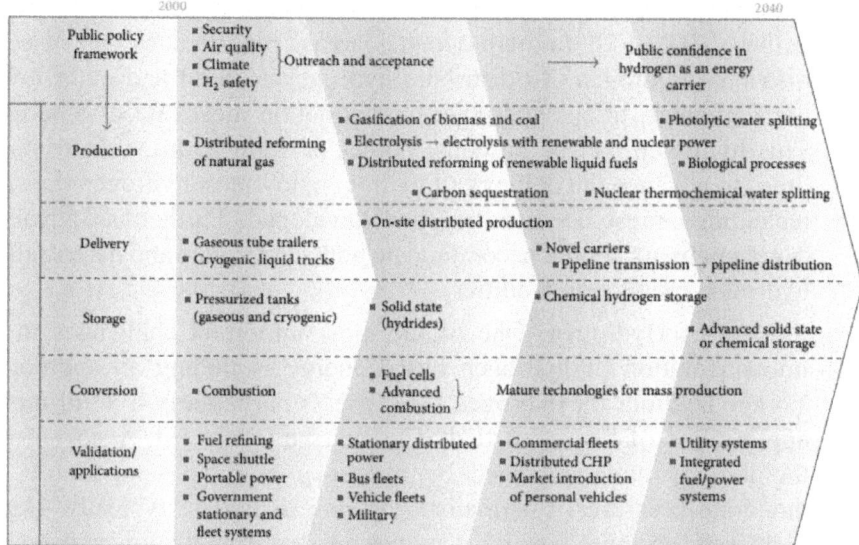

Figure 8: The present status of various elements of hydrogen energy infrastructure. The predicted mature technologies by the end of 2040, envisioned by United States Department of Energy, are indicated on the right side [36].

- Conversion. The conversion of hydrogen fuel into energy can be accomplished in either an internal combustion engine or fuel cells. The former produces energy by burning hydrogen in the presence of oxygen, while the latter use the chemical energy of hydrogen molecules. Combustion method comprises two technologies: gas turbines and reciprocating engines [57]. At present, the hydrogen fuel based gas turbines are exclusively used in aviation industry [72, 73]. The latter techniques, that is, the use of fuel cells to generate energy from hydrogen, are envisaged as attractive alternative to combustion-based engines due to formers higher efficiency [74]. Consequently, the maturing of hydrogen based energy system depends on cost-competitive mass production of fuel cells [58, 75]. This is hindered as none of the present fuel cell technology has met all criteria for performance, durability, and cost [58]. Additionally, standalone conversion devices such as combustion turbines and reciprocating engines need to be developed to off-board applications.

- Applications. In the application sector, hydrogen should be available for every end-user energy need in the economy. This includes the onboard applications, such as transportation and mobile utility systems, and stationary applications, such as central and distributed electric power and combined heat and power for buildings and industrial processes. For short terms, the most important sector that will be using hydrogen will be the transportation sector. For mobile applications hydrogen-based fuel cell is the most appealing option. Most of the major automobile makers presently have hydrogen-fueled vehicle programs [76]. The acceptance of the hydrogen fueled vehicles for end-user applications can be enhanced by performing successful field tests and demonstrations. Additionally, supportive public policies should be developed to stimulate the acceptance of infrastructure and market readiness. For stationary applications, combustion-based process, such as gas turbines and reciprocating engines, can be designed to use hydrogen.

- Codes, Standards, and Education. The final element in the implementation of hydrogen energy infrastructure is the codes and standards. Families of model building codes should be available for adoption. These deal with comprehensive references to equipment standards for both hydrogen and fuel cell technologies for the commercial and residential applications. In addition, published safety standard for the certification of fuel cell vehicles and published fuel gas code should be available. Increased public awareness on the benefits and merits of hydrogen economy can be highlighted by education and outreach programs. It is also important to establish regional, local, and national networks that commit resources for long-term education of students at all levels. Such training and certification program finally intensify the acceptance of the hydrogen based storage and application devices.

A schematic relation of the various hydrogen energy infrastructures is envisaged by the Department of Energy, USA (Figure 8). This incorporates the present status and the expected matured technology by the year 2040 [36].

HYDROGEN STORAGE: TARGETS AND OPPORTUNITIES

The establishment of a hydrogen fuel in the transportation applications depends largely on the availability of novel hydrogen storage media that satisfies a set of selection criteria. This includes parameters, such as high hydrogen content per unit mass and unit volume, limited energy loss during operation, fast kinetics during charging, low self-discharge during stoppage, high stability with cycling, cost of recycling and charging infrastructures, and safety concerns in regular service or accidents. The technical limits of above parameters depend on the region of operation and are proposed by agencies, such as Department of Energy (DOEs, US) [DOE], world energy network (WE-NET, Japan) [WENET], and International Energy Agency (IEA) [IEA].

Targets to Be Met and Their Explanations

The technical targets of a hydrogen storage medium that can be successfully used for mobile applications are time-bound and the upper limit in the energy characteristics of a storage medium is proposed by the partnership between US DOE and FreedomCAR [36]. Amongst several parameters mentioned above, the most significant ones are the gravimetric storage capacity and the volumetric storage capacity. The respective DOE target for the year 2015 is 5.5 wt% and 40 g/L at 233–358 K and 3–100 bars. The ultimate target set by the DOE is 7.5 wt% and 70 g/L [77–80]. The targets are designed so as to enable a refueling distance of 500 km for light motor vehicles. Another important factor in selecting a storage medium is the safety associated with the system during its regular use and accidents. The storage techniques such as the liquefied hydrogen and compressed gas which are presently being used are high risk methods of hydrogen storage [81]. Another factor of central importance is the cost of the storage system. The proposed target for the cost of storage system is about 4 $/kWh [82, 83]. Also, there are stringent requirements regarding the response time and maximum and minimum operating temperature at which hydrogen storage and release can be performed. The former (response time) controls the vehicle performance and proposed technical limit is 0.75 s [36]. In order to gain a complete acceptance for the transportation use, the storage system must track the needs of the fuel cell closely to provide the adequate power and a suitable driving experience. This asymmetric parameter affects the system performance, fuel cell durability, and vehicle stability. The acceptable maximum and minimum temperature conditions for the storage system are −40 and 85°C, respectively. The temperature limit depends also on the working conditions of the fuel cells, which is presently 85°C [70, 80].

As mentioned previously, the most important challenge in storing hydrogen is achieving the specific energy density target of FreedomCAR. In order to do this, hydrogen needed to be compressed into a smaller volume. This can be accomplished using few methods. One method is to mechanically change the pressure and temperature so that the energy per unit volume of hydrogen increases [11]. This method currently being used in all hydrogen powered vehicles uses either compressed gaseous hydrogen or liquefied hydrogen [67]. Liquid hydrogen storage at low pressures and cryogenic temperature is light and compact. Also, it has high hydrogen content that satisfies both energy characteristics (Figure 6). However, it suffers a very high disadvantage that nearly 33% of overall energy content is lost during the liquefaction process [84]. Further, the evaporation during off-periods can only be partially reduced by recooling systems [76]. Both the compressed gaseous and the liquid hydrogen storage are considered as huge security risks due to potential explosions in case of accidents [77–80]. Other promising solid state materials for high-density hydrogen storage are metallic or intermetallic hydrides and complex chemical hydrides [85, 86]. Additionally, the state-of-the-art methods of storing hydrogen in nanostructured materials such as carbon nanotubes, metal-doped carbon nanotubes, metal and covalent-organic frameworks (MOFs) and (COFs), and clathrate hydrates have gained increased attention. The latter two options will prove to be of larger importance to the portable fuel cell industry because energy density is a much more important consideration for portable systems when compared to the off-board applications. Next subsections of this review are dedicated to these novel solid state hydrogen storage techniques. Only if these methods obtain extremely high efficiency will they become realistic candidate for off-board scale energy storage.

In Figure 6, the specific energy (gravimetric capacity) and the energy density (volumetric capacity) of hydrogen in various physical states are provided [43]. From Figure 6, it is clear that only liquefied hydrogen exceeds the aforementioned criteria at present. As evident from Figure 6, any other forms of its storage, hydrogen possesses low volumetric energy density. This suggests that to store hydrogen in any useful form, even in the stationary context where the size of the storage is of smaller importance, it must be altered to achieve a higher energy density. This constitutes one of the big challenges in implementing the hydrogen fuel based economy.

Metals and Intermetallic Hydrides

Hydrogen reacts with many transition metals and metallic alloys at elevated temperatures to form hydrides [87,88]. Lanthanide, actinides, and the members of Ti and V groups are the most reactive transition elements that readily form the hydrides. The binary hydrides of transition metals are metallic in character and are commonly referred to as metallic hydrides. Their compositions are generally expressed as MH_x, where x can vary from ideal stoichiometry and forms multiphase systems depending on the pressure and temperature [89]. Most interesting metallic hydrides belong to intermetallic hydrides with general composition of AB_nH_x. Here, element A is a rare earth or an alkaline earth metal and B is a transition metal. Since the hydriding and dehydriding properties of intermetallic hydride depend on the ratio of A and B, it is possible to tailor these properties by appropriate selection of metals and their combinations. A typical example of intermetallic ternary system is $LaNiH_6$, where A is the lanthanum and B is nickel. One of most attractive features of metallic hydrides is the extremely high volumetric density of hydrogen. $LaNi_5$, for instance, can reach a volumetric hydrogen density of $115\,kg/m^3$ [16, 86]. In Table 2, most important families of hydride form intermetallic alloys and their prototype and respective structure are provided.

Table 2: The families of hydride forming intermetallic compounds, their prototypes, and structure, adapted from [16]

Intermetallic compound	Prototype	Structure
AB_5	$LaNi_5$	Haucke phase, hexagonal
AB_2	$ZrMn_2$	Laves phase, hexagonal, cubic
AB_3	$CeNi_3$	Hexagonal
A_2B_7	Y_2Ni_7	Hexagonal
A_6B_{23}	Y_6Fe_{23}	Cubic
AB	$TiFe$	Cubic
A_2B	Mg_2Ni	Cubic

Metal hydrides are very effective in storing large amount of hydrogen in a safe and compact way. However, transition metal hydrides, those working reversibly at room temperature and moderate pressure, store a maximum of only 3 wt% of hydrogen (gravimetric storage capacity) [16, 86]. Therefore, the possibilities of lightweight metal hydrides which can reversibly store more than 5 wt% of hydrogen are needed to be explored.

Complex Chemical Hydrides

Lighter elements of groups 1, 2, and 3, such as Li, Mg, and B, have been found to form a series of complex hydrides which can be used a chemical method to store hydrogen [85, 90–92]. They are particularly interesting for the mobile hydrogen storage application because of their low density and high hydrogen-to-metal ratio [90–92]. Unlike metallic hydrides the complex hydrides are ionic or covalent in character. Sodium borohydride (NaBH$_4$) is typical example of complex metal hydride. Most of complex metal hydrides have tetrahedral structure as seen in Figure 9(a) with hydrogen occupying the corners of the tetrahedron [91, 92]. Also these materials are stable compounds decomposing only above their melting points.

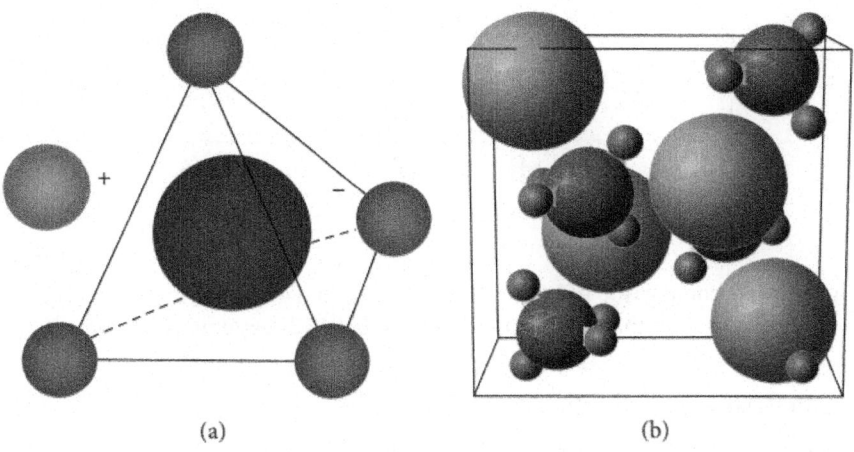

(a)   (b)

Figure 9: Molecular structure of NaBH$_4$ (a). Sodium, boron, and hydrogen are represented by green, blue, and red spheres. (b) Showing the crystal structure of the same.

Complex chemical hydrides usually possess enormous gravimetric storage capacities [91, 92]. The complete release of hydrogen from LiBH$_4$ is equivalent to 45.5 wt% of hydrogen [91, 92]. At room temperature, the amount of hydrogen released from LiBH$_4$, for instance, is equivalent to 18.5 wt% [91, 92]. The challenge in utilizing these materials includes reducing their decomposition temperature. This high decomposition temperature is related to their stability which in turn is related to the percentage ionic character and steric effects. Therefore, to use these materials for mobile hydrogen storage, it is essential to find out the conditions under which their decomposition temperature can be reduced. In Figure 10, a comparison of the gravimetric and volumetric storage capacities of metal hydrides with the proposed 2010 and 2015 FreedomCAR target is provided [93]. A new set of 2020 and ultimate target of gravimetric and volumetric capacity for onboard fuel cell vehicles is 1.8 kWh/kg, 1.3 kWh/L and 2.5 kWh/kg, 2.3 kWh/L [94].

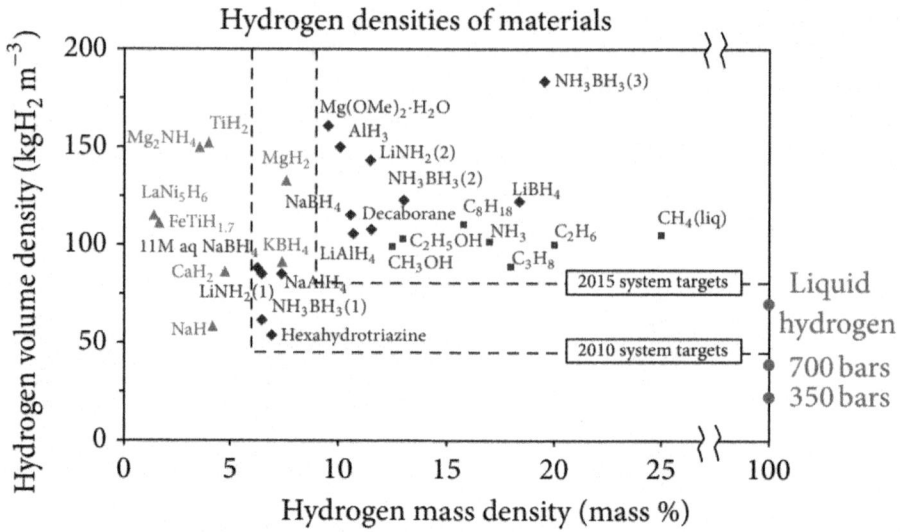

Figure 10: A comparison of hydrogen volumetric and gravimetric densities of complex metal hydrides with that of proposed 2010 and 2015 FreedomCAR targets [93].

Nanostructured Carbon Materials

Amongst all emerging materials for the solid state hydrogen storage, nanostructure carbon materials, especially carbon nanotubes (CNTs) and carbon nanofibers (CNFs), received the most attention. Also, the hydrogen storage in carbon nanotubes is probably the single most controversial topic as far as the solid state hydrogen storage is concerned [95]. Single-wall carbon nanotubes (SWCNTs) can be assumed to form by rolling a single graphene sheet [96]. Owing to several unique properties of CNTs, they are envisaged as a very good medium for solid state hydrogen storage. For instance, they are typically inert to surface contaminants and therefore less severe activation conditions are only required. Also, in SWCNTs, all carbon atoms are exposed to the surface. This makes them the material with highest surface-to-bulk atom ratio and therefore highly surface active. The densities of CNTs are considerably lower than that of metals, intermetallic alloys, COFs, and MOFs. Also, these nanomaterials are known to possess large amount of void spaces in the form of pores which can accommodate large quantities of hydrogen. Hydrogen can be physisorbed in SWCNT bundles on various sites such as external wall surface, grooves, and interstitial channels [97]. Therefore, it can also have large energy density as required for the mobile applications. It is also found that by tuning the adsorption conditions hydrogen can be either chemisorbed or physisorbed on carbon nanotubes.

The studies on the hydrogen storage properties of carbon nanotubes were triggered after the pioneering work by Dillon et al. [98]. They reported a gravimetric storage capacity between 10 and 20 wt%. The experiments on hydrogen adsorption in CNTs can be broadly divided into two, depending on the method of storage: gas-phase hydrogen storage and electrochemical hydrogen storage [95]. In gas-phase hydrogen storage techniques, a macroscopic sample of nanotubes, typically weighing less than 1 g, is exposed to pure hydrogen gas under various temperature and equilibrium pressure conditions. The amount of hydrogen adsorbed by nanotubes is then measured gravimetrically using microbalance [99]. A more popular method of determining the amount of stored hydrogen is by volumetrically using Sieverts type volumetric apparatus [100]. The latter technique involves the exposure of carbon nanotubes to hydrogen gas in a known volume and determining the storage capacity from the change in the free volume of the system upon exposure. With smaller quantities of sample, the gravimetric storage capacities can also be determined using temperature programmed desorption (TPD) or thermogravimetric analysis (TGA) [101]. The second method of storing hydrogen in carbon nanotubes is electrochemical. To do this, an electrochemical cell is constructed with CNTs as the working electrode, Pt as the counter electrode, and an appropriate electrolyte [102]. In such a system, hydrogen is stored in the CNT electrode by the reduction of water at a suitable potential [102].

The earliest experiments on hydrogen storage capacity of CNTs performed by Dillon's group obtained 10–20 wt% of capacity. In their study, they assumed that only CNTs contribute to the gas adsorption. Due to a very dilute CNT mixture they used the large mass correction (99.8% of mass) of the material introduced considerable error into their storage capacity [98]. The volumetric measurement of the CNT-hydrogen storage capacity was initially performed by Ye et al. in 1999 [100]. The experiment performed using a Sieverts type volumetric instrument suggested that, at cryogenic temperatures and an equilibrium pressure of 120 bars, SWCNTs store nearly 8.5 wt%. Hydrogen was stored in CNT electrochemically for the first time by Nützenadel et al. [103]. They found that storage equivalent of a discharge capacity of 110 mAh/g can be obtained when hydrogen is stored electrochemically. Though pristine-carbon nanotubes exhibit remarkable hydrogen storage capacity at cryogenic temperatures, its storage capacity diminishes to less than 1.0 wt% at room temperature [95].

It is worth noticing that, quite recently, there have been some efforts to tailor the surface characteristics of pristine nanotubes so as to increase their hydrogen storage capacities [104, 105]. These surface modifications include high-energy atomic bombardment, reactive ball milling, high temperature annealing, and doping with transition metals [104–108]. Both high-energy atomic bombardment and reactive ball billing result in partial destruction of carbon nanotube structure by creating high density of defects [106–108]. These defect sites serve as additional hydrogen binding sites and up to X wt% of hydrogen can be stored in defect enriched carbon nanotubes [107]. However, here hydrogen binds irreversibly via chemisorption and therefore, to release the stored hydrogen, the samples are required to heat over 500°C [109].

Metal-Doped Nanostructured Carbon Materials

The enhanced hydrogen storage properties of transition metal-doped carbon nanotubes have gained wide attention recently [19, 110–113]. The transition metals that are typically used for doping CNTs are the ones which form hydride at ambient conditions including palladium, platinum, and vanadium [110] and are doped by incipient wetness or electron-beam evaporation technique [113–116]. The transition metal doping is found to increase the hydrogen storage capacity of CNTs by around 30% without adversely affecting the fast desorption kinetics [19, 110]. The enhanced hydrogen storage capacities of doped nanotubes are explained using the spillover phenomenon, where hydrogen molecules were initially adsorbed by the transition metal particles and were subsequently spilled onto different adsorption sites of carbon nanotubes (see the schematic of spillover phenomenon in Figure 11) [111, 112, 116]. Readsorption data of Pd- and V-doped CNTs have indicated that more than 70% of hydrogen is spilled onto the low-energy binding sites such as external wall or groove sites of nanotubes [19, 110]. These sites are associated with considerable low desorption barrier ($\sim k_B T$) and therefore explain why the doping does not affect the observed kinetics of desorption.

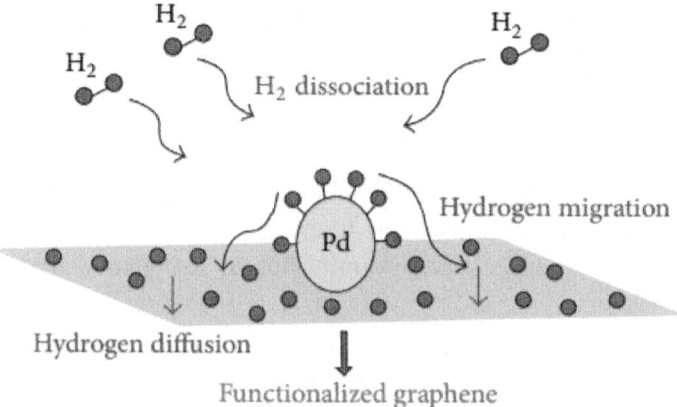

Figure 11: A schematic representation of the spillover phenomenon.

Hydrogen storage behavior of Ti-doped CNTs was predicted using first principle calculations [117]. Each Ti atom in the doped CNT can bind up to four hydrogen molecules which is equivalent to a gravimetric storage capacity of 8.0 wt% (Figure 12).

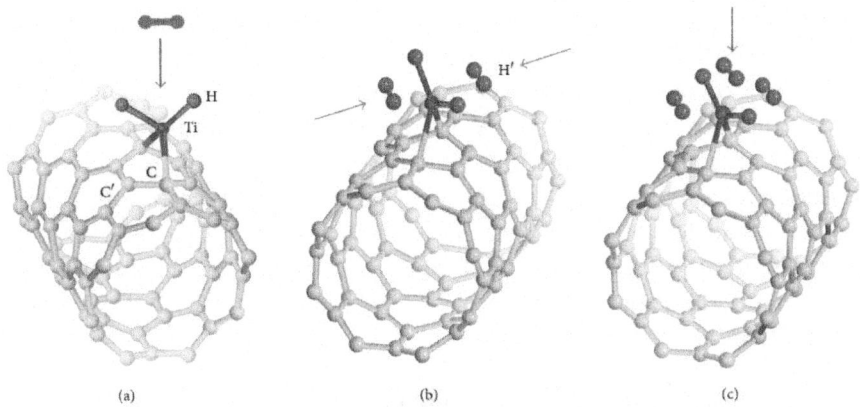

(a)   (b)   (c)

Figure 12: The configurations of hydrogen on Ti-doped carbon nanotubes [117].

The bonding of hydrogen molecules in Ti-doped CNTs occurs via stepwise processes as indicated in Figure 12. In the first step, a hydrogen molecule undergoes dissociative chemisorption and binds to a single titanium atom (Figure 12(a)). Remarkably, this occurs without an energy barrier. In the second step, two hydrogen molecules are physisorbed as shown in Figure 12(b). In the final configuration, the fourth hydrogen molecule is physisorbed to Ti atom in a direction parallel to C-Ti plane (Figure 12(c)). This unexpected bonding is explained using unique hybridization between Ti-d, hydrogen antibonding, and SWCNT C-p orbitals [117]. The experimental studies of the hydrogen storage remain as a controversial topic due to the large disparity in the storage capacities measured in various experiments [118]. This is primarily associated with the uncertainties in the sample purity used in various experiments. It is generally understood that CNTs contain some amount of metallic impurities, which are incorporated during their synthesis [119]. These metallic impurities adsorb hydrogen; therefore the unknown hydrogen storage contributions of metals are sometimes wrongly assigned to the storage capacity of nanotubes.

Other complications arise due to erroneous measurements, typically observed, in the case of volumetric determination of storage capacity [120]. The samples are exposed to hydrogen gas in a constant volume setup. For this type of study, the accurate dead volume of the instruments should be known. Though the precise volume of each of the components is known prior to the experiments, there will be some uncertainty in the total volume of the system when the components are fixed together. This uncertainty is not accounted for in many volumetric studies and constitutes some error in the measured capacity. Another factor that leads to erroneous measurement is the change in the storage capacity due to the change in the ambient temperature, during the experiment [95]. For an adsorption experiment performed at room temperature and 10 MPa pressure, each degree change in the ambient temperature results in 33 kPa of pressure which is equivalent to a storage capacity of 2.6 wt% [95]. This becomes very crucial for the measurements involving very small amount of nanotubes. Thus, it is vital to keep the experimental setup strictly in isothermal conditions. Additionally errors due to temperature fluctuation due to the expansion of the hydrogen gas and leaks in system affect the reliable measurement of gravimetric storage capacity [95]. Therefore, accurate, consistent, and reliable hydrogen storage data are obtained only if necessary precautions are taken during the storage experiments.

Metal-Organic Frameworks (MOFs)

Metal-organic frameworks (MOFs) also known as porous crystalline coordination polymers formed by metallic polyhedra and organic molecules [121–123]. Inorganic building units, metal ions or clusters, and organic units carboxylates or other organic anions such as phosphonate, sulfonate, and heterocyclic compounds are linked together to form MOFs. Structure of MOFs is dependent on connectivity and geometry of organic linkers [124]. These frameworks have large cubic cavities of uniform sizes. Different structures of cavities formed in MOFs make them highly porous material. BET surface area of MOFs is quite high as compared to CNT, hydrides, zeolites, and clathrates and PCN-521 is considered as the most porous MOF among all the MOFs formed from tetrahedral linkers [124]. Some typical examples of MOFs are MOF-5, IRMOF-6, IRMOF-8, and so forth shown in Figure 13. The synthesis method for MOFs is simple, inexpensive, and straight forward. For instance, the MOF-5 can be synthesized from an alkaline zinc solution and 1,4-benzene dicarboxylic acid [125]. Different routes of synthesis method such as conventional heating, electrochemistry, microwave-assisted heating, mechanochemistry, and sonochemistry are used to prepare variety of MOFs [17, 126].

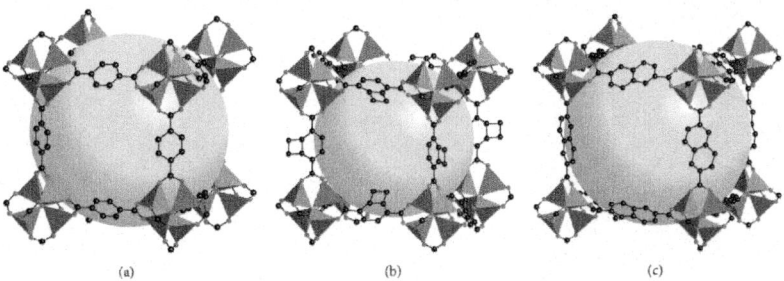

(a)   (b)   (c)

Figure 13: Single crystal X-ray structure of MOF-5 (a), IRMOF-6 (b), and IRMOF-8 (c) on each of the eight corners is a cluster of tetrahedral $[OZn_4(CO_2)_6]$. The large yellow sphere represents the largest sphere that would fit into the cavities without touching the van der Waals atoms of the frameworks [17].

Microporous, open metal sites, high surface area, chemically tunable structures, and different types of cavities of MOFs labelled them as most promising candidate for hydrogen storage material [127, 128]. Hydrogen sorption behavior of MOFs indicates that they reversibly store hydrogen at ambient temperature and moderate pressures [129]. The typical storage capacity of MOF-5 under aforementioned conditions is 1.0 wt%. The storage capacity of the MOFs can easily be increased by introducing larger organic moieties into the frameworks. Maximum hydrogen storage capacity at room temperature of different kinds of microporous MOFs by varying organic linkers and metal sites reaches up to 1.65 wt% and hydrogen pressure 48 bars [17]. Hydrogen binding sites of MOFs are obtained from the neutron diffraction data [130]. The superimposition of the scattering intensity with molecular structure indicates the cup sites (green-yellow-red regions in the left panel of Figure 14), where hydrogen molecules are attached. Covalent-organic frameworks (COFs) are crystalline, highly porous, and large surface area. COFs possess the same properties like MOFs such as structure rigidity, surface area, pore volume, and different kinds of cavities but have low density due to the presence of C, Si, B, and O as main clusters with organic unit carboxylates. Recently COFs are also emerging as the promising candidate for the hydrogen storage material due to the above-mentioned properties [131–133].

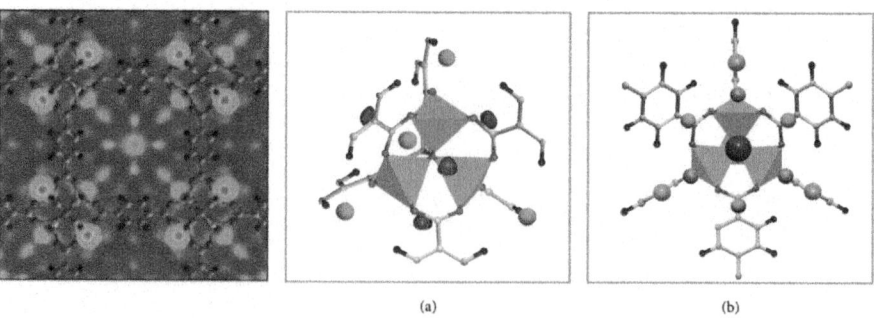

(a)   (b)

Figure 14: The super position of scattering intensity from neutron diffraction data with molecular structure of MOF-5 suggests the binding sites [130].

Metal-Doped Metal-Organic Frameworks (MOFs)

Hydrogen storage capacity in pristine MOFs at room temperature is nowhere near the ultimate target (7.5 wt% and 70 g/L) set by the Department of Energy (DOE) for onboard hydrogen storage systems [128]. However, at 77 K, MOFs show very high hydrogen storage capacity which reaches close to the ultimate US DOE hydrogen storage target for vehicular applications; but for practical onboard hydrogen storage purposes, 77 K is not feasible. Recently different kinds of routes were implemented to enhance hydrogen storage capacity in MOFs at room temperature and one of the methods is doping. Different kinds of metals with and without support (carbon materials) were doped into micro-porous MOFs [134–137]. Different techniques of metal doping into MOFs were used; the most common is incipient wetness and ball milling method [135, 137, 138]. Liu et al. reported that hydrogen uptake capacity of micropo-rous CuBTC was increased 3.5-fold at 298 K and 20 bars by addition of Pt/C catalyst [135]. Furthermore, structural integrity was maintained upon mixing catalyst with CuBTC. Spillover mechanism in metal-doped MOFs is a prom-ising method to increase the hydrogen storage capacity at RT. In addition, by introducing carbon with the metal into MOFs, carbon acts as a bridge between the source and receptor. Building bridging by carbon with metal in MOFs en-hances the hydrogen storage capacity remarkably [139–141]. Comparison of hydrogen uptake capacity of pure MIL-101 with doped MIL-101 via spillover mechanism and bridging is shown in Figure 15 [139].

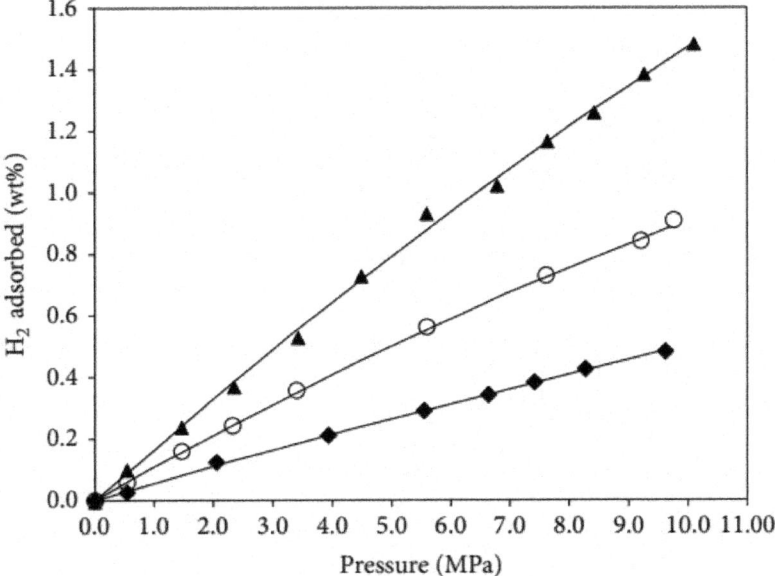

Figure 15: Hydrogen storage capacity of MIL-101, Pt/C-MIL-101, and MIL-101-bridges-Pt/C at 298 K [139].

Clathrate Hydrates

Clathrates are crystalline molecular complexes formed from the mixtures of host molecules and suitably sized gas molecules [142]. They are known also as caged compounds, since in these systems guest molecules are occluded in the cages formed by the host molecules. When the cages are formed by the water molecules, clathrates are known clathrate hydrate. The structure II consists of two types of polyhedra, dodecahedra and hexadecahedra, which are, respectively, termed as S (small) and L (large). In general, clathrates are stable only under extremely high pressure conditions, typically of the order of 120 bars. The hydrogen storage properties of clathrate hydrate have gained attention recently, when Lee et al. found that it is possible to stabilize clathrate hydrate at lower pressures by cooccluding large organic molecules, such as tetrahydrofuran (THF) in these caged materials [21]. Complete occupation of all large and small cages of the structure II clathrate hydrate corresponds to 5 wt% of gravimetric storage capacity (Figure 16) [143].

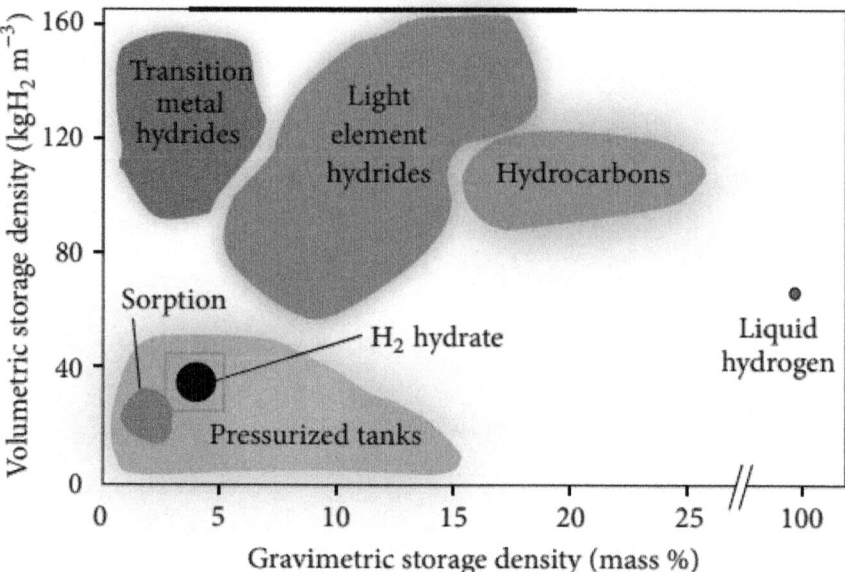

Figure 16: The comparison of gravimetric and volumetric storage capacities of clathrate hydrates and other fuels [143].

It can be seen from Figure 16 that the storage capacities of clathrate hydrates presently do not satisfy the 2010 R&D targets of FreedomCAR. However, if suitable pressure conditions are obtained to stabilize the hydrogen inside the cages, they can offer intermediate solution for the hydrogen storage problem. One of biggest merits of clathrates is that all chemicals required for their formation are inexpensive and easily available.

CONCLUSION

In this review we have presented an overview of the implementation of the hydrogen fuel based energy economy. In the first part we exclusively dealt with the trends of current fossil-fuel based energy economy. It was clear that the transition from the present energy economy to a sustainable and cleaner energy such as hydrogen energy is motivated by both economical and environmental factors. Amongst various forms of alternate energies, hydrogen occupies a distinct status, owing to its universal abundance and the highest specific energy. However, the energy density of hydrogen at ambient pressure and temperature conditions is nearly three times lower than that of the conventional fuels. Therefore, to use hydrogen as a fuel, especially in the mobile context, it is vital to improve this storage attribute by several manifolds.

In the second part of the review, we presented the novel solid state hydrogen storage techniques employing carbon nanotubes, metal-doped carbon nanotubes, intermetallic hydrides, metal-organic frameworks, metal-doped metal-organic frameworks, and clathrates. Carbon nanotubes and metal-organic frameworks, in particular the ones doped with transition metal particles, are considered to offer far-reaching solution as a hydrogen storage medium for transportation applications. Other materials discussed in the review are mostly in their preliminary phase and much intensive works are required, for their applications in the mobile hydrogen storage sector.

REFERENCES

1. J. Yang, A. Sudik, C. Wolverton, and J. S. Donald, "High capacity hydrogen storage materials: attributes for automotive applications and techniques for materials discovery," Chemical Society Reviews, vol. 39, no. 2, pp. 656–675, 2010.

2. T. Nakata, "Energy-economic models and the environment," Progress in Energy and Combustion Science, vol. 30, no. 4, pp. 417–475, 2004.

3. U. Jørgensen, "Energy sector in transition—technologies and regulatory policies in flux," Technological Forecasting and Social Change, vol. 72, no. 6, pp. 719–731, 2005.

4. E. Martinot, "Renewable energy investment by the World Bank," Energy Policy, vol. 29, no. 9, pp. 689–699, 2001.

5. A. Masini and E. Menichetti, "Investment decisions in the renewable energy sector: an analysis of non-financial drivers," Technological Forecasting & Social Change, vol. 80, no. 3, pp. 510–524, 2013.

6. G. D. Berry, A. D. Pasternak, G. D. Rambach, J. R. Smith, and R. N. Schock, "Hydrogen as a future transportation fuel," Energy, vol. 21, no. 4, pp. 289–303, 1996.

7. B. Johnston, M. C. Mayo, and A. Khare, "Hydrogen: the energy source for the 21st century,"Technovation, vol. 25, no. 6, pp. 569–585, 2005.

8. P. Tseng, J. Lee, and P. Friley, "A hydrogen economy: opportunities and challenges," Energy, vol. 30, no. 14, pp. 2703–2720, 2005.

9. A. Midilli, M. Ay, I. Dincer, and M. A. Rosen, "On hydrogen and hydrogen energy strategies I : current status and needs," Renewable and Sustainable Energy Reviews, vol. 9, no. 3, pp. 255–271, 2005.

10. M. Momirlan and T. N. Veziroglu, "Current status of hydrogen energy," Renewable and Sustainable Energy Reviews, vol. 6, no. 1-2, pp. 141–179, 2002.

11. L. Zhou, "Progress and problems in hydrogen storage methods," Renewable and Sustainable Energy Reviews, vol. 9, no. 4, pp. 395–408, 2005.

12. P. Chen and M. Zhu, "Recent progress in hydrogen storage," Materials Today, vol. 11, no. 12, pp. 36–43, 2008.

13. G. Principi, F. Agresti, A. Maddalena, and S. Lo Russo, "The problem of solid state hydrogen storage," Energy, vol. 34, no. 12, pp. 2087–2091, 2009.

14. W. J. Piel, "Transportation fuels of the future?" Fuel Processing Technology, vol. 71, no. 1–3, pp. 167–179, 2001.

15. A. E. Farrell, D. W. Keith, and J. Corbett, "A strategy for introducing hydrogen into transportation," Energy Policy, vol. 31, no. 13, pp. 1357–1367, 2003.

16. A. Züttel, "Materials for hydrogen storage," Materials Today, vol. 6, no. 9, pp. 24–33, 2003.

17. N. L. Rosi, J. Eckert, M. Eddaoudi et al., "Hydrogen storage in microporous metal-organic frameworks," Science, vol. 300, no. 5622, pp. 1127–1129, 2003.

18. E. Klontzas, E. Tylianakis, and G. E. Froudakis, "Hydrogen storage in 3D covalent organic frameworks. A multiscale theoretical investigation," Journal of Physical Chemistry C, vol. 112, no. 24, pp. 9095–9098, 2008.

19. R. Zacharia, S.-U. Rather, S. W. Hwang, and K. S. Nahm, "Spillover of physisorbed hydrogen from sputter-deposited arrays of platinum nanoparticles to multi-walled carbon nanotubes," Chemical Physics Letters, vol. 434, no. 4–6, pp. 286–291, 2007.

20. J. Weitkamp, M. Fritz, and S. Ernst, "Zeolites as media for hydrogen storage," International Journal of Hydrogen Energy, vol. 20, no. 12, pp. 967–970, 1995.

21. H. Lee, J.-W. Lee, D. Y. Kim et al., "Tuning clathrate hydrates for hydrogen storage," Nature, vol. 434, no. 7034, pp. 743–746, 2005.

22. H. Reardon, J. M. Hanlon, R. W. Hughes, A. Godula-Jopek, T. K. Mandal, and D. H. Gregory, "Emerging concepts in solid-state hydrogen storage: the role of nanomaterials design," Energy & Environmental Science, vol. 5, no. 3, pp. 5951–5979, 2012.

23. S. H. Sang, H. Furukawa, O. M. Yaghi, and W. A. Goddard III, "Covalent organic frameworks as exceptional hydrogen storage materials," Journal of the American Chemical Society, vol. 130, no. 35, pp. 11580–11581, 2008.

24. bp, "Putting energy in the spotlight-BP statistical review of World Energy,"   2005,http://www.nioclibrary.ir/free-e-resources/BP%20 Statistical%20Review%20of%20World%20Energy/statistical_review_ of_world_energy_full_report_2005.pdf.

25. P. Moriarty and D. Honnery, "Hydrogen's role in an uncertain energy future," International Journal of Hydrogen Energy, vol. 34, no. 1, pp. 31–39, 2009.

26. K. Mazloomi and C. Gomes, "Hydrogen as an energy carrier: prospects and challenges," Renewable and Sustainable Energy Reviews, vol. 16, no. 5, pp. 3024–3033, 2012.

27. Annual energy outlook 2015, US Energy Information Administration, 2015,http://www.eia.gov/forecasts/aeo/.

28. K. Mukhopadhyay and O. Forssell, "An empirical investigation of air pollution from fossil fuel combustion and its impact on health in India during 1973–1974 to 1996–1997," Ecological Economics, vol. 55, no. 2, pp. 235–250, 2005.

29. D. Mage, G. Ozolins, P. Peterson et al., "Urban air pollution in megacities of the world," Atmospheric Environment, vol. 30, no. 5, pp. 681–686, 1996.

30. J. Alcamo and E. Kreileman, "Emission scenarios and global climate protection," Global Environmental Change, vol. 6, no. 4, pp. 305–334, 1996.

31. D.   G.   Kessel,   "Global   warming—facts,   assessment, countermeasures," Journal of Petroleum Science and Engineering, vol. 26, no. 1–4, pp. 157–168, 2000.

32. O. Kaarstad, "Fossil fuels and responses to global warming," Energy Conversion and Management, vol. 36, no. 6–9, pp. 869–872, 1995.

33. A. D. Sagar, "Automobiles and global warming: alternative fuels and other options for carbon dioxide emissions reduction," Environmental Impact Assessment Review, vol. 15, no. 3, pp. 241–274, 1995.

34. W. R. Moomaw, "Industrial emissions of greenhouse gases," Energy Policy, vol. 24, no. 10-11, pp. 951–968, 1996.

35. National Energy Policy, Report of the National Energy Policy Development Group, Department of Energy, 2001, https://www. whitehouse.gov/energy.

36. Hydrogen posture plan, An Integrated Integrated Research, Development and Demonstration Plan, United States Department of Energy, 2006, http://www.hydrogen.energy.gov.

37. N. Matsuo, "Key elements related to the emissions trading for the Kyoto protocol," Energy Policy, vol. 26, no. 3, pp. 263–273, 1998.

38. J. Leggett, "A guide to the Kyoto protocol: a treaty with potentially vital strategic implications for the renewables industry," Renewable and Sustainable Energy Reviews, vol. 2, no. 4, pp. 345–351, 1998.

39. Y. Hu and C. R. Monroy, "Chinese energy and climate policies after Durban: save the Kyoto Protocol,"Renewable and Sustainable Energy Reviews, vol. 16, no. 5, pp. 3243–3250, 2012.

40. K. Vatopoulos, D. Andrews, J. Carlsson, I. Papaioannou, and G. Zubi, Study on the State of Play of Energy Efficiency of Heat and Electricity Production Technologies, European Commission, 2012.

41. M. Specht, F. Staiss, A. Bandi, and T. Weimer, "Comparison of the renewable transportation fuels, liquid hydrogen and methanol, with gasoline—energetic and economic aspects," International Journal of Hydrogen Energy, vol. 23, no. 5, pp. 387–396, 1998.

42. J. M. Ogden, M. M. Steinbugler, and T. G. Kreutz, "A comparison of hydrogen, methanol and gasoline as fuels for fuel cell vehicles: implications for vehicle design and infrastructure development," Journal of Power Sources, vol. 79, no. 2, pp. 143–168, 1999.

43. R. L. David, CRC Handbook of Chemistry and Physics, CRC Press, Boca Raton, Fla, USA, 73rd edition, 1992.

44. L. M. Das, "On-board hydrogen storage systems for automotive application," International Journal of Hydrogen Energy, vol. 21, no. 9, pp. 789–800, 1996.

45. D. J. Durbin and C. Malardier-Jugroot, "Review of hydrogen storage techniques for on board vehicle applications," International Journal of Hydrogen Energy, vol. 38, no. 34, pp. 14595–14617, 2013.

46. Comparative Properties of Hydrogen and Other Fuels, US Department of Energy, 2015,http://hydrogen.pnl.gov/hydrogen-data/comparative-properties-hydrogen-and-other-fuels.

47. G. A. Karim, "Hydrogen as a spark ignition engine fuel," International Journal of Hydrogen Energy, vol. 28, no. 5, pp. 569–577, 2003.

48. K.-A. Adamson, "Hydrogen from renewable resources-the hundred year commitment," Energy Policy, vol. 32, no. 10, pp. 1231–1242, 2004.

49. A. Steinfeld, "Solar thermochemical production of hydrogen—a review," Solar Energy, vol. 78, no. 5, pp. 603–615, 2005.

50. M. Momirlan and T. Veziro□lu, "Recent directions of world hydrogen

production," Renewable & sustainable energy reviews, vol. 3, no. 2, pp. 219–231, 1999. ·

51. A. A. van Benthem, G. J. Kramer, and R. Ramer, "An options approach to investment in a hydrogen infrastructure," Energy Policy, vol. 34, no. 17, pp. 2949–2963, 2006.

52. J. Tomei, "Planning for a transition to a hydrogen economy: a review of roadmaps," UKSHEC Working Paper 2, 2009.

53. B. D. Solomon and A. Banerjee, "A global survey of hydrogen energy research, development and policy,"Energy Policy, vol. 34, no. 7, pp. 781–792, 2006.

54. C.-J. Winter, "Hydrogen energy—abundant, efficient, clean: a debate over the energy-system-of-change,"International Journal of Hydrogen Energy, vol. 34, no. 14, pp. S1–S2, 2009.

55. S. O. Akansu, Z. Dulger, N. Kahraman, and T. N. Veziro☐lu, "Internal combustion engines fueled by natural gas—hydrogen mixtures," International Journal of Hydrogen Energy, vol. 29, no. 14, pp. 1527–1539, 2004.

56. J. W. Heffel, "NOx emission and performance data for a hydrogen fueled internal combustion engine at 1500 rpm using exhaust gas recirculation," International Journal of Hydrogen Energy, vol. 28, no. 8, pp. 901–908, 2003.

57. United States Department of Energy, National Hydrogen Energy Roadmap, 2002,http://www.hydrogen.energy.gov/pdfs/national_h2_roadmap.pdf.

58. A. J. Appleby, "Issues in fuel cell commercialization," Journal of Power Sources, vol. 58, no. 2, pp. 153–176, 1996.

59. C.-J. Winter, "Into the hydrogen energy economy—milestones," International Journal of Hydrogen Energy, vol. 30, no. 7, pp. 681–685, 2005.

60. G. Marbán and T. Valdés-Solís, "Towards the hydrogen economy?" International Journal of Hydrogen Energy, vol. 32, no. 12, pp. 1625–1637, 2007.

61. D. Day, R. J. Evans, J. W. Lee, and D. Reicosky, "Economical CO_2, SO_x, and NO_x capture from fossil-fuel utilization with combined renewable hydrogen production and large-scale carbon sequestration," Energy, vol. 30, no. 14, pp. 2558–2579, 2005.

62. G. Karagiannakis, C. Kokkofitis, S. Zisekas, and M. Stoukides, "Catalytic and electrocatalytic production of H_2 from propane decomposition over

Pt and Pd in a proton-conducting membrane-reactor," Catalysis Today, vol. 104, no. 2–4, pp. 219–224, 2005.

63. A. I. Miller and R. B. Duffey, "Sustainable and economic hydrogen cogeneration from nuclear energy in competitive power markets," Energy, vol. 30, no. 14, pp. 2690–2702, 2005.

64. A. Domashenko, A. Golovchenko, Y. Gorbatsky, V. Nelidov, and B. Skorodumov, "Production, storage and transportation of liquid hydrogen. Experience of infrastructure development and operation,"International Journal of Hydrogen Energy, vol. 27, no. 7-8, pp. 753–755, 2002.

65. Air Products and Chemicals Inc, "Hydrogen distribution options,"http://www.airproducts.com/products/Gases/Hydrogen.aspx.

66. United States Department of Energy, Targets for On-Board Hydrogen Storage Systems for Light Duty Vehicles, United States Department of Energy, 2009,http://www.eere.energy.gov/hydrogenandfuelcells/pdfs/.

67. K. Pehr, P. Sauermann, O. Traeger, and M. Bracha, "Liquid hydrogen for motor vehicles—the world›s first public LH2 filling station," International Journal of Hydrogen Energy, vol. 26, no. 7, pp. 777–782, 2001.

68. S. A. Sherif, N. Zeytinoglu, and T. N. Veziroᐁlu, "Liquid hydrogen: potential, problems, and a proposed research program," International Journal of Hydrogen Energy, vol. 22, no. 7, pp. 683–688, 1997.

69. M. R. Swain, P. Filoso, E. S. Grilliot, and M. N. Swain, "Hydrogen leakage into simple geometric enclosures," International Journal of Hydrogen Energy, vol. 28, no. 2, pp. 229–248, 2003.

70. K. Verfondern and B. Dienhart, "Experimental and theoretical investigation of liquid hydrogen pool spreading and vaporization," International Journal of Hydrogen Energy, vol. 22, no. 7, pp. 649–660, 1997.

71. F. Rigas and S. Sklavounos, "Evaluation of hazards associated with hydrogen storage facilities,"International Journal of Hydrogen Energy, vol. 30, no. 13-14, pp. 1501–1510, 2005.

72. H. W. Pohl and V. V. Malychev, "Hydrogen in future civil aviation," International Journal of Hydrogen Energy, vol. 22, no. 10-11, pp. 1061–1069, 1997.

73. A. Contreras, S. Yiğit, K. Özay, and T. N. Veziroğlu, "Hydrogen as aviation fuel: a comparison with hydrocarbon fuels," International Journal of Hydrogen Energy, vol. 22, no. 10-11, pp. 1053–1060, 1997.

74. C. E. Borroni-Bird, "Fuel cell commercialization issues for light-duty vehicle applications," Journal of Power Sources, vol. 61, no. 1-2, pp. 33–48, 1996.

75.　T. E. Lipman, J. L. Edwards, and D. M. Kammen, "Fuel cell system economics: comparing the costs of generating power with stationary and motor vehicle PEM fuel cell systems," Energy Policy, vol. 32, no. 1, pp. 101–125, 2004.

76.　California Hydrogen Highway Network Initiative, http://www.hydrogenhighway.ca.gov/.

77.　Department of Energy, Target Explanation Document: Onboard Hydrogen Storage for Light-Duty Fuel Cell Vehicles, Department of Energy, 2015,http://energy.gov/sites/prod/files/2015/05/f22/fcto_targets_onboard_hydro_storage_explanation.pdf.

78.　Multi-Year Research, Development and Demonstration Plan, 2012,http://energy.gov/sites/prod/files/2014/03/f12/introduction.pdf.

79.　S. Satyapal, J. Petrovic, C. Read, G. Thomas, and G. Ordaz, "The U.S. Department of Energy›s National Hydrogen Storage Project: progress towards meeting hydrogen-powered vehicle requirements," Catalysis Today, vol. 120, no. 3-4, pp. 246–256, 2007.

80.　J. L. Mendoza-Cortes, W. A. Goddard, H. Furukawa, and O. M. Yaghi, "A covalent organic framework that exceeds the DOE 2015 volumetric target for H_2 uptake at 298 K," The Journal of Physical Chemistry Letters, vol. 3, no. 18, pp. 2671–2675, 2012.

81.　T. N. Veziroglu, S. Y. Zaginaichenko, D. V. Schur et al., Hydrogen Materials Science and Chemistry of Carbon Nanomaterials, Springer, Dordrecht, The Netherlands, 2005.

82.　D. Stolten, "Hydrogen and fuel cells: fundamentals, technologies and applications," Angewandte Chemie, vol. 50, no. 42, p. 9787, 2011.

83.　R. Ewald, "Requirements for advanced mobile storage systems," International Journal of Hydrogen Energy, vol. 23, no. 9, pp. 803–814, 1998.

84.　M. T. Syed, S. A. Sherif, T. N. Veziroglu, and J. W. Sheffield, "An economic analysis of three hydrogen liquefaction systems," International Journal of Hydrogen Energy, vol. 23, no. 7, pp. 565–576, 1998.

85.　M. B. Ley, L. H. Jepsen, Y.-S. Lee et al., "Complex hydrides for hydrogen storage—new perspectives,"Materials Today, vol. 17, no. 3, pp. 122–128, 2014.

86.　B. Sakintuna, F. Lamari-Darkrim, and M. Hirscher, "Metal hydride materials for solid hydrogen storage: a review," International Journal of Hydrogen Energy, vol. 32, no. 9, pp. 1121–1140, 2007.

87.　S.-U. Rather, R. Zacharia, S. W. Hwang, M.-U. Naik, and K. S. Nahm,

"Hyperstoichiometric hydrogen storage in monodispersed palladium nanoparticles," Chemical Physics Letters, vol. 438, no. 1–3, pp. 78–84, 2007.

88. A. Züttel, A. Borgschulte, and L. Schlapbach, Hydrogen as a Future Energy Carrier, Wiley, Chichester, UK, 2008.

89. M. Morinaga, H. Yukawa, K. Nakatsuka, and M. Takagi, "Roles of constituent elements and design of hydrogen storage alloys," Journal of Alloys and Compounds, vol. 330–332, pp. 20–24, 2002.

90. M. E. Arroyo y de Dompablo and G. Ceder, "First principles investigations of complex hydrides AMH_4 and A_3MH_6 (A=Li, Na, K, M=B, Al, Ga) as hydrogen storage systems," Journal of Alloys and Compounds, vol. 364, no. 1-2, pp. 6–12, 2004.

91. A. Züttel, S. Rentsch, P. Fischer et al., "Hydrogen storage properties of LiBH4," Journal of Alloys and Compounds, vol. 356-357, pp. 515–520, 2003.

92. S.-I. Orimo, Y. Nakamori, J. R. Eliseo, A. Züttel, and C. M. Jensen, "Complex hydrides for hydrogen storage," Chemical Reviews, vol. 107, no. 10, pp. 4111–4132, 2007.

93. IPHE International Hydrogen Storage Technology Conference, Lucca, Italy, June 2005.

94. Technical System Targets: Onboard Hydrogen Storage for Light-Duty Fuel Cell Vehicles, 2012,http://www.energy.gov.

95. R. G. Ding, J. J. Finnerty, Z. H. Zhu, Z. F. Yan, and G. Q. Lu, Encyclopedia of Nanoscience and Nanotechnology, American Scientific Publishers, 2004.

96. R. Ghosh, T. Maruyama, H. Kondo, K. Kimoto, T. Nagai, and S. Iijima, "Synthesis of single-walled carbon nanotubes on graphene layers," Chemical Communications, vol. 51, no. 43, pp. 8974–8977, 2015.

97. K. A. Williams and P. C. Eklund, "Monte Carlo simulations of H_2 physisorption in finite-diameter carbon nanotube ropes," Chemical Physics Letters, vol. 320, no. 3-4, pp. 352–358, 2000.

98. A. C. Dillon, K. M. Jones, T. A. Bekkedahl, C. H. Kiang, D. S. Bethune, and M. J. Heben, "Storage of hydrogen in single-walled carbon nanotubes," Nature, vol. 386, no. 6623, pp. 377–379, 1997.

99. K. Shindo, T. Kondo, and Y. Sakurai, "Hydrogen physisorption capacities of mechanically milled activated carbon powders in a H2 atmosphere using a gravimetric method," Journal of Alloys and Compounds, vol. 379, no. 1-2, pp. 252–255, 2004.

100. Y. Ye, C. C. Ahn, C. Witham et al., "Hydrogen adsorption and cohesive energy of single-walled carbon nanotubes," Applied Physics Letters, vol. 74, no. 16, pp. 2307–2309, 1999.

101. A. Badzian, T. Badzian, E. Breval, and A. Piotrowski, "Nanostructured, nitrogen-doped carbon materials for hydrogen storage," Thin Solid Films, vol. 398-399, pp. 170–174, 2001.

102. N. Rajalakshmi, K. S. Dhathathreyan, A. Govindaraj, and B. C. Satishkumar, "Electrochemical investigation of single-walled carbon nanotubes for hydrogen storage," Electrochimica Acta, vol. 45, no. 27, pp. 4511–4515, 2000.

103. C. Nützenadel, A. Züttel, D. Chartouni, and L. Schlapbach, "Electrochemical storage of hydrogen in nanotube materials," Electrochemical and Solid-State Letters, vol. 2, no. 1, pp. 30–32, 1999.

104. P. F. Weck, E. Kim, N. Balakrishnan, H. Cheng, and B. I. Yakobson, "Designing carbon nanoframeworks tailored for hydrogen storage," Chemical Physics Letters, vol. 439, no. 4–6, pp. 354–359, 2007.

105. B. Adeniran and R. Mokaya, "Low temperature synthesized carbon nanotube superstructures with superior CO_2 and hydrogen storage capacity," Journal of Materials Chemistry A, vol. 3, no. 9, pp. 5148–5161, 2015.

106. W. Z. Huang, X. B. Zhang, J. P. Tu et al., "The effect of pretreatments on hydrogen adsorption of multi-walled carbon nanotubes," Materials Chemistry and Physics, vol. 78, no. 1, pp. 144–148, 2003.

107. F. Liu, X. Zhang, J. Cheng et al., "Preparation of short carbon nanotubes by mechanical ball milling and their hydrogen adsorption behavior," Carbon, vol. 41, no. 13, pp. 2527–2532, 2003.

108. Y. Xia, J. Z. Zhu, M. Zhao et al., "Enhancement of hydrogen physisorption on single-walled carbon nanotubes resulting from defects created by carbon bombardment," Physical Review B—Condensed Matter and Materials Physics, vol. 71, no. 7–15, Article ID 075412, 2005.

109. S. Orimo, T. Matsushima, H. Fujii, T. Fukunaga, and G. Majer, "Hydrogen desorption property of mechanically prepared nanostructured graphite," Journal of Applied Physics, vol. 90, no. 3, pp. 1545–1549, 2001.

110. R. Zacharia, K. Y. Kim, A. K. M. Fazle Kibria, and K. S. Nahm, "Enhancement of hydrogen storage capacity of carbon nanotubes via spill-over from vanadium and palladium nanoparticles," Chemical Physics Letters, vol. 412, no. 4–6, pp. 369–375, 2005.

111. A. Lueking and R. T. Yang, "Hydrogen storage in carbon nanotubes: residual metal content and pretreatment temperature," AIChE Journal, vol. 49, no. 6, pp. 1556–1568, 2003.

112. A. Lueking and R. T. Yang, "Evidence of hydrogen spillover onto carbon nanotubes in the presence of nickel," Journal of Catalysis, vol. 165, p. 206, 2002.

113. R. Bhowmick, S. Rajasekaran, D. Friebel et al., "Hydrogen spillover in Pt-single-walled carbon nanotube composites: formation of stable C-H bonds," Journal of the American Chemical Society, vol. 133, no. 14, pp. 5580–5586, 2011.

114. J. A. Schwarz, C. I. Contescu, and K. Putyera, Dekker Encyclopedia of Nanoscience and Nanotechnology, vol. 1, CRC Press, New York, NY, USA, 2004.

115. Y. Zhang, N. W. Franklin, R. J. Chen, and H. Dai, "Metal coating on suspended carbon nanotubes and its implication to metal-tube interaction," Chemical Physics Letters, vol. 331, no. 1, pp. 35–41, 2000.

116. V. B. Parambhath, R. Nagar, K. Sethupathi, and S. Ramaprabhu, "Investigation of spillover mechanism in palladium decorated hydrogen exfoliated functionalized graphene," Journal of Physical Chemistry C, vol. 115, no. 31, pp. 15679–15685, 2011.

117. T. Yildirim and S. Ciraci, "Titanium-decorated carbon nanotubes as a potential high-capacity hydrogen storage medium," Physical Review Letters, vol. 94, no. 17, Article ID 175501, 2005.

118. C. Liu, Y. Chen, C.-Z. Wu, S.-T. Xu, and H.-M. Cheng, "Hydrogen storage in carbon nanotubes revisited,"Carbon, vol. 48, no. 2, pp. 452–455, 2010.

119. A. C. Dillon, K. E. H. Gilbert, J. L. Alleman et al., "Carbon nanotube materials for hydrogen storage," inProceedings of the Hydrogen Program Review, National Renewable Energy Laboratory, San Ramon, Calif, USA, May 2000.

120. C. Zhang, X. Lu, and A. Gu, "How to accurately determine the uptake of hydrogen in carbonaceous materials," International Journal of Hydrogen Energy, vol. 29, no. 12, pp. 1271–1276, 2004.

121. M. Eddaoudi, J. Kim, N. Rosi et al., "Systematic design of pore size and functionality in isoreticular MOFs and their application in methane storage," Science, vol. 295, no. 5554, pp. 469–472, 2002.

122. M. Kurmoo, "Magnetic metal-organic frameworks," Chemical Society Reviews, vol. 38, no. 5, pp. 1353–1379, 2009.

123. J. L. C. Rowsell and O. M. Yaghi, "Metal—organic frameworks: a new

class of porous materials,"Microporous and Mesoporous Materials, vol. 73, no. 1-2, pp. 3–14, 2004.

124. W. Lu, Z. Wei, Z.-Y. Gu et al., "Tuning the structure and function of metal-organic frameworks via linker design," Chemical Society Reviews, vol. 43, no. 16, pp. 5561–5593, 2014.

125. B. Panella and M. Hirscher, "Hydrogen physisorption in metal-organic porous crystals," Advanced Materials, vol. 17, no. 5, pp. 538–541, 2005.

126. N. Stock and S. Biswas, "Synthesis of metal-organic frameworks (MOFs): routes to various MOF topologies, morphologies, and composites," Chemical Reviews, vol. 112, no. 2, pp. 933–969, 2012.

127. H. W. Langmi, J. Ren, B. North, M. Mathe, and D. Bessarabov, "Hydrogen storage in metal-organic frameworks: a review," Electrochimica Acta, vol. 128, pp. 368–392, 2014.

128. L. J. Murray, M. Dinc, and J. R. Long, "Hydrogen storage in metal–organic frameworks," Chemical Society Reviews, vol. 38, no. 5, pp. 1294–1314, 2009.

129. L. Pan, M. B. Sander, X. Huang et al., "Microporous metal organic materials: promising candidates as sorbents for hydrogen storage," Journal of the American Chemical Society, vol. 126, no. 5, pp. 1308–1309, 2004.

130. T. Yildirim and M. R. Hartman, "Direct observation of hydrogen adsorption sites and nanocage formation in metal-organic frameworks," Physical Review Letters, vol. 95, no. 21, Article ID 215504, 2005.

131. S. B. Kalidindi and R. A. Fischer, "Covalent organic frameworks and their metal nanoparticle composites: prospects for hydrogen storage," Physica Status Solidi B, vol. 250, no. 6, pp. 1119–1127, 2013.

132. S. S. Han, H. Furukawa, O. M. Yaghi, and W. A. Goddard III, "Covalent organic frameworks as exceptional hydrogen storage materials," Journal of the American Chemical Society, vol. 130, no. 35, pp. 11580–11581, 2008.

133. S.-Y. Ding and W. Wang, "Covalent organic frameworks (COFs): from design to applications," Chemical Society Reviews, vol. 42, no. 2, pp. 548–568, 2013.

134. T. Stergiannakos, E. Tylianakis, E. Klontzas, P. N. Trikalitis, and G. E. Froudakis, "Hydrogen storage in novel li-doped corrole metal-organic frameworks," The Journal of Physical Chemistry C, vol. 116, no. 15, pp. 8359–8363, 2012.

135. X. M. Liu, S.-U. Rather, Q. Li, A. Lueking, Y. Zhao, and J. Li, "Hydrogenation of CuBTC framework with the introduction of a PtC

hydrogen spillover catalyst," The Journal of Physical Chemistry C, vol. 116, no. 5, pp. 3477–3485, 2012.

136. W. Qin, W. Cao, H. Liu, Z. Li, and Y. Li, "Metal-organic framework MIL-101 doped with palladium for toluene adsorption and hydrogen storage," Royal Society of Chemistry Advances, vol. 4, pp. 2414–2420, 2014.

137. N. R. Stuckert, L. Wang, and R. T. Yang, "Characteristics of hydrogen storage by spillover on Pt-doped carbon and catalyst-bridged metal organic framework," Langmuir, vol. 26, no. 14, pp. 11963–11971, 2010.

138. F. Schröder and R. A. Fischer, "Doping of metal-organic frameworks with functional guest molecules and nanoparticles," in Functional Metal-Organic Frameworks: Gas Storage, Separation and Catalysis, vol. 293 of Topics in Current Chemistry, pp. 77–113, Springer, Berlin, Germany, 2010. ·

139. Y. Li and R. T. Yang, "Hydrogen storage in metal-organic and covalent-organic frameworks by spillover,"AIChE Journal, vol. 54, no. 1, pp. 269–279, 2008.

140. Y. Li and R. T. Yang, "Significantly enhanced hydrogen storage in metal-organic frameworks via spillover," Journal of the American Chemical Society, vol. 128, no. 3, pp. 726–727, 2006.

141. A. J. Lachawiec Jr., G. Qi, and R. T. Yang, "Hydrogen storage in nanostructured carbons by spillover: bridge-building enhancement," Langmuir, vol. 21, no. 24, pp. 11418–11424, 2005.

142. F. H. Herbstein, Crystalline Molecular Complexes and Compounds: and Principles, vol. 1, 2005.

143. F. Schüth, "Technology: hydrogen and hydrates," Nature, vol. 434, no. 7034, pp. 712–713, 2005.

CITATION

CHAPTER 1

Liga Grinberga and Janis Kleperis (2011). Composite Nanomaterials for Hydrogen Technologies, Advances in Composite Materials for Medicine and Nanotechnology, Dr. Brahim Attaf (Ed.), ISBN: 978-953-307-235-7, InTech, DOI: 10.5772/14231.

CHAPTER 2

Wawrousek K, Noble S, Korlach J, Chen J, Eckert C, Yu J, et al. (2014) Genome Annotation Provides Insight into Carbon Monoxide and Hydrogen Metabolism in Rubrivivax gelatinosus. PLoS ONE 9(12): e114551. doi:10.1371/journal.pone.0114551.

CHAPTER 3

Michail Obolensky, Andrew Basteev, Vladimir Beletsky, Andrew Kravchenko, Yuri Petrusenko, Valeriy Borysenko, Sergey Lavrynenko, Oleg Kravchenko, Irina Suvorova, Vladimir Golovanevskiy and Leonid Bazyma (2012). Postsynthesis Treatment Influence on Hydrogen Sorption Properties of Carbon Nanotubes, Hydrogen Storage, Prof. Jianjun Liu (Ed.), ISBN: 978-953-51-0731-6, InTech, DOI: 10.5772/50151.

CHAPTER 4

Aal, S. and Soliman, A. (2015) High Capacity Hydrogen Storage in Ni Decorated Carbon Nanocone: A First-Principles Study. Journal of Quantum Information Science, 5, 134-149. doi: 10.4236/jqis.2015.54016.

CHAPTER 5

C. Mota, M. Culebras, A. Cantarero, A. Madroñero, C. Gómez, J. Amo and J. Robla, "Effects of Gamma Irradiation on the Kinetics of the Adsorption and Desorption of Hydrogen in Carbon Microfibres," Advances in Materials Physics and Chemistry, Vol. 3 No. 2, 2013, pp. 153-160. doi: 10.4236/ampc.2013.32021.

CHAPTER 6

Vladimir A. Blagojević, Dejan G. Minić, Jasmina Grbović Novaković and Dragica M. Minic (2012). Hydrogen Economy: Modern Concepts, Challenges and Perspectives, Hydrogen Energy - Challenges and Perspectives, Prof. Dragica Minic (Ed.), ISBN: 978-953-51-0812-2, InTech, DOI: 10.5772/46098.

CHAPTER 7

Doki Yamaguchi, Liangguang Tang, Nick Burke, Ken Chiang, Lucas Rye, Trevor Hadley and Seng Lim (2012). Small Scale Hydrogen Production from Metal-Metal Oxide Redox Cycles, Hydrogen Energy - Challenges and Perspectives, Prof. Dragica Minic (Ed.), ISBN: 978-953-51-0812-2, InTech, DOI: 10.5772/50030.

CHAPTER 8

Michael U. Niemann, Sesha S. Srinivasan, Ayala R. Phani, Ashok Kumar, D. Yogi Goswami, and Elias K. Stefanakos, "Nanomaterials for Hydrogen Storage Applications: A Review," Journal of Nanomaterials, vol. 2008, Article ID 950967, 9 pages, 2008. doi:10.1155/2008/950967

CHAPTER 9

Ong, Yit Thai, Ahmad, Abdul Latif, Zein, Sharif Hussein Sharif, & Tan, Soon Huat. (2010). A review on carbon nanotubes in an environmental protection and green engineering perspective. Brazilian Journal of Chemical Engineering, 27(2), 227-242. https://dx.doi.org/10.1590/S0104-66322010000200002

CHAPTER 10

Peter A Schultz and Clark S Snow, Mechanical properties of metal dihydrides: doi:10.1088/0965-0393/24/3/035005

CHAPTER 11

Renju Zacharia and Sami ullah Rather, "Review of Solid State Hydrogen Storage Methods Adopting Different Kinds of Novel Materials," Journal of Nanomaterials, vol. 2015, Article ID 914845, 18 pages, 2015. doi:10.1155/2015/914845

INDEX